Dragons with Clay Feet?

RURAL ECONOMIES IN TRANSITION

Series Editors

Zvi Lerman
The Hebrew University of Jerusalem

Csaba Csaki
The World Bank and the Budapest Economics University

Johan Swinnen
Katholieke Universiteit Leuven

Much of the world's rural population lives in transition countries that are attempting to transform their economies from the centrally planned socialist system to a market-oriented system. The Rural Economies in Transition series evaluates the deep impact these transformations are having in Central and Eastern Europe, the former Soviet Union, and Asia. Based on original empirical work by international teams at the cutting edge of research, each volume presents the results of one complete empirical study devoted to a single country or region. The editors have designed the series to provide agricultural, development, and political economists, as well as scholars in other disciplines, with authoritative and up-to-date factual information about rural transition. The editors welcome manuscript submissions from scholars engaged in original empirical research in the field.

Private Agriculture in Armenia, by Zvi Lerman and Astghik Mirzakhanian

Romanian Agriculture and Transition toward the EU, edited by Sophia Davidova and Kenneth J. Thomson

Transition, Institutions, and the Rural Sector, edited by Max Spoor

Greening Industrialization in Asian Transitional Economies: China and Vietnam, edited by Arthur P. J. Mol and Joost C. L. van Buuren

Agriculture in Transition: Land Policies and Evolving Farm Structures in Post-Soviet Countries, by Zvi Lerman, Csaba Csaki, and Gershon Feder

Building Market Institutions in Post-Communist Agriculture: Land, Credit, and Assistance, edited by David A. J. Macey, William Pyle, and Stephen K. Wegren

Russia's Food Policies and Globalization, by Stephen K. Wegren

Cooperation in the Romanian Countryside: An Insight into Post-Soviet Agriculture, by Rachel Sabates-Wheeler

Reinventing the Cuban Sugar Agroindustry, edited by Jorge F. Pérez-López and José Alvarez

Integrated Development of Agriculture and Rural Areas in Central European Countries, edited by Sophia Davidova, Kai Bauer, and Michael Cuddy

Measuring Social and Economic Change in Rural Russia: Surveys from 1991 to 2003, by David J. O'Brien and Valery V. Patsiorkovsky

Dragons with Clay Feet?

Transition, Sustainable Land Use, and Rural Environment in China and Vietnam

Edited by
Max Spoor, Nico Heerink,
and Futian Qu

LEXINGTON BOOKS

A division of
ROWMAN & LITTLEFIELD PUBLISHERS, INC.
Lanham • Boulder • New York • Toronto • Plymouth, UK

LEXINGTON BOOKS

A division of Rowman & Littlefield Publishers, Inc.
A wholly owned subsidiary of The Rowman & Littlefield Publishing Group, Inc.
4501 Forbes Boulevard, Suite 200
Lanham, MD 20706

Estover Road
Plymouth PL6 7PY
United Kingdom

British Library Cataloguing in Publication Information Available

Library of Congress Cataloging-in-Publication Data

Dragons with clay feet? : transition, sustainable land use, and rural environment in
China and Vietnam / edited by Max Spoor, Nico Heerink, and Futian Qu.
    p. cm.
  Includes index.
  ISBN-13: 978-0-7391-1369-1 (cloth : alk. paper)
  ISBN-10: 0-7391-1369-0 (cloth : alk. paper)
  1. Agriculture and state—China. 2. Agriculture and state—Vietnam.
3. Agriculture—Economic aspects—China. 4. Agriculture—Economic aspects—
Vietnam. 5. Sustainable development—China. 6. Sustainable development—
Vietnam. I. Spoor, Max. II. Heerink, Nico, 1955– III. Qu, Futian.

  HD2098.D73 2007
  338.1'851—dc22   2006024443

Printed in the United States of America

Contents

List of Tables

List of Figures

Introduction

This book presents state-of-the-art research on the impact of ongoing and anticipated economic policy and institutional reforms on agricultural development and sustainable rural resource use in two East-Asian transition (and developing) economies, China and Vietnam. The contributions to this edited volume focus on (a) the regional and sectoral impact of transformational policies, (b) farm household decision making under a changing economic and institutional environment, and (c) potential trade-offs between agricultural growth and sustainable land management in the two countries.

Economic reforms and institutional change since the emergence in the early 1980s of the "household responsibility" system in China and its "output contract system" counterpart in Vietnam several years later, have stimulated an impressive growth of agricultural production and marketed surplus. In both cases this growth was at the heart of the success of the transition from a stagnating planned economy to a thriving market economy. These countries were not only Asian "dragons," but they became the counter-examples for the dramatic supply shocks that struck all of the transition countries of the former Soviet Union. Nevertheless, agricultural growth, in particular in the paddy-rice sector, was obtained in many regions at the expense of substantial (and most often uncontrolled) degradation of rural resources, such as land, forests, and water. Future agricultural growth, which is essential for the feeding of growing populations of these densely-populated countries, and the supply of primary commodities for the expanding agro-industry is seriously threatened by this process of unsustainable resource use. In that sense, the dragons have "clay feet."

Government policies aimed at controlling the negative effects of rapid agricultural modernization are largely based on direct regulation and large-scale projects. With the economic reforms of the 1980s and 1990s, however, resource use decisions have to a large extent been transferred to households. To support policy development within this changing socioeconomic context, there is an urgent need to understand household responses to economic policies and changing institutions (particularly prevailing land tenure systems), and their implications for agricultural production and sustainable resource use. This volume intends to contribute to an improved understanding of such household responses and their implications by analyzing (1) how the institutional and economic environment of households is changing, (2) how households respond to these changes, and (3) what the implications are for environmental degradation and sustainable land use. The contributions in the volume focus on different aspects of these relationships, with most contributions using detailed empirical material and suitable statistical analysis methods.

Major macroeconomic and institutional changes during "transition" in China and Vietnam, have also caused increased differentiation within the rural economy and population, and between rural and urban areas. Access to, and degradation of, land (and water) resources, have become crucially interlinked with income differentiation, marginalization, and poverty. Especially in poorer regions, rural households are still dependent on the agricultural sector as a mass generator of income, food, and employment. Sustainable land use is therefore of crucial importance for rural dwellers in both countries, in spite of rapid economic development and urbanization.

The volume consists of three parts, preceded by an introductory chapter written by the editors. Heerink, Spoor, and Qu, open the volume with an analytical chapter on "Transition, Economic Policy and Institutional Reforms in China and Vietnam" and their impact on sustainable land use. They argue that the transitional and institutional reforms in agriculture that were implemented in China and Vietnam since the 1980s show substantial similarities. The reforms have led to rapid agricultural growth in both countries, but also to major land degradation problems, such as erosion, salinization, reduction of soil organic matter content, and water and soil pollution. Policies addressing these land degradation problems have a strong top-down character in both countries, focusing on direct regulation and state-mandated technological improvements, with little attention for the use of economic incentives. Economic and institutional policy reforms, however, have significant effects on household decision making. Economic instruments such as taxes or subsidies and institutional policies promoting security of land rights can therefore be effective tools in influencing household decision making and thereby promoting sustainable land use. The effects of economic and institutional policies on the sustainability of land use, however, are not easy to disentangle. The purpose of this chapter is to contribute to an improved understanding of such effects. To this end, the authors first present an overview of the main macroeconomic and agricultural policy reforms and institutional changes in both countries since the end of the 1970s, which is followed by a partial analytical framework of the impact of economic reforms on sustainable land use. They then provide a brief overview of available empirical studies that examine the impact of economic and/or institutional reforms on sustainable land use in China and Vietnam, and use the analytical framework to draw initial conclusions and suggest avenues for further research.

After this introduction, the volume starts with Part I, entitled "Regional and Sectoral Impact of Economic Reforms." Fan and Chan-Kang proceed, in Chapter 2, with analyzing regional inequality in China. The increased regional inequality in China over the past two decades has alarmed the policymakers, researchers, and public opinion at large. A more balanced regional growth is actually called for and this is not only itself a social development objective, but also is a necessary condition for future overall economic growth. The authors synthesize the recent literature on the measures and sources of regional inequality and offer insights in development strategies to reduce it. The main sources of increased regional inequality in China are identified as the increased rural/urban divide and large regional differences in non-farm employment, both of which in turn are the results of regional disparity in infrastructure, human capital, and natural resource endowments. The recent rather biased fiscal policies further exacerbated the already large regional disparity. The authors conclude that in order to promote a more balanced regional growth, the government needs to correct urban and regional biased policies and to promote both agricultural and non-farm growth in western China.

Zhang and Fan, in Chapter 3, follow suit with discussing in more detail the fiscal decentralization and spatial inequality in rural China. Neoclassical theory predicts a convergence of economy as marginal returns to capital and labor equalize across regions. Although migration has become much easier, regional inequality in rural China has not narrowed but widened over the past several decades. The growing disparity has become one of the top concerns of China's leaders and posed great challenge for scholars to study this phenomenon. Using a data set at county level from 1986 to 1996, the author find that the uneven development of non-farm activities, has been the major contributing factor to the widening rural regional inequality. The results of a regression-based inequality decomposition show that the uneven regional distribution of infrastructure and educational levels are a key factor explaining the differential growth of the non-farm economy. Finally, it is argued that fiscal decentralization intensifies the difference in initial conditions and enlarges the gap in public finance, which in turn lead to the uneven regional development of infrastructure and education.

Zhong, in Chapter 4, examines the contribution of diversification to the growth and sustainability of Chinese agriculture. Feeding a growing population exceeding 1.3 billion is a big challenge to the Chinese government and has partly led to the self-sufficiency policy in grain production and supply. However, the efforts to push grain production have not only resulted in a deterioration of the environment but also to a stagnation or even reduction in farmers' income as production costs continue to increase. This situation might be worsening with China's accession to the WTO (in 2001) that provided market access for bulk commodity imports. A sustainable development of Chinese agriculture depends on diversification, or "structural adjustment," that allows Chinese farmers to fully utilize their comparative advantage in production of labor-intensive goods. Past experience has shown that diversification has contributed more than a half of the growth in Chinese agriculture during the reform period and reduced stresses on the environment at the same time. It is likely to contribute even more to Chinese agriculture in the future and in a sustainable manner.

Spoor, in Chapter 5, returns to the issue of growing regional (and spatial inequality), but now for both Asian dragons. While in China spatial inequality is very much between the fast-growing Eastern provinces and inland China, in Vietnam it is between the two river deltas (the Red River and the Mekong), and the Central and Northern highlands, where poverty rates are higher, and resource deterioration more severe. The author applies the analytical framework, first developed by Kanbur and Zhang (2005), who use the importance of heavy industry in the pre-reform period, the degree of decentralization, and the increased trade openness, as explanatory factors of spatial inequality. The conclusion of the chapter is that in spite of the fact that both Asian dragons have done impressively well in terms of growth and poverty reduction, they are at a point where inequality (in income and assets), in particular spatial inequality, can become an obstacle to growth and a cause of social instability, which could have negative consequences for foreign direct investment and macroeconomic growth perspectives.

Tuan, in Chapter 6, analyzes land policy in the case of Vietnam's agrarian transition, using a case study of *Namđinh* Province, in the Northern part of the country. *Namđinh* province was selected because it has typical characteristics of Vietnam's agriculture. The author considers agricultural commercialization as a way to develop the agricultural sector, and to improve living standards of peasant households. This process needs land accumulation and labor specialization among peasant households. There is, however, a concern with the leadership of the Vietnamese Communist Party that land accumulation may lead to rural class differentiation. In addition, because of the supposed inverse relationship between farm size and productivity, it seems that the existing equal distribution of land has contributed to both the efficiency and equity objectives. The study shows that land concentration within *Namđinh* province is not related to income inequality or to the level of production efficiency. Growing inequality in Vietnam mainly stems from the widening urban-rural gap. Besides rural-urban transfer of labor surplus, accelerating agricultural growth is expected to reduce rural poverty and to curtail the urban-rural gap. The main reason for low land productivity is land fragmentation. Land consolidation will therefore lead to higher land productivity and further agricultural commercialization.

Chen, Heerink, and Zhu, in Chapter 7, present an analysis of the effects of economic policy reforms on the "economic environment," that is the prices of major agricultural products and inputs, of farm households in one of the poorer regions in China, *Jiangxi* province. Time-series data from 1981 to 2003 for *Jiangxi* Province, China as a whole and the world market are used for this purpose. The results show that domestic supply and demand factors had a significant impact on grain and pork prices in *Jiangxi* Province during the period 1985–93. As a result of the opening up of China's agricultural sector to the world market and the inability of the Chinese government to control domestic grain trade, the world market price has become the main determinant of the prices of grain in *Jiangxi* Province and in China since 1994. The world market prices for pork, on the other hand, do not significantly affect domestic pork prices due to the negligible size of international trade in pork in China. The reform of the fertilizer market has only been partial; to protect farmers from excessive price fluctuations, the state successfully intervenes in fertilizer prices. Although world market prices do affect domestic prices of fertilizer, domestic supply of fertilizers plays a role as well. Likewise, local supply and demand factors within *Jiangxi* province affect fertilizer prices in this province since the mid 1990s. As a result of these policies, the fertilizer—grain price ration has declined more in *Jiangxi* Province than in China as a whole, meaning that farmers have benefited relatively more from the economic reforms than farmers elsewhere in China.

Part II of this volume is focusing on "Farm Household Decision Making under a Changing Economic and Institutional Environment." Heerink, Kuiper and Shi, in Chapter 8, analyze the impact of direct income payments to grain farmers and agricultural tax abolition—two major elements of the rural income support policy that started in 2004—on grain production and chemical inputs use by farm households. A village-level general equilibrium (CGE) model is used to separate the impact of the income support measures from recent price trends in grains and inputs, and to account for differences in household responses to the income support measures. The model is applied to two villages with different degrees of market access in Northeast *Jiangxi* Province. Simulation results with the two models show that the income support policy does not reach its goal of promoting grain production; the large increase in grain production in 2004 was not caused by the income support policy but by the rapid price increases in 2003–2004. The increased incomes allow farm households to buy more inputs for livestock production or for other activities that are more profitable than grain farming. Moreover, because leisure is valued higher with increasing incomes, farmers tend to switch to less intensive rice production. The simulation results further indicate that the income support policy has a negligible effect on total input use, although tax abolitions seem to have a minor positive impact on manure application (and hence improves soil quality) in villages with good market access. It is, however, overshadowed by the rapid grain price increase since the second half of 2003, which has caused a large increase in the use of chemical inputs and a more unbalanced use of inputs.

Ruben, Lu, and Kuiper, in Chapter 9, analyze in detail the marketing chains and transaction costs in managing tomato production and sales in Nanjing City (*Jiangsu* Province). Delivery conditions, trust relationships and quality demands widely differ amongst the different outlets (direct selling, selling to traders, and selling on the wholesale market), offering farmers different prices and occasioning various types of transaction costs. The degree of efficiency of the supply chain provides specific incentives to local producers for enhancing the use of chemical inputs to improve yield and product quality. The chapter analyses the effect of farmers' resource endowments and household characteristics for the selected typical market outlets, the implications of market channel choice for the occurrence of different types of transaction costs (information, negotiation, monitoring and transport costs), and the impact of transaction costs on input use decisions and production-specific investments for a sample survey of farmers. Based on these results, conclusions are drawn on the importance of transaction costs for scale and specialization in production management.

Shi, Heerink, and Qu, in Chapter 10, analyze the impact of off-farm employment on village factor market development and input use in farm production using a village case study from *Jiangxi* province in China. Four household groups are distinguished, based on the resources they have for earning incomes from off-farm employment and/or agricultural production. They use a village SAM multiplier analysis to examine the effects of migration, local off-farm employment and investments in infrastructure. Government investment in infrastructure gives the best results for stimulating all types of agricultural production, but it has the smallest impact on household incomes. Migration of low- and high-educated labor is found to have the largest impact on household incomes. The development of factor markets (agricultural labor, land and oxen).within the village is stimulated most by infrastructure investment. The simulation results further reveal that off-farm employment tends to have a small positive effect on land productivity and environmental quality, because manure application increases more than chemical inputs use.

Tan, Kruseman, and Heerink, in Chapter 11, examine the impact of land fragmentation on the variable costs of household rice production, using empirical material from the same Chinese region as the preceding chapter. The main finding is that land fragmentation—as measured by farm size, the Simpson index and average distance to plots—has a significant impact on production costs. Keeping other factors constant, an increase in farm size and a reduction of the average distance to plots decrease the total production costs per ton. Changes in the average plot size and its distribution do not affect total production costs but cause a shift in input use. Farmers with more and smaller plots tend to switch to more labor-intensive methods and use fewer modern technologies, making the net impact on total production costs not significant. The authors argue that major reforms are needed in the system of land allocation in China to deal with the problem of land fragmentation in order to promote the adoption of new agricultural technologies.

Dung and Spoor, in chapter 12, analyze the intensification of rice production and its negative health effects for farmers in the Mekong Delta during Vietnam's transition. As in many parts of Eastern China, both deltas in Vietnam represent a rice production system which is highly external input intensive. The authors examine the response of rice farming households to changes in the economic and institutional environment, with a focus on the usage of agro-chemicals and its impact on farm profitability and farmer's health. Their work is based on panel data from two household surveys in the Mekong Delta, the main rice-growing area of Vietnam. The chapter concludes that policy changes towards land property rights' security and market liberalization of agro-chemicals (and rice itself) have offered incentives for farm households to increase their land productivity and family income via intensification. Although more and more households implement integrated pest management (IPM) methods, the increased cropping intensity still outweighs the positive effects of diminishing use of fertilizers and pesticides per hectare. There is also evidence that this form of intensification has severe health costs for the farmers, due to the toxicity of agro-chemicals that are used.

Zhu and Li, in Chapter 13, analyze investment in agricultural research and the use of agro-chemicals in grain production in China. The excessive application of fertilizer and pesticides in grain production in China poses a threat to the environmental sustainability of agricultural production. One of the major reasons of farmers' increasing the use of these chemical inputs is to achieve higher yield, especially so in the absence of adequate public inputs, such as rural development infrastructure and technology. By using provincial data for the past 20 years, the authors examine the effects of public investment in agricultural research on the reduction of farmers' private material inputs of major grain crops in China. It is concluded that public investment in agricultural research and development (R&D) have contributed to reducing private cost, and that the impact on wheat is larger than on rice and, to a lesser extent, on corn. However, compared to the total use of fertilizer and pesticides, the impact of R&D investment is marginal.

The third part of the volume deals with "Agriculture and Sustainable Land Management" issues. In Chapter 14, Van Keulen provides a fundamental discussion on issues of sustainability in East Asian rice cultivation. The chapter reviews the available evidence on the environmental impact of the Green Revolution in lowland rice production, with a focus on water use, nutrient supply (soil organic matter, nitrogen, and other nutrients) and the occurrence of weeds and pests and diseases. Next, the impact of these environmental changes on the sustainability of rice cropping systems is discussed. Van Keulen presents a number of options for sustaining soil quality, including green manure planting, crop diversification and site-specific nutrient management, and discusses opportunities for short-term measures to promote environmental sustainability as well as some critical management issues in transforming lowland rice production.

Van den Berg, Wang, and Roetter, in Chapter 15, take the discussion on sustainability a step further by analyzing "agricultural technology and nitrogen pollution in Eastern China." The chapter looks at the potential role of new agro-technologies in decreasing agricultural pollution for *Pujiang*, a county in the intensively cultivated coastal zone of China where rice and vegetables are important cropping systems. The adopted case-study approach is forward looking: it allows assessing the potential of on-the-shelf technologies and feasible new technological developments. The approach is exemplified for newly developed rice technologies that have shown a large potential of reducing nitrogen pollution and for future technologies for rice and vegetables. The results show that substantial reductions in nitrogen pollution are feasible for both types of crops.

Finally, in Chapter 16, Lu, Li, and Tan, provide a scenario and trend analysis of the developments in China's farmland acreage. Arable land is perceived as the most important resource for China to produce the food for its huge population. In recent years, China's limited farmland has been threatened by the rapid expansion of nonagricultural land use as driven by economic growth and urbanization. The authors review China's land use changes in the last 50 years, and explore the trend in farmland for the next 50 years using a scenario analysis approach. The results indicate a decreasing trend of the farmland area, particularly in the coastal zone. Farmland conversion into "built-up" land (for construction purposes) can be reduced by an improvement of land use efficiency. However, the national aim to maintain a balance in farmland losses and gains may not be realistic for most provinces, because the limited cultivable land cannot compensate the farmland losses.

The analysis of household responses to economic policies and changing institutions, and their implications for agricultural production and sustainable resource use in East-Asian transition economies, is a relatively new research field. Much of the research presented in this volume was carried out with financial support from the Dutch Government (SAIL-SPP program) and the European Union (INCO-DC program). We very much appreciate there critical support for initiating research in this field. Preliminary versions of the contributions in this volume were presented at an international seminar on "Economic Transition and Sustainable Agricultural Development in East Asia" that was held in Nanjing, China from 20–22 October, 2003. During that seminar, a group of Chinese, Vietnamese and international researchers working in this field gathered for the first time to discuss mutual research efforts, detailed analysis of survey-based data, and their policy implications. The contributions in this volume, which were subsequently revised and updated, reflect the rapid progress that is being made in this important research field.

The Hague/Beijing/Nanjing, June 2006

1

Transition, Economic Policy, and Institutional Reforms in China and Vietnam

Impact on Sustainable Land Use

Nico Heerink, Max Spoor, and Futian Qu

During the past two decades of transition in China and Vietnam, economic growth has been spectacular, agricultural exports boomed and rural poverty strongly declined. However, beyond this apparent success story of transition, attention should be given to the environmental impact of the growth model both countries have followed in their rather successful transition from planned to market economies. Taking into account the large difference in size of their economies, this process has been comparable in terms of length, sequencing, and profoundness of reforms, while resource degradation caused by (shifts in) agricultural activities is also substantial in both countries. In this introductory chapter a detailed analysis will be provided of the initial conditions of "pre-reform" China and Vietnam, and the consequences of economic and institutional reforms for (un)sustainable land use in both countries.

In the first section, an overview of main macroeconomic and agricultural policy reforms and institutional changes in both countries since the late 1970s is presented, with a discussion of similarities and differences in the process of economic transition. The focus in this part is on changes in crop-mix and land use and on opening of agricultural markets. In the second section, a partial analytical framework is developed to help in understanding the impact of economic (and institutional) reforms on sustainable land use, focusing on shifts in sets of relative prices and their impact, in a context of changing institutions and markets. Household decision-making, regarding crop-mix, input use and investment, and its consequences for land quality have a central position in this framework. In the third section, a brief overview of available empirical studies is presented that examine the impact of economic and/or institutional reforms on sustainable land use in China and Vietnam. The final section uses the analytical framework and the overview of available empirical studies to draw initial conclusions and suggest avenues for further research.

INITIAL CONDITIONS AND IMPACT OF AGRARIAN REFORMS

China: Cracks in the "Iron Rice Bowl"

Market and Price Adjustments

Replacing the collective agricultural production system by the household responsibility system (HRS) in the late 1970s/early 1980s formed the start of a series of significant economic reforms in China that brought profound change in the rural economic system. The reforms were characterized by a relaxation of direct planning control, a movement toward greater decentralization of decision-making and an increased reliance on the market to allocate resources. The introduction of the HRS also implied a major upheaval in the land tenure system. While the government maintained the ownership of collective land, rural households were given private land use rights. Farmers still had to deliver compulsory procurement quota at fixed prices for a number of strategic crops, particularly grains. Quota and surplus production was to be sold to the state at fixed prices, which were gradually increased over time, with higher prices paid for the "above-quota" surplus. For other crops, agricultural producers were free to make planting decisions and were allowed to sell (a part or all of) their output in noncontrolled markets. By 1985, the first phase of the agricultural reform was completed (Chen 2000).

The second phase of agricultural reform, which started in the mid-1980s, aimed at a gradual liberalization of grain prices and marketing. Compulsory sales of above-quota production were abolished and replaced by the "contracted purchase system." Remaining production could freely be sold. Contrary to expectations, the policy changes were followed by a dramatic decline in grain production in the second half of the 1980s. To stimulate farmers to increase productivity and sell to the government, contract prices were raised over time and input price increases were restricted (Findlay and Chen 1999). As a result, grain production recovered to new high levels in the beginning of the 1990s. An attempt to further reform was made in 1993, when procurement quotas were reduced and in some regions fully eliminated. This in turn led to sharp food price increases by the end of 1993.

The third phase of agricultural reform, starting in 1994, was characterized by renewed government intervention, combined with decentralization, and market-oriented prices. To deal with the problem of food price inflation and the danger of food supply shortages, the state grain procurement system was reintroduced. Procurement prices were raised above world market prices in order to stimulate farmers to shift production from cotton and oil seed crops to grain, whereas the above-quota grain surplus could be disposed of at market prices, although these were usually lower (Fan and Chen 1999; Gale et al. 2005). In 1995 the so-called Governor's grain bag responsibility system was introduced, which intended to promote regional food security and stabilize food prices by making governors and regional governments responsible for balancing grain supply and demand in their own provinces (Huang 1997). The central government still controlled grain imports and exports and managed a central grain stock that could be released in case of natural disasters. The policy was successful in increasing grain output from 445 million tons in 1994 to 512 million tons in 1998 and in reducing short-term price fluctuations and lowering prices. However, this result was probably achieved at the expense in efficiency of resource allocation, diversification of agricultural production, and farmers' incomes (Findlay and Chen 1999).

A process to homogenize government procurement prices with market prices started in 1997 (Fan and Chen 1999). In June 1998 the "Grain Purchasing Regulations" were announced. These allowed only state-owned grain enterprises to purchase surplus grain in rural areas and prohibited private merchants and private enterprises from doing so. In addition, measures were taken to improve the efficiency and profitability of the state-owned grain purchasing and storage enterprises (Huang 2002). This policy helped to keep prices stable, but deterred efficiency improvements in grain marketing (Fan and Cohen 1999).

In early 2001, addressing the challenges posed by China's accession to the World Trade Organization (WTO), the Chinese government initiated a policy of strategic adjustment of agriculture and the rural economy (Du 2002; Chen 2004). Its main purpose was to shift the focus of China's agriculture away from quantity increases towards quality improvements, and to stimulate labor productivity in agriculture. The main foci of this policy were: optimization of product varieties and improvement of product quality, development of processing industries, adjustment of regional production patterns based on regional comparative advantages, promotion of non-farm employment and labor migration, and enhancing the rural environment by converting grain land into forests and meadows. Administrated prices for grain procurement were eliminated in many parts of the country. Grain prices are mainly set in free markets, with government procurement prices following market prices (Gale et al. 2005).

From 1998 to 2003, total grain production in China declined from 512 to 431 million tons while the income gap between urban and rural households continued to widen. To deal with this undesirable situation, at the beginning of 2004 the Chinese government replaced its centuries-old policy of taxing rural households by a policy aimed at stimulating rural incomes. This marked the start of a new phase of agricultural policy (Gale et al. 2005; OECD 2005). Under this policy—which is more in line with WTO's regulations than the costly policy of subsidizing grain procurement, storage, and exports—state pricing and procurement have been abolished, except for tobacco. Agricultural taxes have been cut, and will be fully abolished throughout China in the course of 2006.

To stimulate grain production, farmers growing grains receive a small direct income subsidy based on the land area cultivated with rice, wheat, or maize. In addition, modern inputs are subsidized and public investments in agriculture, rural infrastructure and rural services are raised significantly. New regulations were adopted in 2004 to reduce the dominant role of government-sponsored enterprises in domestic grain trade. However, they also stipulate that government departments still remain responsible for ensuring the balance between grain supply and demand, similar to the "Governor's grain bag responsibility system" which was mentioned above (Gale et al. 2005)

Input Markets and Prices

A number of agricultural input markets and prices had been liberalized since the 1980s, but the reform of the fertilizer market had only been partial (Huang 1997). Attempts to lift fertilizer price controls at the retail level led to rapid price increases. To deal with this problem, in 1995 a provincial responsibility system aimed at meeting excess input demand and stabilizing prices was implemented, parallel to the above mentioned regional food self-sufficiency policy.

In addition, measures were taken to control fertilizer imports and to limit private enterprise involvement in domestic fertilizer distribution. China is the world's largest producer of fertilizers, but still needs to import about a quarter of its domestic consumption. In particular, domestic potassium production is limited, with imports accounting for 90–98 percent of consumption. Comparisons of international fertilizer prices with domestic prices at the retail level indicate that China again protected its domestic fertilizer industries since 1993 (Huang 1997; Wu 2000). In the course of 2005, the Chinese government imposed an export tax and abolished the added value tax on urea. Moreover, it introduced a chemical fertilizers' reserve system at provincial and national level in order to ensure domestic market supply and price stability of this strategic commodity. Since 2004, high-quality seeds and farm machinery are subsidized by the Chinese government as part of the new policy to stimulate rural incomes (Gale et al. 2005).

Land Tenure Reform

The introduction of the HRS implied a major restructuring of the land tenure system. Whereas the state maintained the ownership of "collective" land, agricultural land within each production team was distributed among the households according to the number of household members and workers, with land use contracts that initially had a length of not more than fifteen years. Nevertheless, (partial) redistribution of land frequently occurred within villages to correct for demographic changes.

In the second half of the 1990s, in order to stimulate investments in agricultural production, land use contracts were extended to a maximum period of thirty years. In addition, rural households obtained the right to sell their use rights (Huang 1997). Land rental markets were gradually introduced in different parts of China, but still had a minor impact. According to the Agricultural Census conducted in early 1997, only 5 percent of the rural households rented land in 1996, while the size of the rented land accounts for less than 3 percent of the total cultivated land (Diao et al. 2000). Scattered evidence presented in recent studies by Deininger and Jin (2002), Kung (2002), and Feng et al. (2004), however, points to much larger household participation in rural land rental markets in recent years.

Despite the HRS and its recent adjustments, farmers' land use rights are frequently under attack: Contract periods are shortened at will, contracted farmland is taken back or adjusted, contract fees are increased, and freedom of farmland management is disregarded. The lower-level of government (village committee and township government), which are responsible for protecting land resources are also to blame for this. Farmers often face ambiguous rights and duties and are powerless when their land use rights and interests are invaded. Similarly, agricultural production scale is very small, farmland is highly fragmented, and it is very difficult to effectively transfer farmland. This has become a major "persistent disease" restricting agricultural development, limiting effective allocation of agricultural resources, and restricting the modernization process in agriculture.

In order to protect farmers' legal rights and promote the development of the land transfer market, China accelerated the process of land tenure reform and farmland legislation, and implemented the "Rural Land Contract Law" (RLCL) on March 1, 2003. The core of this law is to endow farmers with longer and guaranteed land use rights and to confirm the legitimacy of the land transfer market. The law stipulates the contractors' rights and obligations, which include the right of land use, usufruct, product choice and management, and land transfer.

They are to receive compensation when their farmland is confiscated. Land use rights can be transferred through subcontracts, lease, exchange, transfer, and inheritance. In this law, contractors of land have the right to decide on land transfers, and the form in which they are realized. When land use rights are exchanged or transferred, it is necessary to register the change of ownership, while it is not necessary to register for the subcontracting or subleasing of land use rights. Lower-level governments cannot take back and adjust land during the contracted period.

In addition, the RLCL contains regulations to promote sustainable use of agricultural resources and environmental protection. Farmland contractors have the responsibility to properly protect and use this land and avoid permanent damage to it by unsustainable farm practices. Contractors of wasteland (barren mountains, canals, hills and bottomlands) should obey related laws and regulations and avoid soil erosion and protect the environment (Research Group MoA 2003). Lower-levels of government have the responsibility to uphold and implement this new law.

China's External Agricultural Trade

Macroeconomic and foreign trade policies are another major factor that determines the development of the agricultural sector. Since the start of the economic reforms, China's foreign trade regime has gradually changed from a highly centralized, plan-based, import-substitution regime to a more decentralized, market-based, export-promotion regime. Due to their strategic importance, however, trade in food grains, textiles, fiber, and chemical fertilizers still remain largely monopolized by the state. China's average import tariff was lowered from 47 percent in 1991 to 17 percent in 1997, while on agricultural products this was still 24 percent in 1997. Huang (2002) concludes that China's agricultural sector has been relatively less open to international trade than the rest of its economy, although China already made significant steps in "opening up" to international trade before it finally became a member of the WTO in 2001.

Historically, export incentives were reduced by the overvaluation of China's currency, the *yuan*. Since the 1980s, the degree and extent of this distortion decreased as a result of exchange rate adjustments. Between 1980 and 1994 the exchange rate changed from 1.50 *yuan* per USD to 8.62 *yuan* per USD, remaining at a level of around 8.30 *yuan* per USD until the summer of 2005, when the *yuan* was appreciated by 2 percent. Public control of foreign exchange has been relaxed, and by December 1996 the *yuan* became convertible on the current account. Since 1994, the system of managed floating is used to maintain the exchange rate at its current level. By increasing export competitiveness, falling exchange rates have made an important contribution to China's impressive export growth and spectacular economic performance (Huang 2002).

China's increasing participation in the world agricultural market is mainly in the export market, with the share of exports in agricultural GDP increasing from 4.6 percent in 1980 to 9.1 percent in 1997 (and the import ratio increasing from 5.6 percent to 5.8 percent during that period).[1] Figure 1.1 shows the composition of agricultural exports over the period 1980–1997, making a distinction between land-intensive products (mainly grains, oil seeds, cotton), labor-intensive products (live animals, dairy products, coffee, tea), and labor- and capital-intensive products (meat, sugar, cocoa). Huang and Chen (1999) inferred from these data that the shares of labor-intensive and, particularly, labor/capital-intensive products in agricultural exports, have increased since 1985, at the expense of land-intensive products. They conclude that the pattern of China's agricultural trade since 1985, particularly exports, has moved increasingly in line with China's resource endowments and comparative advantage (figure 1.1).[2]

Using the data available for specific products from FAOSTAT (2005), and analyzing time series between 1980 and 2003, similar tendencies can be spotted. Since the late 1990s, with the continued high growth rates that the country has managed to keep, the trends have changed somewhat. Domestic demand for labor/capital-intensive products, such as meat and cotton, has increased substantially, leading to fewer exports and more imports. The main expansion in terms of agricultural exports can now be found in fruit and vegetables (see figure 1.2).

Imports of oil seeds, soybeans, cotton lint (and since the early 2000s also cereals) grew strongly, to satisfy domestic demand, and overcome the shift in crop-mix toward less land-intensive production.

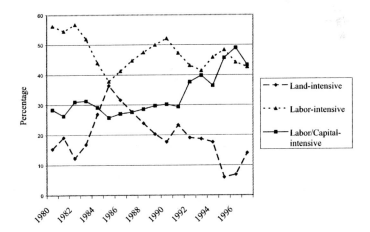

Figure 1.1.   China's Agricultural Exports (1980–1997)

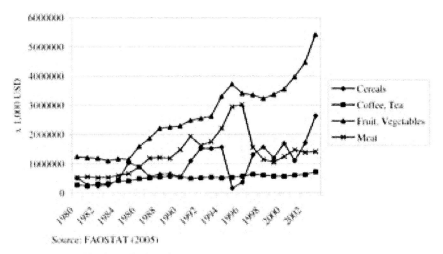

Source: FAOSTAT (2005)

Figure 1.2.   China's Agricultural Exports (1980–2003)

In December 2001, China became a member of the WTO. The most important commitments made with respect to agriculture were (see Lin 2000; Du 2002; Yu and Frandsen 2002; Anderson et al. 2004; Huang et al. 2004):

- Elimination of nontariff measures and adoption of a tariff-rate quota (TRQ) system for imports of grain and other strategic products, with low tariffs for within-quota imports and high tariffs for imports outside the quota.
- Reduction of import tariffs and a gradual increase in TRQ volumes; after a phase-in period; within-quota tariffs will range between 1 percent (for grains and cotton) and 15 percent by 2007.
- Elimination of export subsidies on farm products; rebates of value-added tax, however, can still be given to exporters.
- Facilitate competition from private firms in imports and exports of farm products (except tobacco), at least within the tariff-rate quotas.
- Limit nondistorting domestic support for farmers to 8.5 percent of the value of production (which is the agreed level of the Aggregate Measurement of Support, AMS for China).

Because the agricultural sector has been least involved in international trade and is considered relatively inefficient, accession into the WTO is often believed to cause major changes in this sector. Several studies have argued, however, that the impact of WTO accession on agricultural production and rural welfare is not as dramatic as is often feared. Lin (2000) argues that even if the TRQ on grain would be fully used, the resulting grain imports will be less than 5 percent of China's grain production in the 1990s. The impact on domestic grain prices (and farm incomes) is therefore limited. Tariff reduction on other agricultural products mainly relates to labor-intensive products, in which China has a comparative advantage. China's net exports of these products will not be hurt, which was confirmed in recent years. Furthermore, the liberalization of the markets for agricultural exports and imports is expected to benefit China's producers and consumers, because the state monopoly is rather inefficient. Huang et al. (2000) argues that the changes in agriculture caused by WTO are more of a continuation of thirty years of reform rather than a radical policy change. Domestic price changes resulting from WTO accession have a limited impact on production, because agricultural markets are still underdeveloped in large parts of China, with buyers and sellers facing high transaction costs. Moreover, lower consumption prices and greater off-farm employment possibilities cushion negative effects on rural welfare. In the long-run, agriculture may profit from domestic infrastructure investments and other policies aimed at promoting food security, from falling trade barriers in China's trading partners, and from improved access to imported technologies.

Model simulations of the impact of WTO accession on the impact on agricultural prices and production, based on the Global Trade Analysis Project (GTAP) database, confirm that the impact is modest (Anderson et al. 2004). Price changes for farm products over the period 2002–2007 range from +0.1 percent (for cotton) to –2.8 percent (for oilseeds). Production of cotton (+16 percent) and, to a lesser extent, livestock and meat (+1 percent) increases, while grain production (–2 percent) and production of sugar (–6.5 percent) and oilseeds (–8 percent) declines. National self-sufficiency in farm products decreases slightly, particularly for feed grains (due to higher demand for livestock products) and cotton (due to increased textiles and clothing production). However, the magnitude of these effects is small if compared to the changes caused by normal economic growth.

With respect to income effects, the model simulations show that poverty may increase in remote rural areas and regions with little infrastructure needed to attract investments in textiles and clothing or other activities. Rural non-farm poverty, however, declines as a result of WTO accession due to higher off-farm wages for unskilled labor (Anderson et al. 2004).

Economic Transition in Vietnam: "A Dragon Emerges from the Clay"

The far-reaching economic reforms that were introduced in Vietnam in 1986, also known under the name *Đoi Moi* (reconstruction), were actually preceded by several partial reforms affecting the institutional framework of collectivized agriculture. Actually, it will be shown in this section that the legalization of subcontracting of land to families (rather than to work brigades) in the period 1979–1981, was comparable with the household responsibility system introduced in China at the same time. This parallel development is even more interesting to note in view of the fact that there was great tension between the two countries at that moment.[3] Agricultural production was stagnating within the Chinese and Vietnamese collective systems in nearly the same period, and institutional reforms were called for. The bureaucratic management system and the stifling cooperative setting, provided disincentives to rural households in terms of output performance. As in China, many Vietnamese peasant families saw the collective institutions as an interface between them and the state, in which the latter had ample opportunities to control surplus and transfer it to other sectors.

Apart from these comparable aspects, there were additional reasons for the Vietnamese government to reform. Since the end of the devastating war, and the reunification of North and South Vietnam in 1976, economic development had been problematic. The collective agricultural system in the North stagnated and the forced collectivization move that was tried in the South (particularly the large Mekong Delta) had failed. The collective agrarian system was disintegrating, and tax evasion, illegal renting of land, and "spontaneous privatization" at microlevel became the rule, rather than the exception. After the border war with China in 1979, and with the elimination of the substantial food-aid that used to be received from its neighbor, in 1980 a severe food shortage emerged, in particular in the north of Vietnam. Hence, there were compelling reasons to introduce reforms in the agricultural sector (Spoor 1988: 107–8). In this section, the Vietnamese land reform and institutional changes that have taken place will be discussed in near chronological order. In that way the similarities and differences between the Chinese and Vietnamese cases can be visualized, focusing on the changing role of the agricultural households and the development of rural markets.

Land Reform in the DRV and South Vietnam

While in the wake of the *Doi Moi* reforms since 1986, a gradual but widespread de-collectivization of agriculture came about, Vietnam has known other types of land reforms. First of all there was a land reform very similar to the Chinese one of the 1950s. In the Vietnamese case this took place in the Northern (and partly Central) regions of the country. It started during the anti-French war and was completed in the mid-1950s within the territory of the Democratic Republic of Vietnam, or DRV (Spoor 1987). The land reform was directed against rich (and sometimes middle) peasants, although these were only categories of relative importance, and can be categorized as radical and often even violent in its implementation. By 1956, the redistribution of (the very scarce) agricultural land in the DRV had been completed, and land reform was followed by a forced collectivization campaign, to be largely completed in 1965, around the beginning of the first American bombing on the country.

The Vietnamese government had chosen a model that neither included very large scale units, such as the Chinese People's Communes, nor had it gone through enormous upheavals such as took place in China in that period. Actually, as Fforde and de Vylder (1996) pointed out, the Vietnamese production cooperatives were mostly formed on the basis of the traditional administrative hamlet. Furthermore, when the land reform and collectivization had been completed, war broke out, and the collective units became first and foremost defense and social protection units in the rural areas. In South Vietnam, governed by a pro-American government, a "land to the tiller" land reform was introduced in the early 1970s, comparable to the Philippines. However, in the large and fertile Mekong Delta the medium and rich peasants remained the powerful groups, while there was a large mass of landless (or land-deficit) peasants who did not benefit from any redistributive reform. This difference is important to note, as the "land question" was crucial for the Vietnamese government in a reunified Vietnam, hence after 1975. The leadership first tried to transplant the Northern model to the South, which failed. As Nghiep (1994: 148) noted, by 1979–1980, not more than 6.9 percent of households in the Mekong Delta were registered as members of a cooperative, whereas in the Red River Delta this was nearly 100 percent. The failure of the collectivization in the Mekong Delta coincided with a stagnation of the food production in the North. This led to a generalized food crisis in the whole country, aggravated by the cutting of food aid from China in the aftermath of the border war of early 1979. In response, by the early 1980s, the first steps were taken toward the contemporary set of economic policies and institutional reforms, which were later formalized in the *Doi Moi* reforms of 1986.

The first real reforms date indeed much earlier than *Doi Moi*. During the period 1979–1980, at the same time as China started with the household responsibility system, land within the cooperatives was rented out to individual households. Production was therefore, albeit still illegally, shifting into private hands. Even earlier, in the early 1970s, experiments have been reported in cooperatives, in which land was leased or rented to households. However, because of the fear that some farmers would enrich themselves these experiments were repressed. Forced by the economic crisis and the food shortages of 1979–1980, the government started to legalize a practice that had been implemented illegally. The contracting of land to individual households, made possible by the Decree #100 of January 1981, also gave them the opportunity to sell their "above-procurement quota." This caused an initial boost in rice production, similarly to China. While in 1979 paddy production was in total 11.4 million tons, in 1982 this became 14.4 million tons. It resolved the food deficit that had been structural, and was for years supplemented by food aid from the USSR and China. Nevertheless, as Fforde and de Vylder (1996) correctly noted, this was a short-run effect, as much of the administrative market structure had remained untouched. Furthermore, land tax was very high (40 percent of average production per hectare), which actually strengthened surplus transfer and dampened the long-term effects of the reform of 1981. It can be observed that on a per capita basis, production stagnated afterward, and a second food crisis emerged, which was triggered by extremely bad weather in 1987.

At first there was still a temporary retreat from reform after 1982, in particular concerning market liberalization. As the agricultural sector grew rapidly, and some farm households were doing quite well, the government (and particularly the Vietnamese Communist Party, the VCP) became worried about growing income disparity and "emerging capitalism" in the countryside. It would only be after a disastrous money reform in September 1985, when a renewed and now more comprehensive decision was made to reform the Vietnamese economy. However, this reform package was largely focused on macroeconomic stabilization, liberalization of prices, and diminishing of the omnipotent state influence on the economy. In 1987, again in times of severe food crisis, a decision was made which abolished the existing policy of district food self-sufficiency, which had been at the root of the forced rice cultivation, to the detriment of an alternative crop-mix. This policy decision also created a more nationwide market for rice in Vietnam (Men 2003). However, the most important step in agricultural reform, namely the dismantling of the cooperative structures, making the individual household the cornerstone of agricultural development, was made possible with Resolution #10 of the VCP's Politburo regarding the "Renewal of Agricultural Economic Management" (Men 2003). This decree focused on improving economic incentives to households, by reducing the "coercive impact of local collectives by improving their democratic content," and substantially diminishing the agricultural bureaucracy at district level. This combination of policy measures "finally destroyed the rural basis of the command economy" (Fforde and de Vylder 1996: 157).

Market and Price Adjustments

Contracts between cooperatives and individual households had to be such that the share of output left for the household to dispose of freely was no less than 40 percent. This had also been the case in the partial reform of 1981, but seems to have never been implemented. Furthermore, land allocated to households was minimally fixed for fifteen years, and the contract terms for five years. Land tax, which had been a major point of contention, was reduced to only 10 percent of "normal output/hectare." After these problematic years, during the following season of 1989 peasants indeed responded to the new opportunities and opening of markets, and the second bumper crop (19.0 million tons) of the decade was produced, providing the country for the first time in contemporary history with a substantial rice surplus to export. Finally, in 1993, a new Land Law was introduced (Barker 1994; Do and Iyer 2002), which formalized land use rights (or usufruct rights), lengthened the period of land lease, while at the same time the state procurement system was practically abolished. With the issuing of Land Use Certificates (LUC), which "gave households the power to exchange, transfer, lease, inherit and mortgage their land-use rights," practically all resource-use decisions was transferred to the individual household. The implementation of the new Land Law also meant the completion of the de-collectivization of Vietnamese agriculture, although in the Red River Delta remnants of the cooperative structures remained in existence. These still performed certain tasks of collective interest, such as the maintenance of the water-management system at microlevel. In the southern regions of Vietnam, by the late 1980s no residues were left from the ill-fated collectivization wave of the late 1970s. By the end of the decade, some modifications to the Land Law specified that land could be subleased (1998), and land could be used as equity capital in joint ventures with foreign companies. In 2001 land registration and rights of urban land were simplified (Do and Iyer 2002: 6). Rice production rose steadily during the 1990s, and by 2000 the output had risen to 32.5 million tons.

As already touched upon, Vietnam is very short in arable land, and depends for food supplies and other agricultural commodities largely on two fertile river deltas, from the Red River in the North and the Mekong in the South, where most agricultural production, in particular paddy cultivation, takes place. The population pressure in the Red River Delta is enormous, and arable land availability currently is less than 0.10 ha/capita. There are very intensive production systems in the Delta, with most often two rice crops and one dry one (such as peanuts, sweet potatoes, or sesame seed) being grown on the same plot. This intensive system could (and can) only exist with a well-functioning integral water management system, with an important role for local-level organizations and institutions that implement protective and conservation measures (such as weeding the canals, taking care of the dikes, etc.). The other regions of North and Central Vietnam are quite different, with a combination of forestry areas, upland rice systems, agro-forestry, and other cash crops, such as tea, coffee, and sugar cane. The Southern Mekong River Delta is the largest delta of the two, known as the "rice basket" of Southeast Asia, and specifically for the rich soils fed by fertile silt, but also threatened by enormous floods, which regularly destroy crops and displace many thousands of rural dwellers. Arable land that can be used for agriculture is not more than 17.7 percent of the total land area.

This causes continuous pressure on abundant forestry resources, which have been diminishing at an impressive rate of 2.5 percent per annum over the past two decades.

Vietnam's External Agricultural Trade

With the gradual opening of export and import markets in Vietnam, although with remaining important influence of state companies and systems of restricted licensing for the private sector, external trade has boomed since the 1980s. There is also a change in composition of external trade, overall and in the sector of agricultural exports. In figure 1.3, it can be noted that there were several tendencies at first in terms of the export-mix. By the end of the 1980s, in response to the first waves of reforms, rice production was boosted and exports became a very important part of agricultural (and total) exports for the country. Since the mid-1990s, in line with the development that was noted for Chinese agricultural exports, labor and labor-capital intensive agricultural commodities substantially increased their share in exports, in particular coffee and tea, fruit and vegetables, and also natural rubber.

The difference with China is that the changes that took place in the late 1990s, with a move toward larger imports of land-intensive products, and the growing importance of domestic demand (because of agro-industrial growth and increased incomes), are not yet visible in Vietnam. The latter country is clearly becoming dependent on labor intensive exports, searching to use its comparative advantage, but rice remains important.[4]

At the time of this writing Vietnam had not completed the accession negotiations for the WTO, although initially they had planned to be ready by the end of the Doha Round (December 2005). It is yet unclear under which

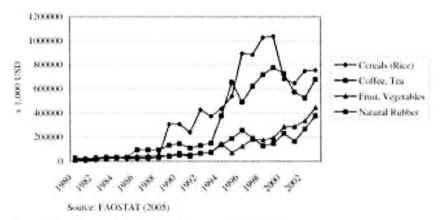

Source: FAOSTAT (2005)

Figure 1.3.   Vietnam's Agricultural Exports (1980–2003)

conditions the country will become a member and how these will affect its crucial agricultural sector. Oxfam (2004) critically stated:

> Vietnam is particularly concerned to be able to use tariff rate quotas (TRQs) and Special Safeguards (SSGs) against import surges. With average holdings of 0.7 hectare, farmers are extremely vulnerable to price falls. Most Working Party members have asked Viet Nam not to apply TRQs and SSG, though Viet Nam's proposal to apply SSGs for only pork, beef, and poultry, and to apply TRQs for eight other products was much more modest than that of China. Those who did not ask Vietnam to remove TRQs and SSGs asked Vietnam to reduce tariffs.

The report wondered whether Vietnam would access the WTO on "pro-development terms," and not by coincidence entitled it *Extortion at the Gate*.

A PARTIAL ANALYTICAL FRAMEWORK: CHANGING RELATIVE PRICES AND SUSTAINABLE LAND USE

The transfer of agricultural decision-making to the household level, and the subsequent policy-induced changes in the economic environment of rural households in both China and Vietnam, not only resulted in impressive production growth rates and income increases for large shares of the rural population. They also had a major impact on the quality of the natural resources used in achieving these growth rates. These effects on the sustainability of resource use, however, are not easy to disentangle. In this section, we intend to develop a partial analytical framework that may be used in understanding the impact of past and anticipated future economic reforms on the sustainability of the major resource used in agricultural production, i.e., land.

The focus in this section is on four types of household decisions that are likely to affect the quality of the soil:

- between traded and non-traded crops
- between organic and chemical fertilizers
- on the cultivation of marginal land
- on investment in soil conservation measures

Economic and institutional policy changes have a major impact on these four categories of household decisions, by changing the (relative) prices of agricultural products and inputs, the availability of markets for output and inputs, the involved institutions and their efficiency, and the transaction costs involved in using such institutions and markets. Our analysis concentrates primarily on the impact of (relative) price changes as perceived by farm households. These so-called effective prices also comprise the transaction costs (costs of transport, obtaining information, monitoring) made by farmers in using markets for selling output or buying inputs.

Public investments in infrastructure and public maintenance expenditures evidently affect market prices as well as transaction costs of both agricultural products and inputs, whereas public expenditure on rural infrastructure, agricultural research, and extension affects the accessibility to modern agricultural techniques, and the prices paid for them by farm households. For reasons of simplicity, the impact of such public expenditures on (relative) price changes is disregarded in the discussion below. Neither is the availability or absence of markets included in the analysis, although in the empirical chapters that will follow some of these gaps are filled.

Crop Choice (Traded vs. Non-Traded Crops)

Liberalization of international trade in agricultural products, reduction of the import (and export) tariffs, and exchange rate adjustments change the prices of traded crops relative to non-traded crops. Positive effects on export crop prices may be partly offset, when domestic food market liberalization would result in higher prices for (non-traded) food crops. Furthermore, a perverse supply response may occur because of the mixture of risk aversion and profit maximization that is followed to protect their livelihood (Ellis 2000), in particular in a situation when markets are fragmented, work inefficiently and have high transaction costs.

When the relative price of tradable commodities increases, a switch is expected to take place among farm households away from subsistence crops toward commercial crops (and between different commercial crops, when their relative prices also change). Individual crops generally react more strongly to price changes than aggregate agricultural production as a whole, because it is easier to shift resources from one crop to another than it is to draw more resources into agriculture. It is however clear, that in the specific context of transition, fundamental changes take place in domestic agricultural markets (and their link with external ones). Therefore, the following analytical framework, focusing exclusively on relative prices, and abstracting from the institutional changes and differential access to markets, can only be seen as a partial instrument to analyze the impact of reforms on household decisions and consequences for sustainable land use. This is the main issue at hand in this volume.

The resulting changes in cropping patterns may have important consequences for soil quality, as depicted in figure 1.4. Some crops provide more protection against erosion than others. Tree crops, like coffee, tea, bamboo, oranges, and tobacco provide a continuous root structure and canopy cover, and are more suitable for sloping terrain. Other commercial crops such as cotton and groundnuts, and nonirrigated staple food crops like maize, soybeans, sweet potatoes, and cassava, leave the soil more susceptible to erosion.

Soil nutrient and organic matter balances can also be affected by changes in crop choice. Some crops (e.g., maize, rice, banana, cassava) remove more nutrients from the soil than others (e.g., coffee, cocoa), and therefore need higher inputs of fertilizers to maintain nutrient balances (Rehm and Espig 1991). In general, nutrient balances for areas under cultivation of commercial export crops are better maintained compared to local staple food crops. The amounts of crop residues that remain after harvesting also differ largely from crop to crop. The decision whether or not to leave these residues on the fields or to use them for other purposes has important implications for both soil nutrient and organic matter balances. Pesticides that are used to annihilate the plagues that easily emerge in monocultures (which is often the case with commercial export crops) may also eliminate useful organisms that contribute to a good physical and organic soil structure.

Fertilizer Use

As in many other countries, a major purpose of the economic reforms in China in the 1990s and in Vietnam was to reduce the "bias against agriculture" that resulted from direct price discrimination (particularly from cheap food policies) as well as indirect price discrimination (overvalued exchange rate, protection of manufacturing industries). The higher (relative) prices of agricultural products as an outcome of economic reforms are expected to stimulate agricultural production either through the use of more inputs and new technologies (intensification) or through the cultivation of more land (extensification), or both. Hence, a shift is expected to occur from "low value, high volume" (mostly land-intensive) crops, towards "high value, low volume" (labor/capital-intensive) ones.

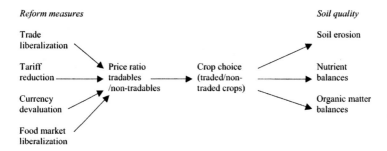

Figure 1.4. Crop Choice and Soil Quality Changes

Due to the very high land pressure in both China and Vietnam, most production growth in these countries comes from increased use of chemical fertilizers and other inputs (see also tables 1.1 and 1.2). The price ratio of chemical fertilizers to crop prices is therefore of great importance for the quantity of fertilizers used in crop production. Economic policies may influence this ratio in a number of ways. Cuts in fertilizer subsidies to farmers, or in subsidies to fertilizer industries, and privatization of fertilizer supply influence the prices that farmers pay to buy these inputs. For imported types of fertilizers, an adjustment of the exchange rate will generally influence the domestic prices for these types as well. For traded crops, however, the latter effect may be offset by the corresponding price change in domestic crop prices after exchange rate adjustment. In addition, domestic crop price levels are affected by food market liberalization, export liberalization, and/or reduction of export taxation.

Increases in the fertilizer-crop price ratio will commonly reduce the use of chemical fertilizers in agricultural production. In areas with negative soil nutrient balances, this could further reduce the soil nutrient stock and hence aggravate the problem of soil nutrient depletion. However, in regions characterized by high levels of chemical fertilizer use, such as China, Vietnam, and several other countries in Asia, an increase in the fertilizer-crop price ratio may help to reduce the environmental problems related to overuse of fertilizers and other chemical inputs (e.g., water pollution, soil pollution). In addition, it may induce a change in fertilizer application from chemical fertilizers toward organic fertilizers (application of manure, green manure planting, and the use of crop residues). This may reduce problems associated with low organic matter content of soils in these areas. See figure 1.5. Furthermore,

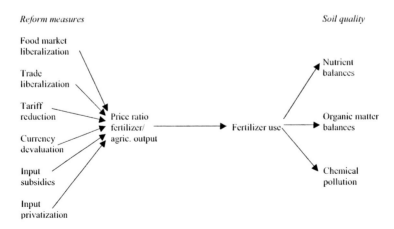

Figure 1.5.   Fertilizer Use and Soil Quality Changes

Table 1.1.   Land Use Intensification in China (1980–2003)

	1980	1985	1990	1995	2000	2003
Arable Land (x 1,000 Ha)	96,924	120,805	123,678	124,055	137,124	142,615
Perennial Crops (x 1,000 Ha)	3,295	5,091	7,719	10,645	11,533	12,235
Rice (x 1,000 Ha)	34,482	32,634	33,519	31,107	30,301	26,780
Rice Area/Arable Land	0.36	0.27	0.27	0.25	0.22	0.19
Chemical Fertilizer (x 1,000 Tons)	15,335	17,279	27,274	35,580	34,218	39,605
Chem. Fertilizer/Arable Land (Kg/Ha)	158.2	143.0	220.5	286.8	249.5	277.7

Source: FAOSTAT (2005)

Table 1.2.   Land Use Intensification in Vietnam (1980–2003)

	1980	1985	1990	1995	2000	2003
Arable Land (x 1,000 Ha)	5,940	5,616	5,339	5,403	6,200	6,680
Perennial Crops (x 1,000 Ha)	630	805	1,045	1,348	1,938	2,300
Rice (x 1,000 Ha)	5,600	5,718	6,043	6,766	7,666	7,452
Rice Area/Arable Land	0.94	1.02	1.13	1.25	1.24	1.12
Chemical Fertilizer (x 1,000 Tons)	155.2	469.2	560.3	1,223.7	2,267.0	1,975.2
Chem. Fertilizer/Arable Land (Kg/Ha)	26.1	83.5	104.9	226.5	365.6	295.7

Source: FAOSTAT (2005)

increased pesticide prices are expected to reduce their use, and promote the embracing of integrated plague management (IPM) by farmers.

It should be noted that the net effect of changing input and output prices depends on the size of the marketed surplus. When farmers produce mainly for their own consumption, that is, because of small landholding sizes, high transaction costs or risk aversion, higher agricultural prices are unlikely to have much impact on the use of external inputs. The effect of price liberalization on farm household's real income is dependent on their net supply or demand position in the food market. Higher farm-gate prices will be favorable for net food-surplus households, but negative expenditure effects arise for net food-deficit households. Furthermore, although ignored for the purpose of this partial analytical framework, the efficiency of markets, the access to them, and accompanying institutions, are crucial in determining the farm household responses to relative price changes. It is well known that in case of "missing markets and institutions," farm households might make other decisions, which combine risk aversion and profit maximization, which may even lead to perverse supply responses.

Extensification

In regions where farm households have access to forestland or wasteland, part of the production increase may come from taking such ecologically fragile areas into cultivation. When no adequate countermeasures (terraces,

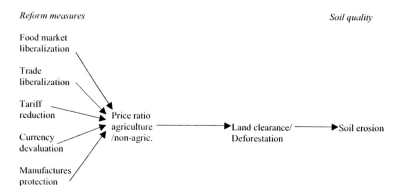

Reform measures   *Soil quality*

Figure 1.6.   Extensification and Soil Quality Changes

ridges, and so on) against soil degradation are taken, soil erosion may be
the result. (See figure 1.6).

Soil Conservation Investments

An important objective of removing price discrimination against agricultural products, besides raising farmers' incentives for increased land and
input use, is to improve prospects for different types of investments in level-
ling, terracing, draining, irrigating, and other forms of land improvement.
When prices are depressed, lower profitability restricts the use of labor and
other yield-increasing inputs and/or reduces the demand for farmland.
Because farmland cannot be massively shifted to other uses, land prices
will be lower than they would be without price control. As a result, returns
on investments in soil fertility and land conservation are depressed. An
increase in relative prices of agricultural products is therefore expected to
stimulate antierosion investments and investments in soil fertility (through
fertilizer, green manure, and crop residues). (See figure 1.7).

A few caveats should be taken into account. Firstly, measures aimed at
soil conservation and land improvement require labor and/or capital inputs. When labor and capital are scarce, conservation measures may not be
undertaken. An increase in the price ratio of agricultural products to nonagricultural products may influence the availability of labor and capital
by attracting more labor to the agricultural sector and by stimulating the
production of cash crops. On the other hand, the level of remittances from
household members employed in the nonagricultural sector is expected to
decline in the long run.

Secondly, security of land use rights is an important precondition for
farm households to consider long-term investments in land improvement.

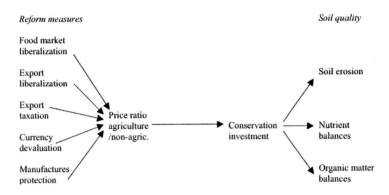

Figure 1.7. Conservation Investment and Soil Quality Changes

When land use rights are frequently redistributed or are ill defined (e.g., no rights of exchange or inheritance), higher prices will hardly enhance household investments in conservation projects, although this depends on how secure use-rights in practice are being seen.[5] Similar effects may occur when price increases are expected to be temporary.

Thirdly, the full benefits of investments in land improvement can only be reaped after several years. This holds especially for investments in soil fertility, the initial effects of which are usually limited. Moreover, farmers need time to learn new technologies and adapt them to local circumstances. Short time horizons, caused by poverty, may also prevent farmers from undertaking such investments.

TRANSITION AND SUSTAINABLE LAND USE: A RESEARCH REVIEW

The Case of China

Land in China consists to a large extent of mountains, deserts, and other land types that are unsuited for agricultural production. Out of the total land area of 960 million hectares, around 130 million hectares are currently used for agricultural production. Around 214 million rural households depend to varying degrees on this land for making a living, implying an average landholding size of 0.6 hectares per household, that is, 0.15 hectare per person (NACO & FASC 1999).

Due to its geographic features, the land available for cultivation is not only scarce but also vulnerable to degradation. The various land, market, and trade reforms implemented since the end of the 1970s have fundamentally changed the economic incentives and institutional setting facing rural households, thereby changing crop-mixes, input use, land use, and investment decisions, with marked affects on the sustainability of land use.

Data from the 2nd National Soil Census indicate that 45.4 million hectares of cultivated land suffer from water and wind erosion, or 34.3 percent of the total cultivated land area. Out of these, 10.5 million hectares (or 8.0 percent of the cultivated land), suffers from intensive erosion, while 34.9 million hectares (26.3 percent) shows signs of light to moderate erosion. The five regions affected most by erosion are the Loess Plateau (where 71 percent of the cultivated land suffers from erosion), the Northeastern hills, the Northern hills and plain, the Southwestern red soil area, and the hills surrounding the middle and lower branches of the *Yangtze* River. More than 90 percent of the eroded cultivated land is located in these five regions (Huang 2000; Zhu 2000). Huang (2000) reviewed official statistics on the areas affected by water erosion over the period 1973 to 1996. The data reveal that the area affected by water erosion remained roughly constant in the 1970s, increased by about 1.5 percent per year on average during the 1980s, and remained constant again during the 1990s.[6]

Soil erosion can have important negative consequences for agricultural productivity in eroded areas. It removes the topsoil layer from cultivated land, which tends to contain relatively large quantities of nutrients and organic matter. It reduces the capacity of the soil for holding water and air, and leads to more water run-off and hard soils. By making the topsoil thinner, it reduces the root depth and thereby the potentials for plant growth. Serious soil erosion can severely damage the ecological environment in affected areas, with summers becoming hot and dry and winters bleak and chilly, natural disasters (droughts, flooding) increasing in frequency, severe shortages of natural fuel, forage, and fertilizer sources (Shi et al. 2000; Tan et al. 2000; Zhu 2000).

To some extent, the impact on agricultural production may be counterbalanced by the cultivation of fertile soil sediments elsewhere within or outside the same region. Not all eroded soil, however, ends up as relatively fertile agricultural land elsewhere. Much of the washed away soil ends up in rivers, lakes, and water reservoirs further down the watershed. An estimated 400 million tons of silt are deposited on the riverbed of the Yellow River every year, causing the riverbed to rise above the surrounding land in some regions. The *Yangtze* River carries an estimated 500 million tons of soil every year, causing the bed of *Poyang* Lake to rise by 2 mm per year, and increasing the danger of flooding (Shi et al. 2000; Zhu 2000). The chemical fertilizers and pesticides that are washed into the rivers and lakes together with the soil add to the pollution of the water. Irrigation systems have been partly or completely damaged by siltation and by reduced storage capacity of water reservoirs (Rozelle et al. 1997; Tan et al. 2000).

Natural factors (geology, climate, parent rock materials, and natural vegetation) are at the root of the soil erosion problem. The thick, finely granulated and undifferentiated soils of the Loess Plateau are highly susceptible to wind erosion, while the shallow and acid clay soils in the Red Soils area are very vulnerable to water erosion. Human activity may aggravate or lessen soil erosion problems. Since 1949, China's population has more than doubled, forcing an expansion of the cultivated area into steep hillsides and other areas previously considered ill-suited for agricultural production.[7] Increasing domestic food demand and improved export opportunities, combined with regulations restraining rural-urban migration, have contributed to the problem, particularly in areas with relatively good market access. On the other hand, increased cash incomes from selling surplus production and from off-farm employment provide additional resources for taking soil conservation measures.

In the Red Soils area of China, farming expanded from the valleys into the easily erodible hills, harvesting trees and brush for fuel and timber. Once the natural protective cover is removed, high rainfall washes away the topsoil unless appropriate conservation measures are taken (Rozelle et al. 1997). The contracting out of jointly managed forestland to individual households during the early reform years caused a rapid increase in the rate of deforestation, thereby aggravating the soil erosion problem. On the other hand, expansion of fruit trees and other perennial crops (from 3.3 million hectares in 1980 to 12.2 million hectares in 2003; see table 1.1) has made a positive contribution to soil conservation. Large-scale government-induced soil conservation projects, such as the "Green for Green" program, are to a large extent responsible for this trend. Private planting of fruit trees is expected to increase with the rising demand for fruit consumption in the near future, provided the new Rural Land Contract Law is indeed able to provide farm households with stable long-term land use, transfer, product choice, and management rights.

Soil fertility decline is an important problem in many parts of China. One-third to one-half of the cultivated land is deficient in P, while one-fourth to one-third is deficient in K (Huang 2000: 4). Long-term trends in soil chemical properties, however, tell a mixed story (Lindert 1999). Examining available data in the 1930s, 1950s, and 1980s, Lindert concludes that organic matter and nitrogen content of soils have declined, whereas other soil nutrients and pH have improved in most regions. Evidence of changes over the period 1984–1997 for different zones in *Jiangsu* Province indicates that organic matter, N and P have increased while K has declined in all zones (Tan et al. 2000: 381–82).

Application of chemical fertilizers has a crucial impact on soil nutrient balances. Profitability of fertilizer application in grain production has increased greatly during the economic reforms. From 1990 to 1994, the fertilizer/grain price ratio declined about 40 percent mainly as a result of increasing grain prices, while it remained relatively stable since then (See chapter 7).[8] Together with the increased decision power of farm households since the start of the reforms, this explains the rapid increase in chemical fertilizer use[9] (table 1.1) and the generally favorable trends in soil fertility mentioned above. The removal of nutrients from the soil, particularly through erosion and crop harvesting, is too small to offset this positive trend. Shortages in K reported by Lindert (2000) and Tan et al. (2000) are the main exception. They are caused by the very limited domestic production and insufficient imports (see overview of policies above).

Recent trends in soil organic matter contents differ greatly between regions. Tan et al. (2000) report steadily rising organic matter contents in *Jiangsu* Province. Although the planting of green manure has declined, the practice of leaving crop residues on the field has extended since the mid-1980s, thereby greatly contributing to the positive soil organic matter balances. Recent research for Northeast *Jiangxi* Province, on the other hand, indicates that improved off-farm employment opportunities lead farmers to apply less green manure and animal manure (Shi 2007). The recent income support policy has a negligible effect on input use in this region, although manure application may slightly increase (and hence improve soil quality) in villages with good market access (Heerink et al. 2007).

Salinization is a major problem in irrigated agriculture in the flat, water-scarce North China Plain. It results from inadequate application of water or from substandard drainage. The area affected by salinity currently amounts to about 7.7 million hectares (5.9 percent of the cultivated land). It increased by about 1 percent per year during the period 1973–1985, and remained more or less constant since then (Rozelle et al. 1997; Huang 2000). Huang and Rozelle (1994) estimate the negative impact on China's national grain production during the reform period at 2 percent per year.

The scale and intensity of rangeland desertification have been on the rise over the past few years, with deserts expanding by an estimated 0.16 million ha per year (Ho 2000; 2001). Rangeland desertification occurs mainly in arid and semiarid areas, especially in areas exploited for both animal husbandry and agriculture (Ning 1997). Agricultural expansion into grasslands, combined with heavy winds and rains on recently ploughed or harvested land, is at the root of the problem. Grassland converted into arable land cannot easily be converted back into productive pastures. Some reforestation projects on fragile cultivated land, however, have been successful in reducing the dust pollution caused by desertification (Rozelle et al. 1997).

Overgrazing is another major factor in rangeland desertification. China's sheep population grew from 37 million in 1952 to 96 million in 1978 and over 110 million in the beginning of the 1990s, while the goat population grew from 16 million in 1949 to 35 million in 1978 and 40 million in the beginning of the 1990s (Rozelle et al. 1997: 233). The institutional setup of rural collectives during the first three decades of communism is generally held responsible for encouraging free riding by pastoralists with no consideration of long-term protection of the rangeland. The Pasture Contract Household Responsibility System introduced in 1985 shifted managerial responsibilities from the state and collectives to individual herdsmen and families. Ho (2000) argues that the high transaction costs for contracting, information and enforcement involved in individual management asks for community-based management of China's pastoral sector, which is still mainly subsistence-oriented. Revitalization of precollectivism institutions or establishment of voluntary herders' and sedimentary livestock farmers' cooperatives that promote internal compliance to regulations are more efficient ways to deal with overgrazing (Ho 2000: chapter 8).

According to Rozelle et al. (1997), China's environmental strategy in rural areas is based on three main tools: direct regulation, targeting and environmental campaigns, and state-mandated technological improvements; economic incentives are rarely used. Financial restraints have precluded the establishment of strong environment protection agencies in rural areas. The State Environmental Protection Agency (SEPA) decided not to become directly involved with most rural natural resource and environmental problems. Instead, the government has delegated responsibility for carrying out rural environmental protection policies and enforcing regulations to the ministries of agriculture and forestry, thereby creating conflicts within these ministries between those who want to exploit resources and those who want to preserve them. Each ministry has established its own environmental management hierarchy, from the central to the township level. Local officials with both environmental and production responsibilities, however, have a much higher incentive for reaching production targets since they receive almost no rewards or punishments for success or failure in executing environmental laws (Rozelle et al. 1997: 239–40). The continued reliance on command-and-control measures and the institutional problems in implementing these measures may explain to a large extent why problems of soil erosion, salinization, and grassland degradation continue to persist in rural China.

The Case of Vietnam

Whereas economic growth and poverty reduction, and more in particular agricultural growth in Vietnam since the late 1980s, has been very impressive, there are environmental limits to the growth model. With very limited arable land (less than 0.10 hectare per capita), a rapidly growing and very large population of around 82 million (in 2003), and often adverse climate conditions (cyclones and flooding), the country has to struggle against increased deterioration of its land resources, which in part is caused by the negative impact of agricultural activities on land and water resources.

The various waves of land and market reform transformed the agrarian structure, the systems of incentives and as a consequence the crop-mix, land use, output, and productivity. As was stated above, Vietnam, in a comparable (and even more rigorous) manner than China, had a regional and district-based food (read rice) self-sufficiency policy, which forced peasants to divert from more remunerative agro-forestry systems, especially in the upland areas. By reducing or eliminating the influence of the bureaucracy and leaving production decisions to individual households, while at the same time opening market outlets, initially there was an output spur of rice, starting with the bumper crop of 1989. By the end of the 1990s, however, it seems that rice production, after a spectacular growth, which made Vietnam one of the world's most important rice exporters, has stagnated. Rice prices at the end of the 1990s and beginning of the new century were very low. If possible, peasant households and commercial farmers moved to crops that were more remunerable, entered into mixed systems, such as shrimp farming in combination with paddy production, in particular in the river deltas and coastal mangrove areas, or intensified rice production by moving toward triple cropping systems, such as in the Mekong Delta (Dung 2003).[10]

Over the past two decades there has been a shift in land use, which expressed itself in the following ways (table 1.2). First, there was an increased intensification of rice production, often in combination with one dry crop, produced on the same field. As there was hardly a possibility to expand the acreage, in particular in the lowland areas, an increased land use ratio was the response to improved economic incentives. This has advantages in terms of land productivity, but as we saw in the previous section, also can cause land quality deterioration, amongst others through salinity. Second, there has been a strong expansion of the area sown with permanent crops (tea, coffee, rubber, citrus trees, and sugarcane). The latter expansion has taken place mostly in the upland areas, at the expense of existing forests. However, according to Vietnamese statistics, the forest cover has actually been expanding, through state-led programs.

Thirdly, the intensification process has caused a strong increase in chemical (and a decrease in the traditional green) fertilizers. In the crisis years 1976–1980 of Vietnamese agriculture, chemical fertilizer use first went down, from 0.27 million tons to 0.15 million tons (FAOSTAT 2003). Thereafter, chemical fertilizer use has increased rapidly, rising to 1.18 million tons in 1994 and 2.27 million tons in the year 2000. By then the use of fertilizers was high and in some cases there is even overuse, such as in the Mekong Delta, as was reported by Dung (2003). In 2003, there was again some decline reported, caused by price shifts (against rice) and the introduction of more sustainable input management, and IPM, but on the whole fertilizer consumption is very high for Southeast Asia, and even higher than in China (see table 1.1).

The shift toward permanent crops is also visible from the growth of the coffee sector in the 1990s. While before that time Vietnam's coffee exports made a negligible contribution to the world market, in 2003 its share was estimated at 8.8 percent. Again, with an initial contraction, coffee acreage went from 15,500 hectares in 1976 to 10,820 hectares in 1980. Thereafter a spectacular growth followed, from an acreage of 61,860 hectares in 1990 to 155,000 hectares in 1995 and 476,900 hectares in 2000 (FAOSTAT 2005). Because of the continuing trough in world market coffee prices in the past few years, a small decrease in the acreage planted with coffee can be noted, followed by a more recent growth. Rice production has been intensified, and with the land shortage Vietnam faces in its densely populated Red River and Mekong deltas, with arable land only slightly expanding, the rice harvest acreage grew faster, by using double and even triple cropping, as was mentioned above.[11]

There are very serious land degradation problems to be tackled in Vietnam, but before the influence of economic transition on environmental (and specifically land) degradation is discussed, a brief inventory of the main problems is presented. Overall, it is estimated that more than 30 percent of the land area of Vietnam is degraded, two-thirds of it even severely. Firstly, soil degradation is particularly due to: (i) continued "slash and burn" cultivation, practiced in upland agriculture (The 2001; for an extensive analysis on Central Vietnam); (ii) cutting of fuel wood for local consumption. Secondly, particularly in upland agricultural systems, there is a rapid expansion of monocropping with coffee, tea, rubber, and sometimes sugarcane. Thirdly, in the lowland areas, because of the longstanding "food = rice first" policy, irrigated rice has been produced for very long periods of time. As water management systems were suffering from underinvestment, water erosion and soil degradation was severe. Finally, the wetlands and mangrove areas have suffered from increased salinity, and more recently from the introduction of shrimp production, and run-off of nutrient pollution of dramatically increased fertilizer and pesticides use (Hue 2004).

Transition and institutional reforms have had very contradictory effects on the sustainability of agricultural production and the use of land in Vietnam. Firstly, with the economic reforms, in particular the introduction of the contracts between cooperatives and individual households since the late 1980s, and in combination with the diminishing of state procurement, the opportunity for households to sell the surplus above the established contracts improved. The decisions on crop-mix that were previously made by the planning system, and mostly by the district bureaucracy, were—since 1993—completely transferred to the household level. This process of de-collectivization has positively contributed to household welfare (Ravallion and van der Walle 2001), and it can be expected that this will translate itself in more investments in the farm. However, the shift from planned agricultural production through the production cooperatives (and some state farms), toward the individual rural household, has also led to the introduction of new crops. These are considered by peasants to be more remunerative than rice, but might well have negative externalities in terms of soil and water resources. The environmental costs of such changing cropping patterns have been estimated by The (2001) for the case of soil erosion in Central Vietnam, taking both on-site and off-site costs into account. However, these environmental costs of resource degradation are mostly an unknown variable, and certainly not a part of private cost-benefit calculations. Another example of environmental damage related to shifts in production are the shrimp production systems that have emerged quite widely in the *Mekong Delta*, contributing to increased salinity of neighboring rice fields, while promoted by the government because of export opportunities (Be and Dung 1999).

Second, market liberalization, both of input and output markets, in combination with the increased decision power of individual households after the process of de-collectivization and improved incomes, has contributed to a modernization of crop technology. Intensification of crop production, under conditions of extreme land scarcity, meant the introduction of more crops, which are produced on the same plot of land. This led to the need for more chemical fertilizer applications. Space limitations have also contributed to the diminished use of green fertilizers, which had always been popular in the Red River Delta. In table 1.2 the large increase in chemical fertilizer use was already pictured. Dung (2003) has shown that this in some cases means overuse, leading to substantial nutrient run-off in water resources and subsequent pollution of water and soil, health implications for farmers, and a non-efficient allocation of resources. On the other hand, the policy of introduction of IPM has been quite successful in the *Mekong Delta*.

Thirdly, the introduction of (more) secure property rights, through the issuing of LUC after the proclamation of the new Land Law of 1993, is a positive factor in increased land productivity (see also Pingali and Xuan 1992), and the increase in long-term investments (Do and Iyer 2002). The latter is also stimulated by the fact that land (with its increased length of the leasehold and secure land use rights), can be used as collateral for credit. An indication of the impact of the issuing of LUC since 1993, is the rapid growth in perennial crops such as coffee in the second half of the 1990s, which need substantial investment while harvest can be reaped only after some years. More secure property rights seem also to have a positive impact on irrigation investment, which had been long abandoned by individual households, as its positive impact did not sufficiently improve private economic benefit. A process of land consolidation has taken place during the 1990s, which improved the economics of scale, reduced transaction costs, and strengthened the above effects on the investment side and improving of technology. Whether increased long-term investment also translates into conservation techniques (soil improvement, avoiding wind and water erosion, pollution control, etc.), is not entirely clear. In his research on land use systems of upland farmers in Central Vietnam, the environmentally most sensitive and poorest area of the country, The (2001) showed that the choice for a particular land use system and crop-mix, whether this is upland rice, sugarcane, fruit tree-based agro-forestry, or forestry, depends on various factors, including policy-related ones. Secure property rights is one of them, but also technological and educational level of the farmer, the specific characteristics of the land, and institutional arrangements, such as access to services and credit, are important.

Fourthly, and following the last point on institutional reforms, the transition in Vietnam has substantially transformed the influence of the state on production decisions. Through a gradual process of deregulation, speeded up by the two most important reforms (1988 and 1993), the previous forced policy of national, regional, and even district food self-sufficiency, which had simply meant that rice was cultivated even when it was not economically sustainable, was eliminated. Since the reforms, individual households take decisions based on a mix of "risk aversion" and "profit maximization," the latter related to (observed) comparative advantage. However, such decision depends very much on the working of input and output markets, access to credit, crop insurance, and agricultural services, which are still relatively underdeveloped in Vietnam. Therefore, individual decisions of peasants might have environmental on-site and off-site costs, which are not observed in the short-run. The availability of credit, in particular for long-term investment, is a key factor to promote investments in land improvement and conservation techniques. Thus far not much empirical work has been done in this respect.

Finally, Vietnam has to cope with the problem of high population density, extreme scarcity of fertile soil, pressure on its forest resources, and the still existing vicious circle of poverty and resource degradation. Public policies will have to be directed to avoid damaging agricultural practices, such as the widespread "slash and burn" cultivation in the upland areas. Through improved agricultural services, the introduction of environmentally friendly crop protection can be promoted. The environmental limits to economic growth, in particular related to the deterioration of land and water resources, are becoming clear in Vietnam's transition to a market economy. Policy-oriented research to the interconnections between economic reform, property rights, the development of land, input, output, and financial markets, and the response of individual households to new economic opportunities, in view of the ongoing process of resource degradation, is needed. The necessity to a real implementation of plans, such as spelled out in the "Pillars of Development" document (World Bank/ADB/UNDP 2000) on improving environmental quality, is greater than ever. The Vietnamese miracle has been very much based on intensified resource use, in particular through the expansion of agricultural production. In that sense, the dragon indeed "emerged from the clay."

CONCLUSION

The transitional and institutional reforms in agriculture that were implemented in China and Vietnam since the 1980s show substantial similarities in many respects: decision-making on crop choice and management was transferred to individual households; land was subcontracted to households but remained collectively owned, with more secure land property rights being assigned in recent years; agricultural input and output markets were gradually liberalized; export and import markets were opened up, while important influences of state companies remained; state grain procurement quota were diminished over time, and regional food self-sufficiency policies were used to ensure sufficient grain supplies. However, there are also important differences. Due to the large size of domestic food demand in China relative to the size of world trade in food, Chinese policy makers continue to put a great emphasis on grain self-sufficiency. For similar reasons, China continues to intervene in the domestic production and trade of chemical fertilizers. The rapid economic growth in China since the start of the reforms has made it possible in recent years to abolish the agricultural tax and to introduce direct income subsidies to grain-producing farmers. Farmers in Vietnam enjoy a larger range of land property rights than Chinese farmers, possessing LUC, that provide more possibilities to obtain credit.

The reforms have led to rapid agricultural growth in both countries, particularly in the first few years after the introduction of private land use rights. Subsequent reforms in domestic markets and foreign trade continued to stimulate agricultural production and boost trade in both countries, with labor and labor-capital intensive agricultural commodities increasing their share in exports. In recent years, however, the growing domestic demand (because of agro-industrial growth and increased incomes) for labor/capital intensive products in China, such as meat and cotton, has led to declining exports and more imports. This trend is not (yet) visible in Vietnam.

The transitional and institutional reforms in agriculture and the resulting impressive growth rates in agricultural production and trade were accompanied in both countries by major changes in the quality of the major resource used in agricultural production, i.e., land. Large areas of land are affected by water erosion and (in China) wind erosion, causing a loss of agricultural productivity and increasing problems with siltation of rivers, reservoirs and irrigation systems, flooding, and (in China) sandstorms. Agricultural intensification has caused a strong increase in the use of chemical fertilizers and a decrease in the use of traditional green fertilizers in both countries, often resulting in improved soil fertility but also in water and soil pollution and in a reduction of soil organic matter content. Salinization is a major problem in irrigated agriculture in the flat, water-scarce North China Plain and in the wetlands and mangrove areas in Vietnam, while large parts of Northwest China are faced with intensified rangeland desertification caused by overgrazing and agricultural expansion into grasslands.

Policies addressing these land degradation problems have a strong top-down character in both countries. Direct regulation (e.g., prohibiting tree cutting), environmental campaigns and targeting (e.g., number of hectares planted with trees), and state-mandated technological improvements (e.g., use of biogas), are the main tools, with little attention for the use of economic incentives. Past and ongoing economic and institutional policy changes, however, have significant effects on household decision-making with respect to choice of crop mix, use of chemical and organic inputs, investments in soil conservation, and the cultivation of marginal land. Economic instruments such as taxes or subsidies and institutional policies promoting, e.g., security of land rights can therefore be effective tools in influencing household decision-making and thereby promoting sustainable land use. The effects of economic and institutional policies on the sustainability of land use, however, are not easy to disentangle. Policy-oriented research into the interconnections between economic reform, property rights, the development of input, output, land, labor and financial markets, and the response of individual households to new economic opportunities, in view of the ongoing process of resource degradation, is needed. The partial analytical framework of the impact of past and anticipated future economic reforms on sustainable land use that we present in this chapter may be helpful in this respect.

The rapid reforms in economic and institutional policy that took place in China and Vietnam and the great similarities in these reforms make these two countries very well suited for policy-oriented research into the responses of rural households and the consequences for the sustainability of land use. The insights gained from such an analysis can also offer important lessons for other countries that embarked on a transition from a centrally planned (or strongly state-intervened) economy to a more market-oriented one, in Asia (such as Cambodia, Mongolia) as well as other parts of the developing world.

NOTES

1. Official statistics do not contain information on agricultural GDP. This makes it unfortunately impossible to examine these trends for the years after 1997.

2. The trends in agricultural imports, however, are less consistent with China's comparative advantage. The share of land-intensive imports has declined, while the shares of labor- and labor/capital-intensive imports increased during most of the 1980s and 1990s (Huang and Chen 1999: table 4.13).

3. In February 1979 there was a short, but very violent and devastating border conflict between the two countries.

4. In fact, rice, as it is produced on very tiny plots of land, with a high labor intensity, and substantial external inputs, such as fertilizers and pesticides, can also be seen as a labor/capital-intensive product.

5. It seems that in China investments made in land also count as some form of security against redistribution (Peter Ho, 9 January 2006, at Seminar "Land and Markets beyond Transition," The Hague: Institute of Social Studies).

6. We do not take the sudden jumps in the observations for 1990 and 1996 into account, as they are probably caused by changes in the method of data collection.

7. The trends in arable land area in table 1.1, seems to confirm the expansion of cultivated land. These data should be interpreted with care, however. They probably overstate the true increase in arable land due to underreporting of arable land in earlier years.

8. Rapid grain price increase in 2003–2004 caused a further decline in the fertilizer/grain price ratio (Gale et al. 2005).

9. Note that if the positive trends in arable land are overstated (see footnote 7) then indeed the positive trend in fertilizer use per hectare is in fact stronger than is shown in table 1.1.

10. Some farmers even produce seven rice harvests in two years (sic!), whereas the margins of profit are very small. This is problematic, as the use of fertilizers and pesticides, in spite of the widespread introduction of IPM and regulatory measures by the government, per hectare use of chemical inputs has only risen since the late 1990s (Dung 2003).

11. This explains also why the paddy harvest area/arable land ratio is greater than one.

BIBLIOGRAPHY

Andersen, H. "The Economic Behaviour of Rural Households in Vietnam." Paper presented at the SPC/CASS/Stockholm School of Economics Conference on "Rural Development: An Exchange of Chinese and Vietnamese Experiences," Hanoi, 28 February–2 March 1995.

Anderson, K., J. Huang, and E. Ianchovichina. "The Impact of WTO Accession on Chinese Agriculture and Rural Poverty." Pp. 101–15 in *China and the WTO: Accession, Policy Reform, and Poverty Reduction Strategies*, edited by D. Bhatatasali, S. Li, and W. Martin. Oxford, U.K.: Oxford University Press, 2004.

Be, T. T. and L. C. Dung. "Economic and Environmental Impacts of Rice–Shrimp Farming in the Mekong Delta." Pp. 221–51 in *Economy & Environment: Case Studies in Vietnam*, edited by D. Glover and H. Francisco. Singapore: EEPSEA, International Development Research Centre, 1999.

Barker, R. "Agricultural Policy Analysis for Transition to a Market-Oriented Economy in Vietnam." *Economic and Social Development Paper*, no. 123, Rome: FAO, 1994.

Chen, J. "Economic Thought on Chinese Agricultural Reform." Pp. 24–36 in *Chinese Economic Analysis on Reform and Open Experience*, edited by Z. Zhang. Beijing: Economic Management Press, 2000.

Chen, X. "China's Agricultural Development and Policy Readjustment after Its WTO Accession." Pp. 69–79 in *China and the WTO: Accession, Policy Reform, and Poverty Reduction Strategies*, edited by D. Bhatatasali, S. Li, and W. Martin. Oxford, U.K.: Oxford University Press, 2004.

Deininger, K. and S. Jin. "Land Rental Market as an Alternative to Government Reallocation? Equity and Efficiency Considerations in the Chinese Land Tenure System." *World Bank Policy Research Working Paper*, no. 2930. Washington, D.C.: World Bank, 2002.

Diao, X., Z. Yi, and A. Somwaru. "Farm Holdings, Crop Planting Structure and Input Usage: An Analysis of China's Agricultural Census." *TMD Discussion Paper*, no. 62. Washington, D.C.: International Food Policy Research Institute, 2000.

Do, Q. T. and L. Iyer. "Land Rights and Economic Development." Mimeo, Cambridge, MA: Massachusetts Institute of Technology, 2002.

Du, Y. "China's Agricultural Restructuring and System Reform under Its Accession to the WTO." *ACIAR China Grain Market Policy Project Paper*, no. 12. Adelaide, South Australia: University of Adelaide, 2002.

Dung, N. H. *Economic and Environmental Consequences of Agro-chemicals Use in Intensive Rice Agriculture in the Mekong Delta, Vietnam*. Draft and unpublished chapter, PhD Thesis. The Hague: Institute of Social Studies, 2003.

Ellis, F. *Rural Livelihoods and Diversity in Developing Countries*. Oxford: Oxford University Press, 2000.

Fan, S. and M. J. Chen. "Critical Choices for China's Agricultural Policy." *2020 Policy Brief*, no. 60. Washington, D.C.: International Food Policy Research Institute, 1999.

FAOSTAT. *Statistical Database*. Rome: Food and Agricultural Organization, 2005. www.fao.org.

Feng, S., N. Heerink, and F. Qu. "Factors Determining Land Rental Market Development in Jiangxi Province, China." Paper presented at 7th European Conference on Agriculture and Rural Development in China (ECARDC), September 8–10, 2004, Greenwich, U.K.

Fforde, A. and S. de Vylder. *From Plan to Market: The Economic Transition in Vietnam*. Boulder CO: Westview Press, 1996.

Findlay, C. and C. Chen. "A review of China's Grain Marketing System Reform." *ACIAR China Grain Market Policy Project Paper*, no. 6. Adelaide, South Australia: University of Adelaide, 1999.

Gale, F., B. Lohmar, and F. Tuan. *China's New Farm Subsidies*. USDA Outlook WRS-05-01. Washington, D.C.: United States Department of Agriculture (USDA), Economic Research Service (ERS), 2005.

Glover, D. and H. Francisco, eds. *Economy & Environment. Case Studies in Vietnam.* Singapore: EEPSEA, International Development Research Centre, 1999.

Green, D. J. and R. W. A. Vokes, "Agriculture and the Transition to the Market in Asia." *Journal of Comparative Economics* 25, no. 2 (October 1997): 256–80.

Heerink, N. F. Qu, M. Kuiper, X. Shi, and S. Tan. "Policy Reforms, Rice Production and Sustainable Land Use in China: A Macro-micro Analysis." *Agricultural Systems* (2007) (forthcoming).

Ho, P. "China's Rangelands under Stress: A Comparative Study of Pasture Commons in the Ningxia Hui Autonomous Region." *Development and Change* 31, no. 2 (March 2000): 385–412.

Ho, P. "Rangeland Degradation in North China: A Preliminary Statistical Analysis to validate Non-equilibrium Range Ecology." *Journal of Development Studies* 37, no. 3 (2001): 99–132.

Huang, J. "Agricultural Development, Policy and Food Security in China." Pp. 3–38 in *Proceedings Workshop Wageningen—China*, edited by P. W. J. Uithol and J. J. R. Groot. Report 84. Wageningen, Netherlands: AB-DLO, 1997.

Huang, J. "Trade Liberalization and China's Food Economy in the 21st Century: Implications to China's National Food Security and International Trade." Paper presented at Mansholt Seminar, September 15, 1999, Wageningen University, The Netherlands, 1999.

Huang, J. and C. Chen. "Effects of Trade Liberalization on Agriculture in China: Institutional and Structural Aspects." *Working paper*, no. 42, Bogor, Indonesia: CGPRT Centre, 1999.

Huang, J. "Land Degradation in China: Erosion and Salinity Component." CCAP's *Working Paper Series*, no. WP-00-E17. Beijing: Centre for Chinese Agricultural Policy, Chinese Academy of Agricultural Sciences, 2000.

Huang, J. and S. Rozelle. "Environmental Stress and Grain Yields in China." *American Journal of Agricultural Economics* 77 (1994): 246–56.

Huang, J., S. Rozelle, and L. Zhang. "WTO and Agriculture: Radical Reforms or the Continuation of Gradual Transition?" *China Economic Review* 11 (2000): 397–401.

Huang, J. "Trade Liberalization and China's Food Economy in the 21st Century: Implications for China's National Food Security." Pp. 83–102 in *Globalization and the Developing Countries: Emerging Strategies for Rural Development and Poverty Alleviation*, edited by D. Bigman. Wallingford, U.K.: CABI Publishing, 2002.

Huang, J., S. Rozelle, and M. Chang. "The Nature of Distortions to Agricultural Incentives in China and Implications of WTO Accession." Pp. 81–97 in *China and the WTO: Accession, Policy Reform, and Poverty Reduction Strategies*, edited by D. Bhatatasali, S. Li, and W. Martin. Oxford, U.K.: Oxford University Press, 2004.

Huang, Y. "China's New Grain Marketing Policy." *ACIAR China Grain Market Policy Project Paper*, no. 5. Adelaide, South Australia: University of Adelaide, 1999.

Hue, L. T. V. *Coastal Resource Use and Management in a Village of Northern Vietnam.* PhD Thesis, Institute of Social Studies. Maastricht: Shaker Publications, 2004.

Kham, T. N. and L. X. Thuy. "An Economic Analysis of Agroforestry Systems in Central Vietnam." Pp. 173–203 in *Economy & Environment: Case Studies in Vietnam*, edited by D. Glover and H. Francisco. Singapore: EEPSEA, International Development Research Centre, 1999.

Kung, J. K.-S. "Off-Farm Labor Markets and the Emergence of Land Rental Markets in Rural China." *Journal of Comparative Economics* 30 (2002): 395–414.

Lin, J. Y. "WTO accession and China's agriculture." *China Economic Review* 11 (2000): 405–8.

Lindert, P. H. "The Bad Earth? China's Soils and Agricultural Development since the 1930s." *Economic Development and Cultural Change* 47, no. 4 (1999): 701–36.

Lindert, P. H. *Shifting Ground: The Changing Agricultural Soils of China and Indonesia.* Cambridge, MA: MIT Press, 2000.

Men, N. T. *Vietnamese Rice Markets in Transition,* Draft Manuscript, PhD Thesis, Mimeo. The Hague: Institute of Social Studies, 2003.

National Agricultural Census Office of China (NACO) & Food and Agricultural Statistics Centre, China (FACS). *Abstract of the First National Agricultural Census in China.* Beijing: China Statistics Press, 1999.

National Bureau of Statistics of China. *China Statistical Yearbook.* Beijing: China Statistics Press, 2002.

Nghiep, L. T. "Agricultural Development in Vietnam: Issues and Proposals for Reform." Pp. 144–56 in *Vietnam's Dilemmas and Options. The Challenge of Economic Transition in the 1990s,* edited by M. Than and J. L. H. Tan. Singapore: ISEAS, 1994.

Ning, D. "An Assessment of the Economic Losses Resulting from Rangeland Degradation in China." Occasional paper. Project on Environmental Scarcities, State Capacity, and Civil Violence. Toronto, Canada: University of Toronto, 1997.

Organization for Economic Cooperation and Development (OECD). *Review of Agricultural Policies—China.* Paris: OECD, 2005.

Otsuka, K. "Enhancing Land Rights and Improving Land Use Efficiency in Marginal Area: Regional Overview and International Context." Tokyo: Foundation for Advanced Studies on International Development, 2002.

Oxfam. "Extortion at the Gate. Will Viet Nam Join the WTO on Pro-Development Terms?" *Briefing Paper,* no. 67. London: Oxfam, 2004.

Pingali, P. and V. T. Xuan. "Vietnam: Decollectivisation and Rice Productivity Growth." *Economic Development and Cultural Change* 40, no. 4 (1992): 697–719.

Ravallion, M. and D. van der Walle. "Breaking up the Collective Farm: Welfare Outcomes of Vietnam's Massive Land Privatization." *Policy Research Working Paper Series,* no. 2710. Washington D.C.: World Bank, 2001.

Rehm, S. and G. Espig. *The Cultivated Plants of the Tropics and Subtropics: Cultivation, Economic Value, Utilization.* Weikersheim, Germany: Margraf Verlag, 1991.

Research Group of Ministry of Agriculture (MoA), Peoples Republic of China, "Reform of Grain Subsidy in China." *Problems of Agricultural Economics* 2003(5): 4–11 (*in Chinese*).

Rozelle, S., J. Huang, and L. Zhang. "Poverty, Population and Environmental Degradation in China" *Food Policy* 22 (1997): 229–51.

Shi, X., N. Heerink, F. Qu, and X. Huang. "Soil Erosion in Jiangxi Province." Pp. 354–71 in *Sustainable Resource Use and Sustainable Economic Development,* edited by F. Qu, N. Heerink, and X. Huang. Beijing: China Forestry Publishing House, 2000.

Shi, X., *Off-Farm Employment, Factor Market Development and Sustainbable Land Use in Rural China.* Unpublished PhD Thesis, Institute of Social Studies (Forthcoming), 2007.

Spoor, M. "Finance in a Socialist Transition: The Case of the Democratic Republic of Vietnam (1955–1964)." *Journal of Contemporary Asia* 17, no. 3 (1987): 339–65.

Spoor, M. "Reforming State Finance in Post–1975 Vietnam." *Journal of Development Studies* 24, no. 4 (1988): 102–14.

Spoor, M. *Transition, Institutions, and the Rural Sector*. Lanham, MD and Boston: Rowman and Littlefield, Lexington Books, 2003.

Tan, S., N. Heerink, F. Qu, and X. Huang. "Soil Degradation in Jiangsu Province: An Inventory of Major Issues." Pp. 372–91 in *Sustainable Resource Use and Sustainable Economic Development*, edited by F. Qu, N. Heerink, and X. Huang. Beijing: China Forestry Publishing House, 2000.

Thach, T. V. A. "The Present State of Environment in Vietnam and the Task Assigned to Our Country's Environmental Protection Policy." Internet Document, Environmental Systems Index, 1997.

The, B. D. "The Economics of Soil Erosion and the Choice of Land Use Systems by Upland Farmers in Central Vietnam." *Research Paper*. Singapore: EEPSEA, 2001.

World Bank/ADB/UNDP. *Vietnam 2010: Entering the 21st Century. Vietnam Development Report: Pillars of Development*. Joint Report of the Consultative Meeting for Vietnam. Hanoi: World Bank, Asian Development Bank, and UNDP, 2000.

World Bank. "Vietnam: Transition to the Market, An Economic Report." *Report*, No. 11902-VN, Washington: World Bank, 1993.

Wu, B. "Fertilizer Buying Behaviour of Farmers on the Edge of Market Liberalization: the Case of Jiangxi Province, P. R. China." MSc Thesis, Wageningen University, The Netherlands, 2000.

Yu, W. and S. E. Frandsen. "China's WTO Commitments in Agriculture: Does the Impact Depend on OECD Policies?" Paper prepared for the Fifth Annual Conference on Global Economic Analysis, June 5–7, 2002, Taipei, Taiwan.

Zhu, P. "Soil Erosion and Conservation in China: An Overview." Pp. 324–37 in *Sustainable Resource Use and Sustainable Economic Development*, edited by F. Qu, N. Heerink, and X. Huang. Beijing: China Forestry Publishing House, 2000.

I

REGIONAL AND SECTORAL IMPACT OF ECONOMIC REFORMS

2

Regional Inequality in China

Scope, Sources, and Strategies to Reduce It

Shenggen Fan and Connie Chan-Kang

Over the past two decades, economic reforms have led to rapid economic growth in China. As a result, the living standard has risen significantly. This is reflected in various indicators of development such as average years of schooling, life expectancy, infant mortality rate, and the percentage of malnourished children, that reached the level of many middle-income countries (World Development Report 2003). Poverty, particularly in rural areas, has declined significantly. However, not all have benefited from this economic growth. The Gini coefficient rose from 0.33 in 1980 to 0.46 in 2000, and the income gap between the coastal areas and western regions of the country widened dramatically over the same period, a reflection of the worsening income distribution and regional inequality. For example, in 2003, per capita gross domestic product (GDP) of Shanghai was thirteen times higher than that of Guizhou.

The income gap between the rural and urban sectors has also increased drastically. In 1984, the average rural income was 60 percent of urban income, but the ratio dropped to 33 percent in 2002. Growing inequality has an adverse effect on poverty reduction and may lead to tensions within a country. Such tensions may constrain the prospects for future growth, through a variety of social, political, and economic mechanisms. With the demand for industrial products reaching a saturated level in affluent coastal areas, future growth will depend on increased demand from the lesser-developed areas of western and central China. The lack of purchasing power of these lesser-developed regions due to income inequality may hinder this process.

The objective of this chapter is to make an in-depth analysis of the causes and consequences of regional inequality in China and propose a strategy for a more equitable development of the economy. The chapter begins with a review of regional development in China and a comparison of development indicators and their periodic changes across different regions. From the existing literature, in the following section we look into the various aspects of regional inequality in China and analyze their sources. The last section offers policy implications and suggests an approach that will help in promoting a more equitable regional development in China.

REGIONAL DEVELOPMENT

Income inequality exists on various grounds. The most common sources are income differences between gender, regions, households, and members of households. Among these, regional inequality is most important in China as it is related to two forms of income disparity between the rural and urban sectors, as well as the agricultural and nonagricultural sectors. Therefore, our chapter will be emphasizing the main causes and problems of regional inequality in China.

In the 1950s, prior to China's economic reforms, the country was broadly divided into seven regions based on historical and administrative boundaries as well as economic levels. These regions were: the Northeast, North, Northwest, Central, Southeast, Southwest, and South. During the late 1980s and early 1990s, these regions were aggregated into three regional groups namely: (i) the Western region consisting of the Southwest and Northwest in the old classification; (ii) the Central region, which includes part of the former Central and Northern regions; and (iii) the Coastal (or Eastern) region, encompassing the former Northeast, Southeast, and South regions.

China's Western region is comprised of eleven provinces and autonomous regions and municipalities under the direct administration of the central government. They are *Chongqing, Gansu, Guizhou,* Inner Mongolia, *Ningxia, Qinghai, Shaanxi, Sichuan,* Tibet, *Yunnan,* and *Xinjiang.* These provinces cover more than half of China's total geographic area and account for 20 percent of the national population in 2001 (table 2.1).

However, Western China is trailing other regions in terms of economic development. For example, per capita GDP in the Southwest and Northwest was only half when compared to the Southeastern and Southern provinces. Moreover, per capita incomes in both rural and urban sectors were much lower than the other regions. The income inequality between rural and urban residents in this region is quite large and is a matter of concern. As table 2.1 shows, the average income of rural residents in Southwest and Northwest China was only 28 percent of their urban counterparts, while this ratio is almost 40 percent in other regions.

Table 2.1. Regional Economic Development (2001)

	Population	Per Capita GDP	Rural Per Capita Income	Urban Per Capita Income	Agricultural GDP
	Million	yuan	yuan	yuan	%
Northeast	104.5	9,328	2,175	5,027	13
North	356.5	7,747	2,592	6,689	15
Northwest	54.1	5,317	1,518	5,413	20
Central	163.6	6,092	2,200	5,535	19
East	189.4	11,716	3,845	9,266	11
Southwest	190.3	4,496	1,662	5,904	23
South	155.5	10,280	2,733	7,097	15
Average		7,986	2,401	6,306	15.2

Source: China Statistical Yearbook (2002).

Table 2.2 shows the regional distribution of the rural poor. In 1996, more than 60 percent of the rural poor in China lived in the Western provinces. Given the low population density in these areas, the poverty incidence was higher than the national average. For example, 23 percent of the rural population in *Gansu* and 27 percent in *Xinjiang* lived under the poverty line in 1996.

MEASURES OF REGIONAL INEQUALITY

Table 2.3 shows a summary of twenty-five studies done on inequality in China. Inequality, unless otherwise indicated, here refers to income inequality. The majority of the reviewed studies used the Gini coefficient (and some the Theil coefficient) to measure income inequality. The table provides us with a useful overview on inequality in China.

Most studies depict an increase in inequality at the household level. According to the study undertaken by Aaberge and Li (1997), there was a notable rise in the Gini coefficient from 0.156 in 1986 to 0.200 in 1990 in *Sichuan* Province, and in *Liaoning* Province the coefficient rose from 0.155 in 1986 to 0.174 in 1990. Likewise, Ravallion and Chen (1990) found that the Gini coefficient had increased from 0.271 to 0.287 between 1985 and 1990 in Southwest China.

Similarly, there was an increase in inequality in both rural and urban China. Coady and Wang (2000) found that in 1986 the average income of the ten percent richest households in urban *Liaoning*, was 2.29 times greater than that of the poorest 10 percent and in 1990, the ratio was 2.56. Gustafsson and Shi (2001) found similar evidence using urban household survey data from ten provinces. According to their study, in 1988 the richest 10 percent of the households had incomes which were 5.43 times more than

Table 2.2. Regional Distribution of Rural Poor

Province	Share of Households (%) 1989	Share of Total National Poor (%) 1989	Official Chinese Statistics Number of Poor Counties 1997	Share of total Provincial Counties (%) 1997	Head Count (%) 1988	Head Count (%) 1995	Change (%)
Beijing	0.2	0	0	0.0	8.7	1.3	−7.4
Tianjin	0.4	0	0	0.0	n.a.	n.a.	n.a.
Hebei	13	7.1	39	28.3	29.9	22.7	−7.2
Shanxi	17.4	4.1	35	34.7	51.9	49.5	−2.4
Inner Mongolia	23.5	3.6	31	36.5	n.a.	n.a.	n.a.
Liaoning	8	1.9	9	20.5	27	21.9	−5.1
Jilin	12.2	1.9	5	12.2	41.5	18.3	−23.2
Heilongjiang	18.3	3.6	11	16.4	n.a.	n.a.	n.a.
Shanghai	0	0	0	0.0	n.a.	n.a.	n.a.
Jiangsu	3.4	1.9	0	0.0	27.8	4.7	−23.1
Zhejiang	2	0.8	3	4.7	5.8	4	−1.8
Anhui	7.7	3.9	17	25.4	35.6	19.8	−15.8
Fujian	1.8	0.5	8	13.1	n.a.	n.a.	n.a.
Jiangxi	5	1.6	18	20.9	25.7	27	1.3
Shandong	6.8	5	10	10.6	28.3	19.3	−9
Henan	16.5	12.7	28	24.6	52.5	20.1	−32.4
Hubei	6	2.6	25	36.8	20.3	25	4.7
Hunan	6.2	3.5	10	11.2	13.1	37.5	24.4
Guangdong	0.9	0.5	8	8.3	4.8	5.2	0.4
Guangxi	15.4	6.1	28	34.6	n.a.	n.a.	n.a.
Sichuan	11.2	11.2	43	24.9	n.a.	n.a.	n a.
Guizhou	17.8	5.4	48	60.8	32.5	43.1	10.6
Yunnan	19	6.5	73	59.8	58.3	61.8	3.5
Tibet			5	6.5	47.3	45.6	−1.7
Shaanxi	20.3	5.8	50	56.2	n.a.	n.a.	n.a.
Gansu	34.2	6.7	41	53.9	59.9	58	−1.9
Qinghai	23.7	0.8	14	35.9	69.7	69	−0.7
Ningxia	18.9	0.7	8	44.4	n.a.	n.a.	n.a.
Xinjiang	18.7	1.6	25	29.4	n.a.	n.a.	n.a.
China	11.1	100	592	27.8	35.1	28.6	−6.5

Source: World Bank (2000) and Khan (1997).

the incomes of the poorest households and this number swelled to 9.11 by 1995. Fang et al. (2002) and Khan et al. (1999) confirmed that urban inequality increased during the 1990s. Chan and Chan (2000), Gustafsson and Shi (2002), and Wan (2001) analyzed inequality in rural China. Chan and Chan used the Gini coefficient to assess the change in rural inequality between provinces from 1991 to 1994. The latter two studies analyzed rural household survey data from the mid-1980s to the mid-1990s using the

Table 2.3. Summary of Selected Studies on Regional Inequality in China

Source	Location	Year Studied	Measure Used	Unit of Analysis	Inequality
Aaberge and Li (1997)	Sichuan	1986	Gini	Household	0.156
		1990			0.200
	Liaoning	1986			0.155
		1990			0.174
Akita and Kawamura (2002)	China	1995	Theil	District	0.230
		1998			0.249
	Western China	1995		District	0.027
		1998			0.020
	Central China	1995		District	0.017
		1998			0.021
	Eastern China	1995		District	0.026
		1998			0.033
Benjamin et al. (2000)	Hebei and Liaoning	1995	Gini	Rural household	0.385
			Theil Log		0.238
			Theil Entropy		0.291
Benjamin and Brandt (1999)	Northeast China	1935	Gini	Rural household	0.420
		1995			0.380
Chan and Chan (2000)	China	1991	Gini	Interprovincial, rural sector	0.420
		1994			0.460
Chowdhury et al. (2000)	China	1981	Theil Entropy	Interprovincial	0.054
		1997			0.045
Coady and Wang (2000)	Liaoning	1986	Gini	Urban household	0.180
		1990			0.200
		1986	Decile ratio		2.29
		1990			2.56
Cook (2000)	Zouping County, Shandong	1992	Quartile	Share of income by poorest quartile, rural sector	11%
				Share of income by richest quartile, rural sector	44%
Démurger et al. (2001)	China	1952–1965	CV	Interprovincial	0.442
		1990–1998			0.659
		1952	Income ratio	Top 5 to bottom 5 provinces	2.55

Table 2.3. (Continued)

Source	Location	Year Studied	Measure Used	Unit of Analysis	Inequality
Fang et al. (2002)	China	1998			3.59
		1992	Gini	Urban household	0.244
		1998			0.312
Gang et al. (1997)	China	1995	Gini	All households	0.45
				Rural household	0.42
				Urban household	0.33
Gustafsson and Shi (2001)	China (10 provinces)	1988	Decile ratio	Urban household	5.43
		1995			9.11
		1988	Gini		0.240
		1995			0.304
		1988	MLD		0.106
		1995			0.184
Gustafsson and Shi (2002)	China	1988	MLD	Rural household	0.178
		1995			0.288
		1988	Theil		0.181
		1995			0.339
Howes and Hussain (1994)	China	1985	Gini	Value of output, rural counties	0.244
					0.335
Kanbur and Zhang (1999)	China	1983	Gini	Per capita consumption expenditure by province	0.220
		1995			0.277
		1983	MLD		0.079
		1995			0.120
Kanbur and Zhang (2005)	China	1952	Gini	Per capita consumption expenditure by province	0.224
		2000			0.372
		1952	MLD		0.09
		2000			0.248
Khan et al. (1999)	China (10 provinces)	1988	Gini	Urban household	0.233
		1995			0.332
Kung and Lee (2001)	Hunan and Sichuan (4 counties)	1993	Gini	All income (Rural households)	0.182
				Off-farm income	0.456

Table 2.3. (Continued)

Source	Location	Year Studied	Measure Used	Unit of Analysis	Inequality
Lee (1995)	China			Cash crop	0.480
				Grain	0.261
				Animal husbandry and fishery	0.387
		1952	CV	Per capita national income by province	0.399
		1990			0.490
		1952	Gini		0.283
		1990			0.274
Ravallion and Chen (1990)	Southern China	1985	Gini	Rural household	0.271
		1990			0.287
		1985	MLD		0.12
		1990			0.134
Tsui (1998)	Guangdong	1985	MLD	Rural household	0.149
		1990			0.157
	Sichuan	1985	MLD	Rural household	0.079
		1990			0.107
Wan (2001)	China	1984	Gini	Rural household	0.106
		1996			0.16
Wei and Kim (2002)	Jiangsu	1953	Gini	Per Capita Gross Value of Ind. and Agric.	0.198
		1995			0.495
	Jiangsu, agriculture	1953	Gini	Output per county	0.188
		1995			0.166
	Jiangsu, industry	1953	Gini		0.555
		1995			0.896
Yang (1999)	Sichuan	1986	Gini	Household survey	0.278
		1994			0.392
	Jiangsu	1986	Gini		0.276
		1994			0.349
Zhang and Yao (2001)	China	1952	MLD	GDP by province	0.038
		1999			0.11

Source: Compiled by authors.
Note: CV denotes Coefficient of Variation; MLD denotes Mean Log Deviation.

Theil index and the Gini coefficient respectively. Despite the differences in the unit of analysis and indicator used, all the sources indicated an increase in inequality in rural China.

Although both rural and urban inequality have increased, the inequality between rural households was much larger than the one between urban households. Gang et al. (1997) computed income inequality in rural and urban China using household level data. The authors found that inequality among rural households in 1995 was 0.45 while their urban counterparts had a Gini coefficient of 0.33 for that same year. However, the increase in urban inequality has been faster than that in rural inequality. Gustafsson and Shi (2001) found that urban inequality registered a 73 percent increase (from 0.106 to 0.184) in the Mean Log Deviation (MLD) value during 1988–1995. In a complementary study, Gustaffson and Shi (2002) showed that in rural areas, for the same time period, the MLD had increased from 0.178 to 0.288 (a 61 percent increase).

Table 2.3 also shows a sharp increase in the interregional inequality over the last several decades. There has been a substantial increase since the introduction of reforms in the late 1970s. The coefficient of variation was reported to have increased from 0.442 in the 1950s and early 1960s to 0.659 in the 1990s (Démurger et al. 2001), and from 0.399 in 1952 to 0.490 in 1990 (Lee 1995).

Kanbur and Zhang (2005) showed that the Gini coefficient for inequality, in terms of per capita consumption expenditures between provinces, increased from 0.224 to 0.372 during the period from 1952 to 2000. The Theil MLD index used by Zhang and Yao (2001) also showed a substantial increase in interprovincial inequality in China from 0.038 to 0.11 during 1952–1999.

Finally, inequality differs among economic sectors and appears to be greater for high value crops as well as for the non-farm sector. Kung and Lee (2001) analyzed inequality by sources of income of four counties in rural *Hunan* and *Sichuan*. Their study revealed that the Gini coefficient was the highest for income derived from cash crops and off-farm work. Wei and Kim (2002), on the other hand relied on county level data in *Jiangsu* Province for their analysis. The authors compared inequality on the basis of the economic sector and found that the Gini coefficient was greater in the industrial sector (0.896) than in the agricultural sector (0.555) in 1995. In 1953 however, the opposite was observed.

SOURCES OF REGIONAL INEQUALITY

Policy makers and economists alike can benefit greatly by understanding the sources of income inequality as they can gather vital information for evaluating, designing, and implementing policy interventions to reduce inequality and poverty. For this purpose we include here a brief review of the literature on the sources of regional inequality in China.

Inequality between—and within—Provinces or Regions

Chowdhury et al. (2000), Akita and Kawamura (2000), Galbraith and Wang (2001), Gustafsson and Shi (2002), and Zhang and Yao (2001) used the Theil index to identify the sources of inequality in China during the 1990s.[1] The use of different geographical units underlying the decomposition of the Theil index yields to different findings. Chowdhury et al. (2000) and Zhang and Yao (2001) decomposed total income inequality into inequality between—and within—regions. Both studies found that increases in inequality *between* regions accounted for most of the rise in inequality during the 1990s. Akita and Kawamura (2002), in contrast, found that *within*-province inequality accounted for a major share (62 percent) of China's overall inequality between 1995 and 1998, followed by interregional inequality (27 percent) and inter-provincial inequality (11 percent). These results conjure that the increase in inequality among coastal, central and western regions, and not among provinces, contributes largely to China's overall regional inequality.

Galbraith and Wang (2001) focused on inequality by economic sector within each province and by province within each sector. The authors scrutinized inequality for sixteen sectors in thirty Chinese provinces from 1987 to 2000. They concluded that the rise of China's total inequality was mainly due to inequality between provinces, rather than between economic sectors. They investigated the provinces and sectors primarily responsible for China's inequality growth. Their analysis revealed that the high-income growth in eastern provinces such as *Guangdong,* Shanghai, and Beijing was the main cause of the surge in inequality in China during the 1990s. Among the economic sectors, banking, transportation, and utilities contributed most to income inequality. Interestingly, the authors found that the source of inequality changed over time and they supported their claim by comparing the years 1987 and 2000; for the former the largest source was inequality between sectors, while the major contributor for the latter was geographical factors. Gustafsson and Shi (2001) corroborated these findings. They collected rural household data from eighteen provinces, and attributed the inequality increase in rural China during the mid-1990s to differences in mean income across Eastern, Central, and Western provinces.

In spite of diverse findings on the sources of inequality, all the studies reviewed above concur that the preferential policies given to Eastern China after the reform contributed to the growth of inequality in the 1990s. Eastern China was successful in developing TVEs (Township and Village

Enterprises) and export-oriented enterprises, attracting foreign investment, increasing productivity, and in reinvesting savings (Chowdhury et al. 2000). Thus, Eastern China enjoyed a rapid economic expansion while a number of constraints including poor infrastructure and transportation facilities, the lack of growth centers, and poor education hampered the economic growth in the West and the Central regions.

Rural-Urban Divide

Since the mid-1980s, the focus of China's economic reforms has shifted to the urban sector. These reforms led to the widening of the income gap between urban and rural households and thus triggered an interest in studying the contribution of the rural-urban income gap to inequality. Kanbur and Zhang (1999) decomposed the overall regional inequality in China into rural-urban and inland-coastal inequality in the 1980s and 1990s. They found the contribution of rural-urban inequality to be much greater than that of inland-coastal inequality. However, the contribution of the latter has increased dramatically in recent years. According to Kanbur and Zhang, the greater ease of rural-to-urban migration within provinces, compared to the difficulties of migrating from inland to coastal provinces, provides a partial explanation for this phenomenon.

Yao (1999) concurred with Kanbur and Zhang. Using secondary and household survey data, Yao showed that the urban-rural divide along with their rising spatial inequality accounted for a major share of income inequality in China. The author also found that the incidence of poverty is very susceptible to changes in per capita income and inequality. While the contribution of the rural-urban income gap to inequality has been established by the above studies, Yang (1999), Guillaumont Jeanneney and Hua (2001), and Kwong (1994) investigated the driving factors behind the widening income gap between rural-urban households in China. Yang assessed whether the increase in income inequality was a consequence of the 1978 institutional reforms that replaced the egalitarian rewards with work incentives, employment contracts, and labor mobility. His analysis relied on household survey data recorded from rural and urban regions of *Sichuan* and *Jiangsu* in 1986, 1988, 1992, and 1994. He noticed that urban biased policies and institutions, including labor mobility restrictions, welfare systems, inflation subsidies, and investment credits, were responsible for the long-term rural-urban divide and the recent increase in inequality between the two sectors.

On the other hand, Guillaumont Jeanneney and Hua (2001) attributed the depreciation of the Chinese currency—which played a crucial role in the opening-up policy—to the increase in income inequality between urban and rural areas during 1982–1986. According to their study, the real depreciation of the *yuan* raised the urban bias because of the higher share of tradable goods produced in urban areas when compared to rural areas. Although the real depreciation led to a greater rise in consumer prices in urban areas, the average urban income per capita rose because a larger share of tradable goods was produced and a smaller share was consumed.

Kwong (1994) studied the influence of markets—during the years 1986 and 1988—on the rural-urban income inequality in China. Using regression analysis, the author found that the introduction of markets in China's economy significantly affected the urban-rural divide. On the basis of these findings, Kwong suggested that rural industrialization, along with increased participation into market activities will help in narrowing the income differences among urban and rural residents.

The Rural Non-Farm Sector

One of the most important characteristics of China's recent economic history is the rapid pace of rural industrialization. The expansion of rural industries after the reforms resulted in a substantial increase in the share of off-farm employment and rural non-farm GDP in the post-reform period (Fan et al. 2002).[2] Nevertheless, the increasing importance of the rural non-farm sector is seen by many as a significant source of inequality. Ke (1996) observed that rural industry grew the fastest in Eastern China where the share of off-farm employment and income is the highest when compared to other regions. Likewise, Benjamin and Brandt (1999) ascribed off-farm employment as a main source of income inequality among households in Northeast China. Similar conclusions were drawn by Tsui (1991), Rozelle (1994), Hussain et al. (1994), and Bramall and Jones (1993).

Two recent household level studies done by Benjamin et al. (2000) and Knight and Shi (1997) substantiate these findings. Econometric methods were employed to formally assess the contribution of the rural non-farm sector to inequality across households within villages. The analyses were based on household level data collected in Northeast China (by Benjamin et al.) and in *Hebei* Province (by Knight and Shi). The econometric results confirmed that income inequality is driven mainly by non-farm income sources.

Interestingly, Benjamin et al. found that commercialization and market integration promoted income equality, but this positive effect is tempered by the unequal access to education. Since the opportunities arising from the non-farm sector are taken by the more educated, the chances of aggravating the income inequality are higher. The authors concluded that simultaneous improvement in educational attainment and off-farm market development would allow more households to gain from the rapid growth in rural China, whereas Knight and Shi emphasized that increased opportunities in the non-farm sector are important for raising household incomes.

Ho (1995) shed more light in his study of higher inequality variation in the non-farm sector. According to him, greater inequality prevails in the non-farm sector because of geographical and technological differences among the regions in which rural industrialization takes place. That is, regions with better infrastructure, resources, developed non-farm activities, and closer proximity to urban areas grew more rapidly than the poorly endowed regions, thereby fueling an increase in interregional inequality.

Tsui (1998), and Howes and Hussain (1994) had alternative views. They made use of household level data from *Guangdong* and *Sichuan* provinces for the period 1985–1990. Tsui found that while nonagricultural activities exerted an increasingly important influence on rural inequality, incomes from agriculture remain a significant source of inequality. Hence, policy makers should not overlook the inequality-reducing potential of policies focused on agriculture. Howes and Aussian (1994) found that the principal driving force behind the growth of inequality has been the unprecedented growth of the nonagricultural sector (relative to the agricultural ones), rather than a sole rise in inequality in the nonagricultural sector from uneven growth. Moreover, they did not find a notable increase in inequality in the distribution of nonagricultural output over their period of study. They concluded that nonagricultural output had always been unequally distributed and it is mainly its growing in importance in rich and poor counties alike that is driving inequality in rural China.

Geography and Infrastructure

China is a vast country with highly diverse topography, climate, and natural resource endowments. Western China, with its amazing mountains and highlands, has been susceptible to water loss and erosion. Central China is landlocked, constraining it from gaining access to overseas markets and foreign investments. In contrast, Eastern China enjoys favorable conditions for agricultural production and has been a pivotal center for industrial development and sea-borne trade. Thus, initial conditions linked to geographical location were a source of regional inequality (Hang Seng Bank 2001; Gang et al. 1997). However, access to infrastructure may help to ease some of the geographical constraints and facilitate economic development particularly in Western and Central China.

Démurger (2001) analyzed the impact of infrastructure investments on economic growth by utilizing the data retrieved from twenty-four Chinese provinces during the period between 1985 and 1998. The author estimated a growth model and found that geographical location and infrastructure endowments (transport infrastructure and telecommunication facilities) accounted significantly for the notable differences in growth across provinces. Moreover, the author found that economic policy together with naturally inherited elements contributed to the difference in interprovincial economic growth. For example, the relatively developed infrastructure network and economic activities oriented toward nonagricultural industries explained the high per capita GDP growth in Eastern China while the lack of the same contributed to the lower growth rates in Northwest China over the same period. It was concluded that economic measures that can boost infrastructure might have a significant impact in reducing the per capita income gap.

Démurger et al. (2001) and Knight and Song (1993) have come up with similar results. The regression results obtained from the study done by Démurger et al. showed that geographical location and preferential economic policies influenced the difference in output growth among Chinese provinces. Likewise, Knight and Song also validated the impact of geographical location and availability of infrastructure on increasing inequality across rural counties in the post-reform period. Thus, counties close to city markets with good transport links, in conjunction with human capital and larger investable surpluses per capita, enjoyed higher growth rates.

THE CHANGING POLICY ENVIRONMENT

Policy and Institutional Reforms

As mentioned previously, China has implemented major policy and institutional reforms over the past several decades. A number of studies have provided insights into the links between the level of regional inequality in China and governmental development strategies and regional policies. Kanbur and Zhang (2005) took a historical approach in analyzing the sources of regional inequality in China. The analysis was based on per capita consumption data taken from twenty-eight provinces for the period from 1952 to 2000. Using the Gini coefficient and the General Entropy index, the authors observed that the peaks of inequality coincided with major structural changes in China, namely the Great Famine of the late 1950s, the Cultural Revolution of the late 1960s and 1970s, and the current phase of openness and decentralization. The regression results demonstrated that three key policy variables caused the increase in regional inequality: (i) the share of heavy industry in gross output value; (ii) the degree of decentralization; and (iii) the degree of openness. Kanbur and Zhang concluded that further liberalizing and investing in the economy of inland regions is an important development strategy for the government to promote China's economic growth and reduce regional inequality. Interestingly, heavy industrialization resulted in the swelling of the rural-urban gap in the pre-reform period, while openness and decentralization contributed to the rapid increase in inland-coastal disparity in the reform during 1980s and 1990s. The latter finding concurs with Lin (2000) who found that the openness and global integration of the late 1970s is a key factor in explaining regional inequality during the post-reform period. Wei and Kim (2002) also took a historical approach in their inequality study. To analyze the sources of inequality, they examined inter-county inequality in *Jiangsu* Province between 1950 and 1995. The authors decomposed the Gini index by economic sector and discovered that inequality across counties in *Jiangsu* did not change much until 1984, the year in which urban reforms were launched. While the decomposition of the Gini index showed that the widening inequality resulted mainly from the industrial sector, the regression analysis indicated that foreign direct investment, nonstate enterprises, agglomeration effects (i.e., the proximity to larger cities and network of markets) and human capital were important factors underlying the divergence of intercounty inequality in *Jiangsu*.

Rozelle (1994), Tsui (1996), and Sun and Chai (1998) sought to understand the sources of inequality trends during the post-reform period in China. Utilizing the county level data from *Jiangsu* Province recorded during 1983–1988, Rozelle decomposed the Gini coefficient by income source (crop, livestock, other agricultural activities, and rural industry) and found that inequality across the counties increased after 1983. Rozelle came to the conclusion that the changing patterns of inequality were attributable to changes in the structure of the rural economy. Specifically, the policies that emphasized the importance of agriculture led to a reduction in inequality while those encouraging the expansion of rural industry gave rise to greater inequality. Tsui attributed the increase of inequality in China during the mid-1980s to the reforms implemented to the industrial sector, which included fiscal decentralization, rapid development of rural industries in coastal regions, and foreign direct investment in richer provinces. According to the author, these conflicting forces will aggravate the gap, as the richer provinces will benefit more than the poorer. Similarly, Sun and Chai found that foreign direct investment, decentralization, and the priority in the regional development strategy given to the coastal regions contributed to widening of the gap between eastern and western regions of China.

In a recent study, Fan et al. (2002) offered a different perspective on the sources of inequality. The authors, with the help of a simultaneous equation model, estimated the effects of different types of government expenditures on regional inequality during 1970–1997. They found that investments made in research and development (R&D), irrigation, rural education, and infrastructure, contributed not only to agricultural production growth but also reduced regional inequality. Expenditure on education, in particular, has the largest impact in reducing regional inequality. Interestingly, all types of government spending in the western region led to reductions in regional inequality, while investment in the coastal or central regions worsened existing regional inequalities. Moreover, the contribution of conventional inputs (capital, labor, and land) to inequality declined over time, while the contribution of most public investments (R&D, electrification, and telephones) increased inequality.

Trade

Does globalization intensify inequality or reduce it? According to Reynolds (1987) China's rapid trade growth has probably increased regional inequality. His analysis showed that the fifteen richest provinces (ranked by income per capita) accounted for 43 percent of the population and generated 75 percent of the exports in 1983, while the nine poorest provinces, with 37 percent of the population, accounted for only 9 percent of exports. A more recent study by Yao and Zhang (2001) sought to understand the increase in regional inequality in China from 1978 to 1995. Differences in regional growth particularly arose due to the degree of openness and transportation. However, Wei and Wu (2001) appear to contradict the above findings. The authors analyzed the impact of globalization on income inequality from 1988 to 1993 using data of more than 100 cities and adjacent rural areas. Their regression results showed that cities with a greater degree of openness in trade (proxied by a trade-to-GDP ratio variable) tend to demonstrate a decline in the urban-rural income inequality. Thus it can be concluded that globalization has helped to reduce income inequality between urban and adjacent rural areas.

Strategies to Reduce Regional Inequality

How China will maintain a high growth, which is much needed to increase per capita income, and concurrently ensure that unemployment is kept at bay, will remain a continuous challenge. Many have argued that China has to sacrifice the overall national growth to achieve this objective. But increasing evidence has shown otherwise. In fact, there are many policy options that may achieve the twin goals of both growth and more balanced regional development.

Promoting Growth in Rural Sectors

As Kanbur and Zhang (2005) have shown, the large regional gap in income comes from the rural-urban disparity. The Western region lags behind mainly because agriculture and the rural sector still accounts for a large share of its GDP. Lower productivity in the rural sector, particularly in the agricultural sector, has caused overall productivity of the economy to lag behind. Empirical evidence (e.g., Fan et al. 2004) has shown that investment

Table 2.4.   Returns of Public Investments to Production and Poverty Reduction in China

	Coastal	Central	Western	Average
Returns to Agricultural GDP	yuan per yuan expenditure			
R&D	8.60	10.02	12.69	9.59
Irrigation	2.39	1.75	1.56	1.88
Roads	1.67	3.84	1.92	2.12
Education	3.53	3.66	3.28	3.71
Electricity	0.55	0.63	0.40	0.54
Telephone	1.58	2.64	1.99	1.91
Returns to Poverty Reduction	No. of poor reduced per 10,000 yuan expenditure			
R&D	1.99	4.40	33.12	6.79
Irrigation	0.55	0.77	4.06	1.33
Roads	0.83	3.61	10.73	3.22
Education	2.73	5.38	28.66	8.80
Electricity	0.76	1.65	6.17	2.27
Telephone	0.60	1.90	8.51	2.21
Poverty loan	0.88	0.75	1.49	1.13

Source: Fan et al. (2002: Table 5.3).
Notes: Marginal returns are calculated for 1997.

in rural areas can yield large returns (see table 2.4). Therefore, based on the above attained results, the government should continue to increase overall investment in rural areas. Rural investment accounted for only 19 percent of total government expenditures in 1997, while rural residents account for 69 percent of China's total population.

Moreover, almost 50 percent of China's GDP was produced by the rural sector (agriculture and rural township and village enterprises) in 1997. Government's total rural spending was equivalent to only about 5 percent of the rural GDP, compared with 11.6 percent for the whole economy. The implementation of an urban- and industry-biased investment policy by the Chinese government for the past several decades has resulted in the gradual swelling of the rural-urban income gap. Policies disfavoring the development of the rural sector should be discontinued, otherwise the existing disparity will aggravate.

Correcting Government Regional Biased Policy

In addition to biased regional investment policy, predatory price policy on natural resources used by the government has been the major culprit in worsening regional inequality. For the past several decades, particularly under the previous centrally planned economy, natural resources such as minerals, oil, gas, and even land have been owned by the central government. These resources were supplied to the Eastern regions at very low prices or even free of charge, transferring rents to the coastal areas. The Western provinces, although rich in these resources, benefited very little from these transactions. The latest reform of the state-owned enterprises, further worsened the situation, leaving millions of laid-off workers and degraded environment under the responsibility of local governments.

In order to develop the economy of Western China, the biased policy needs to be corrected and the central government should invest more of its funds for the development of these regions. From table 2.4, it is evident that the Western region shows the highest returns to all kinds of public investments targeted at reducing rural poverty and regional inequality. This is consistent with the national strategy to develop Western China. Investment in agricultural research, education, and rural infrastructure should be the government's top priorities. Considering China's decentralized fiscal system and the western region's small tax base, fiscal transfers from the richer coastal region are required in order to ensure progress in the West.

Reforming the Fiscal System

China is highly decentralized in its government spending. Local government accounts for more than 70 percent of the total government spending. The central government plays a limited role in equalizing regional development through its financial transfers. Most of the transfers from central government to local governments are tax rebates. This mode of transfer is seriously biased and can be considered as one of the major reasons behind the increased regional inequality in China after the introduction of its new financial responsibility system in 1988. Under this system, each province signs a contract with the government with regard to their obligations and responsibilities, thereby—supposedly—providing each province more incentives to develop its economy, so that they can retain more revenues. However, poor provinces suffered since their tax base was low, which eventually led to a dramatic increase of inequality in per capita government budgetary spending. For example, the gap in per capita budget spending between the richest and poorest province has increased from 6.1 times in 1990 to 19.1 times in 1999 (Wong 2003). In order to reduce regional inequality, the central government has to increase its fiscal transfers to support the development of western China.

CONCLUSION

Over the past few decades, China has experienced rapid economic growth and a significant decline in poverty. However, inequality has increased over the same period, causing concerns among policy makers. This chapter sought to review the measures and sources of inequality, and propose a strategy for a more equitable regional development in China.

The review of literature on inequality in China revealed that major sources of regional inequality were both *between* and *within* regions. But the reviews have shown that interregional inequality has increased at a much faster rate than inequality within regions. When regional inequality is decomposed into different sectors, the literature reveals that inequality within the rural sector is much larger than that of the urban sector. Moreover, the increase in regional inequality is mainly due to a slower growth rate in the rural sector.

Therefore, the strategy to reduce the regional inequality should concentrate on the development of the rural sector in western China, as well as promote public investment in the region. Western China lags far behind other regions in terms of economic development, infrastructure endowments, education attainment, and health care provision. Moreover, investing in the western region has high payoffs, as shown by Fan et al. (2004), who indicate that the Western region yielded the highest returns to all types of public investments aimed at reducing rural poverty and regional inequality.

NOTES

1. The Theil index is appealing as it helps to identify the relative contribution of various subgroups of population to the overall inequality. To form population subgroups, the total population is partitioned according to various geographical units such as regions (Eastern, Central, and Western), provinces, and urban-rural areas. Typically, the Theil inequality index is written as the sum of *between* group and *within* group inequality. In a provincial level analysis for example, the Theil index decomposes the total inequality into inequality *between* provinces and inequality *within* provinces.

2. The share of off-farm employment increased from 7 percent to 29 percent between 1978 and 1997, whereas rural non-farm GDP increased from 4 percent to 28 percent over the same period.

BIBLIOGRAPHY

Aaberge, R. and X. Li. "The Trend in Urban Income Inequality in Two Chinese Provinces, 1986–1990." *Review of Income and Wealth* 43, no. 3 (1997): 335–55.

Akita, T. and K. Kawamura. "Regional Income Inequality in China and Indonesia: A Comparative Analysis." Paper presented at the 42nd Congress of the European Regional Science Association, University of Dortmund, Germany, 27–31 August, 2002.

Benjamin, D., L. Brandt, P. Glewwe, and G. Li. "Markets, Human Capital, and Inequality: Evidence from Rural China." *Working Paper*, no. 298. Ann Arbor: William Davidson Institute, University of Michigan Business School, 2000.

Benjamin, D. and L. Brandt. "Markets and Inequality in Rural China: Parallels with the Past." *American Economic Review* 89, no. 2 (1999): 292–95.

Bramall, C. and M. E. Jones. "Rural Income Inequality in China since 1978." *Journal of Peasant Studies* 21, no. 1 (1993): 41–70.

Chan, H. L. and K. T. Chan. "The Analysis of Rural Regional Disparity in China." *Asian Economic Journal* 14, no. 1 (2000): 23–38.

Chowdhury, K., C. Harvie, and A. Levy. "Regional Income Inequality in China." Pp. 238–61 in *Contemporary Developments and Issues in China's Economic Transition*, edited by C. Harvie. New York and London: St. Martin's Press, Macmillan Press, 2000.

Coady, D. P. and L. Wang. "Equity, Efficiency, and Labor-Market Reforms in Urban China: The Impact of Bonus Wages on the Distribution of Earnings." *China Economic Review* 11, no. 3 (2000): 213–31.

Conceição, P. and J. K. Galbraith. "Constructing Long and Dense Time-Series of Inequality Using the Theil Index." *Working Paper*, no. 259. Austin: Levy Economics Institute, Lyndon B. Johnson School of Public Affairs, University of Texas at Austin, 1998.

Cook, S. "Employment, Enterprise and Inequality in Rural China." Pp. 158–79 in *The Chinese Economy under Transition*, edited by S. Cook. New York and London: St. Martin's Press, Macmillan Press, 2000.

Cowell, F. A. *Measuring Inequality*. Hemel Hempstead: Harvester Wheatsheaf (2nd edition), 1995.

Démurger, S., J. D. Sachs, W. T. Woo, S. Bao, G. Chang, and A. Mellinger. "Geography, Economic Policy, and Regional Development in China." *CID Working Paper*, no. 77. Cambridge, MA: Harvard University, Center for International Development, 2001.

Démurger, S. "Infrastructure Development and Economic Growth: An Explanation for Regional Disparities in China?" *Journal of Comparative Economics* 29, no. 1 (2001): 95–117.

Fan, S., L. Zhang, and X. Zhang. "Reforms, Investment, and Poverty in Rural China." *Economic Development and Cultural Change.* 52, no. 2 (2004): 395–421.

Fan, S., L. Zhang, and X. Zhang. "Growth, Inequality, and Poverty in Rural China: The Role of Public Investments." *Research Report*, no. 125. Washington, D.C.: International Food Policy Research Institute, 2002.

Fang, C., X. Zhang, and S. Fan. "Emergence of Urban Poverty and Inequality in China: Evidence from Household Survey." *China Economic Review* 13 (2002): 430–43.

Galbraith, J. K. and Q. Wang. "Increasing Inequality in China: Further Evidence from Official Sources, 1987–2000." *University of Texas Inequality Project Working Paper*, no. 20. Austin: University of Texas, 2001.

Gang, F, D. H. Perkins, and L. Sabin. "People's Republic of China: Economic Performance and Prospects." *Asian Development Review* 15, no. 2 (1997): 43–85.

Guillaumont Jeanneney, S. and P. Hua. "How Does Real Exchange Rate Influence Income Inequality between Urban and Rural Areas in China?" *Journal of Development Economics* 64, no. 2 (2001): 529–45.

Gustafsson, B. and L. Shi. "The Anatomy of Rising Earnings Inequality in Urban China." *Journal of Comparative Economics* 29, no. 1 (2001): 118–35.

Gustafsson, B. and L. Shi. "Income Inequality within and across Counties in Rural China 1988 and 1995." *Journal of Development Economics* 69, no. 1 (2002): 179–204.

Gustafsson, B. and L. Shi. "Types of Income and Inequality in China at the End of the 1980s." *Review of Income and Wealth* 43, no. 2 (1997): 211–26.

Hang Seng Bank. "Western Development and Regional Disparities in Mainland China." *Hang Seng Economic Monthly.* Hong-Kong: Hang Seng Bank, June 2001.

Hare, D. "Rural Nonagricultural Activities and Their Impact on the Distribution of Income: Evidence from Farm Households in Southern China." *China Economic Review* 5, no. 1 (1994): 59–82.

Ho, S. P. S., X. Y. Dong, P. Bowles, and F. MacPhail. "Privatization and Enterprise Wage Structures during Transition: Evidence from Rural Industry in China." *Economics of Transition* 10, no. 3 (2002): 659–88.

Ho, S. P. S. "Rural Non-Agricultural Development in Post-Reform China: Growth, Development Patterns, and Issues." *Pacific Affairs* 68, no. 3 (1995): 360–91.

Howes, S. and A. Hussain. "Regional Growth and Inequality in Rural China." *Working Paper*, no. EF/11. London: London School of Economics, Suntory-Toyota International Centre for Economics and Related Disciplines, 1994.

Howes, S. "Income Inequality in Urban China in the 1980s: Levels, Trends, and Determinants." *Working Paper*, no. EF/3. London: London School of Economics, Suntory-Toyota International Centre for Economics and Related Disciplines, 1993.

Hussain, A., P. Lanjouw, and N. Stern. "Income Inequalities in China: Evidence from Household Survey Data." *World Development*, 22, no. 12 (1994): 1947–57.

Kanbur, R. and X. Zhang. "Which Regional Inequality? The Evolution of Rural-Urban and Inland-Coastal Inequality in China from 1983 to 1995." *Journal of Comparative Economics* 27, no. 4 (1999): 686–701.

Kanbur, R. and Z. Zhang. "Fifty Years of Regional Inequality in China: A Journey Through Central Planning, Reform and Openness." *Review of Development Economics* 9, no. 1 (2005): 87–106.

Ke, B. "Regional Inequality in Rural Development." Pp. 245–55 in *The Third Revolution in the Chinese Countryside—Trade and Development Series*, edited by R. Garnaut, G. Shutian, and M. Guonan. Cambridge, New York, and Melbourne: Cambridge University Press, 1996.

Khan, A. R., K. Griffin, and C. Riskin. "Income Distribution in Urban China During the Period of Economic Reform and Globalization." *American Economic Review* 89, no. 2 (1999): 296–300.

Khan, A. R., K. Griffin, C. Riskin, and R. Zhao. "Household Income and Its Distribution in China." Pp. 25–73 in *The Distribution of Income in China* edited by K. Griffin and R. Zhao. New York: St. Martin's Press, 1993.

Knight, J. and L. Shi. "Cumulative Causation and Inequality among Villages in China." *Oxford Development Studies* 25, no. 2 (1997): 149–72.

Knight, J. and L. Song. "The Spatial Contribution to Income Inequality in Rural China." *Cambridge Journal of Economics* 17, no. 2 (1993): 195–213.

Kung, J. K. S. and Y. F. Lee. "So What If There Is Income Inequality? The Distributive Consequence of Non-farm Employment in Rural China." *Economic Development and Cultural Change* 50, no. 1 (2001): 19–46.

Kwong, T. M. "Markets and Urban-Rural Inequality in China." *Social Science Quarterly* 75, no. 4 (1994): 820–37.

Lee, J. "Regional Income Inequality Variations in China." *Journal of Economic Development* 20, no. 2 (1995): 99–118.

Li, Y. "Rural Income and Wealth Inequality in China—A Study of Anhui and Sichuan Provinces, 1994-1995." Unpublished MSc Thesis, Virginia Polytechnic Institute, Blacksburg, 2000.

Lin, S. "International Trade, Location and Wage Inequality in China." Paper presented at the Cornell-WIDER-LSE Conference on Spatial Inequality and Development, London School of Economics, United Kingdom, 28-30 June 2000.

Litchfield, J. A. "Inequality: Methods and Tools." Text for the World Bank's Web Site on Inequality, Poverty, and Socioeconomic Performance. www.worldbank.org/poverty/inequal/index.htm, 1999.

Ravallion. M. and S. Chen. "When Economic Reform is Faster Than Statistical Reform: Measuring and Explaining Rural Income Inequality in Rural China." *Oxford Bulletin of Economics and Statistics* 61, no. 1 (1999): 33-56.

Reynolds, B. L. "Trade, Employment, and Inequality in Postreform China." *Journal of Comparative Economics* 11, no. 3 (1987): 479-89.

Rozelle, S. "Rural Industrialization and Increasing Inequality: Emerging Patterns in China's Reforming Economy." *Journal of Comparative Economics* 19, no. 3 (1994): 362-91.

Sun, H. and J. Chai. "Direct Foreign Investment and Inter-Regional Economic Disparity in China." *International Journal of Social Economics* 25, nos. 2-4 (1998): 424-47.

Tsui, K. Y. "China's Regional Inequality, 1952-1985." *Journal of Comparative Economics* 15, no. 1 (1991): 1-21.

Tsui, K. Y. "Economic Reform and Interprovincial Inequalities in China." *Journal of Development Economics* 50, no. 2 (1996): 353-68.

Tsui, K. Y. "Trends and Inequalities of Rural Welfare in China: Evidence from Rural Households in Guangdong and Sichuan." *Journal of Comparative Economics* 26, no. 4 (1998): 783-804.

Wan, G. H. "Changes in Regional Inequality in Rural China: Decomposing the Gini Index by Income Sources." *Australian Journal of Agricultural and Resource Economics* 45, no. 3 (2001): 361-81.

Wei, D. Y., S. Kim. "Widening Inter-county Inequality in Jiangsu Province, China, 1950-1995." *Journal of Development Studies* 38, no. 6 (2002): 142-64.

Wei, S. J. and Y. Wu. "Globalization and Inequality: Evidence from within China." *NBER Working Paper*, no. 8611. Cambridge, MA: National Bureau of Economic Research, 2001.

Wong, C. *China: National Development and Sub-National Finance.* Beijing: CITIC Publishing House, 2003.

World Bank. *World Development Indicators.* Washington, D.C.: World Bank, 2003.

Yang, D. T. "Urban-Biased Policies and Rising Income Inequality in China." *American Economic Review* 89, no. 2 (1999): 306-10.

Yao, S. "Economic Growth, Income Inequality and Poverty in China under Economic Reforms." *Journal of Development Studies* 35, no. 6 (1999): 104-30.

Yao, S. and Z. Zhang. "On Regional Inequality and Diverging Clubs: A Case Study of Contemporary China." *Journal of Comparative Economics* 29, no. 3 (2001): 466-84.

Zhang, Z. and S. Yao. "Regional Inequalities in Contemporary China Measured by GDP and Consumption." *Economic Issues* 6, no. 2 (2001): 13-29.

3

Fiscal Decentralization, Non-farm Development, and Spatial Inequality in Rural China

Xiaobo Zhang and Shenggen Fan

Despite continued labor migration and increased capital flows during the past two decades, income inequality is on the increase in rural China (see chapter 2). Hence, in addition to the fragmentation of factor markets, there should be other underlying causes for this phenomenon. The uneven regional development in non-farm activities in the rural sector has been regarded as another major driving force that accounts for the changes in rural regional inequality (Hare 1994; Rozelle 1994). An important question that arises in this context is: why has growth in rural non-farm economy diverged across regions? To answer this, we will focus on the role of uneven distribution of infrastructure and human capital. After showing their significant impact, we will argue that the increasing regional disparity in public capital is largely a result of fiscal decentralization.

Most previous studies on the evolution of regional inequality in China have used data at a provincial level, because reliable official data at a microlevel for China as a whole, over a long period of time, are usually not available (Lyons 1991; Tsui 1991; Jian et al. 1996; Fleisher and Chen 1997; Kanbur and Zhang 1999). These studies in general find that the rise in interregional inequality, especially the increasing gap between the coastal and inland regions, is a major cause of worsening regional inequality. This suggests the need for an inland-based targeting policy to reduce regional inequality. However, if the within-zoneal inequality has been increasing as fast as interregional inequality, then the simple region-based targeting policy may not be sufficient. In order to examine this issue more in depth, in this chapter we will use a more disaggregate data set at a county level for the period 1985–1996. To our knowledge, this is one of the first attempts to examine the changing patterns of China's regional inequality using county-level data.[1]

By pooling the newly available public input data at the county level from the China Agricultural Census in 1996 with production and output data from the State Statistical Bureau (SSB), the importance of different types of public inputs for both farm and non-farm activities can be evaluated. By applying a regression-based decomposition method, we can quantify the distributional effect of various public inputs. At a time when reducing regional inequality seems to be at the top of the Chinese leadership's policy agenda, this analysis provides a crucial understanding of a broader panel of policy instruments.

Few studies have linked public inputs with inequality despite their relevance regarding distributional policy. The focus of these studies however, is normally on only one type of public investment or input, such as road infrastructure (Martin 1999). The government may nevertheless have various forms of public inputs available for equity purposes. Each public good has its own specific characteristics and externalities, and therefore may have a different impact on growth and distribution. However, both the theoretical and empirical literature have largely ignored the comparison of trade-offs amongst different types of public inputs. Hence, little is known about the size and distribution of benefits of public inputs for farm and non-farm activities. In this chapter, we will attempt to fill this knowledge gap and test two key hypotheses. Firstly, each type of public input has a different distributional impact on farm and non-farm activities. Secondly, the uneven distribution of public inputs is a major contributing factor to regional inequality, in particular in terms of non-farm activities. We furthermore argue that the current fiscal decentralization is one of the primary reasons for the observed uneven regional distribution of public inputs.

The chapter is organized as follows. The second section examines regional inequalities based on different regional groups and sources using a decomposition analysis. The third section discusses the distributional importance of various types of public inputs on farm and non-farm activities and additionally explores the role of fiscal decentralization. The last section presents conclusions and policy implications.

THE EVOLUTION OF RURAL REGIONAL INEQUALITY

Inequality Decomposition by Regions

Tables 3.1, 3.2, and 3.3 report on the inequalities measured by Theil's second, or "population-weighted" entropy index (TE) and decomposition analysis under different regional groupings. Table 3.1 presents the overall inequality and the corresponding decompositions, using total rural output per capita as a measure of economic development level. The TE inequality measure of rural output per capita is calculated for each year, using the rural population of each country as weight.

Table 3.1. Inequality Decomposition: Contribution to Overall Inequality

Year	Overall Inequality	Within Province	Between Province	Within Zone	Between Zone	Polarization
1985	0.108	0.054	0.054	0.081	0.027	0.338
		(50.05)	(49.95)	(74.74)	(25.26)	
1986	0.122	0.064	0.058	0.087	0.035	0.396
		(52.59)	(47.41)	(71.62)	(28.38)	
1988	0.170	0.086	0.083	0.116	0.053	0.461
		(50.87)	(49.13)	(68.45)	(31.55)	
1989	0.179	0.094	0.085	0.125	0.055	0.438
		(52.55)	(47.45)	(69.53)	(30.47)	
1990	0.177	0.088	0.089	0.121	0.056	0.464
		(49.81)	(50.19)	(68.31)	(31.69)	
1991	0.204	0.107	0.097	0.142	0.062	0.433
		(52.60)	(47.40)	(69.78)	(30.22)	
1992	0.256	0.136	0.121	0.170	0.086	0.509
		(52.92)	(47.08)	(66.26)	(33.74)	
1993	0.323	0.185	0.138	0.220	0.102	0.464
		(57.18)	(42.82)	(68.28)	(31.72)	
1994	0.325	0.170	0.155	0.216	0.110	0.509
		(52.27)	(47.73)	(66.29)	(33.71)	
1996	0.290	0.154	0.135	0.199	0.091	0.455
		(53.27)	(46.73)	(68.75)	(31.25)	
Growth (%)	167.5	184.8	150.3	146.1	231.1	34.5
		(6.44)	(–6.45)	(–8.02)	(23.74)	

Notes:
1. The figures above are calculated by the authors. The figures in the parentheses refer to the contributions of each component to the total inequality in the second column. The bottom line in the parentheses represents the changes in percentage of the shares of contributions from 1985 to 1996. Due to lack of data for some variables in 1987 and 1995, these two years are not included in regression.
2. The zonal decomposition in columns 5 and 6 is between coastal provinces and inland provinces.
3. The polarization index is defined as the ratio of *between*-zone inequality and *within*-zone inequality following Zhang and Kanbur (2001).

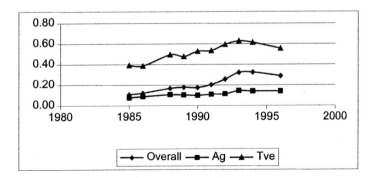

Figure 3.1. Regional Inequality (1986–1996)

Table 3.2.   Inequality Decomposition: Contribution to Inequality in Farm Activities

Year	Inequality	Within Province	Between Province	Within Zone	Between Zone	Polarization
1985	0.077	0.037	0.039	0.057	0.019	0.337
		(48.90)	(51.10)	(74.77)	(25.23)	
1986	0.088	0.046	0.041	0.065	0.023	0.355
		(52.96)	(47.04)	(73.78)	(26.22)	
1988	0.107	0.052	0.055	0.072	0.035	0.480
		(48.88)	(51.12)	(67.55)	(32.45)	
1989	0.103	0.053	0.050	0.075	0.029	0.382
		(51.66)	(48.34)	(72.38)	(27.62)	
1990	0.099	0.050	0.049	0.074	0.025	0.337
		(50.90)	(49.10)	(74.77)	(25.23)	
1991	0.109	0.056	0.054	0.081	0.028	0.345
		(50.83)	(49.17)	(74.37)	(25.63)	
1992	0.114	0.060	0.054	0.081	0.033	0.409
		(52.50)	(47.50)	(70.95)	(29.05)	
1993	0.146	0.088	0.058	0.109	0.037	0.338
		(60.11)	(39.89)	(74.72)	(25.28)	
1994	0.141	0.071	0.070	0.096	0.045	0.464
		(50.39)	(49.61)	(68.31)	(31.69)	
1996	0.141	0.071	0.070	0.096	0.046	0.477
		(50.63)	(49.37)	(67.71)	(32.29)	
Growth (%)	84.5	91.1	78.3	67.1	136.1	41.3
		(3.5)	(–3.4)	(–9.4)	(28.0)	

Note: See table 3.1.

Similarly, the inequalities measured in terms of farm output per capita and non-farm output per capita and their decompositions are presented in tables 3.2 and 3.3, respectively. Figure 3.1 presents the time paths of these inequalities from 1985 to 1996.

Two characterizations are immediately apparent from the above tables and figure 3.1. First, regional inequality shows a sharply increasing trend in China, during this period of fast growth. The overall inequality, farming inequality and non-farm inequality have risen by 167.5 percent, 84.5 percent, and 41.4 percent, respectively, from 1985 to 1996. Interestingly, the rate of change in overall inequality is faster than that in farm inequality and non-farm inequality, reflecting the regional divergence in the composition of rural economies. This result confirms the conclusion of earlier studies based on more aggregate provincial data. Second, there is a wide gap between the levels of inequality in the non-farm and farm sectors. For instance, in 1996, the overall inequality, and the inequalities in the farm and non-farm sectors were 0.290, 0.141, and 0.555, respectively. This suggests that we should pay more attention to the large regional disparity in non-farm activities.

A deeper understanding of the changing patterns of inequalities can be gained through a further investigation of the inequalities from a regional

Table 3.3. Inequality Decomposition: Contribution to Inequality in Non-farm Activities

Year	Inequality	Within Province	Between Province	Within Zone	Between Zone	Polarization
1985	0.393	0.193	0.200	0.244	0.149	0.613
		(49.15)	(50.85)	(62.01)	(37.99)	
1986	0.387	0.172	0.215	0.213	0.174	0.817
		(44.44)	(55.56)	(55.02)	(44.98)	
1988	0.495	0.211	0.285	0.247	0.248	1.004
		(42.55)	(57.45)	(49.89)	(50.11)	
1989	0.477	0.237	0.240	0.283	0.194	0.685
		(49.63)	(50.37)	(59.34)	(40.66)	
1990	0.528	0.207	0.321	0.250	0.278	1.113
		(39.16)	(60.84)	(47.32)	(52.68)	
1991	0.534	0.250	0.284	0.313	0.221	0.705
		(46.76)	(53.24)	(58.65)	(41.35)	
1992	0.595	0.303	0.291	0.363	0.232	0.639
		(50.98)	(49.02)	(61.01)	(38.99)	
1993	0.627	0.305	0.323	0.360	0.267	0.743
		(48.59)	(51.41)	(57.37)	(42.63)	
1994	0.615	0.284	0.330	0.350	0.265	0.758
		(46.23)	(53.77)	(56.89)	(43.11)	
1996	0.555	0.280	0.276	0.337	0.219	0.650
		(50.35)	(49.65)	(60.61)	(39.39)	
Growth (%)	41.4	44.8	38.0	38.2	46.6	6.1
		(2.4)	(–2.4)	(–2.3)	(3.7)	

Note: See table 3.1.

perspective. The third column through the sixth column in tables 3.1, 3.2, and 3.3 present the results of the decomposition analyses for the three inequalities under two kinds of groupings. First, inequalities in per capita total output, farm output, and non-farm output are decomposed into within-province inequality and between-province inequality components, according to the formula proposed by Shorrocks (1984). Second, a similar decomposition is conducted across the coastal-inland divide. The east or coastal zone includes the following provinces: Beijing, *Liaoning, Tianjin, Hebei, Shandong, Jiangsu, Shanghai, Zhejiang, Fujian, Guangdong,* and *Guangxi.* We classify all the remaining provinces as inland for our study.

Three features are apparent from the decomposition analyses in the three tables. First, while the inequalities in each of the components have distinctly increased, the changes in the shares of contributions are less dramatic. The inequalities in all the components have increased by at least 38 percent. The between-zone (coast-inland) overall inequality has increased by 231 percent, the largest amongst all the components. In terms of shares, the contributions of most components to total inequalities have changed less than 10 percent, except for between-zone inequalities in the entire rural sector and farm sector. The coastal-inland contributions to the overall inequality and inequality in farm activities have increased by 24 percent and 28 percent, respectively.

Second, the within-province and within-zone contributions to inequalities were rather high. About half of the total inequalities could be explained by within-province contributions to the three inequalities. The within-zone inequality accounted for about 70 percent of total inequality. This finding suggests that the within-province and within-zone inequalities should also be taken into account when the government implements strategies to reduce overall regional inequality.

Third, there existed a polarization trend between the coastal and inland regions. Following the methodology proposed in Zhang and Kanbur (2001), we use the ratio of between-zone inequality to within-zone inequality to represent the polarization between coastal and inland regions. The within-zone inequality represents the spread of the distributions in a region, while the between-group inequality is a measure of the distance between the zone means. The ratio of between-zone inequality to within-zone inequality can thus be regarded as a scalar polarization index because it captures the average distance between the zones in relation to the differences seen within zones. As income differences within zones diminish, i.e., as the groups become more homogeneous internally, differences across groups are, relatively speaking, magnified and "polarization" is higher. Similarly, as the groups means drift apart, polarization increases, given the within-group differences.

The coastal-inland polarization measured for overall, farm and non-farm sectors have increased by 34.5 percent, 41.3 percent, and 6.1 percent, respectively. Increasing polarization between the coastal and inland regions is a finding consistent with previous studies conducted at the provincial level (Zhang and Kanbur 2001). Interestingly, the trend of increasing polarization in the non-farm sector is much lower than that in the farm sector. This suggests a massive labor transformation from agricultural production to nonagricultural activities within each region. In brief, the coast-inland contributions to overall inequality have increased steadily.

Inequality Decomposition by Activities

Figure 3.1 has shown that inequality in the non-farm sector is much higher than that in the farm sector, and overall inequality is increasing faster than both farming and non-farm inequalities. How much have non-farm activities contributed to the overall increase in inequality? Table 3.4 presents a decomposition analysis by activities. We divided per capita output into per capita agricultural and non-farm outputs. Then overall inequality was decomposed, following the methodology outlined in Shorrocks (1982) and applied in Zhang and Fan (2004).

Table 3.4. Overall Inequality Decomposition by Activities

Year	Overall Inequality	Farm Contribution (%)	Non-farm Contribution (%)
1985	0.108	42	58
1986	0.122	32	68
1988	0.170	25	75
1989	0.179	24	76
1990	0.177	23	77
1991	0.204	22	78
1992	0.256	12	88
1993	0.323	10	90
1994	0.325	9	91
1996	0.290	11	89
Growth (%)	167.5	–73.6	53.5

One feature is evident from the table: the contribution of non-farm activities to overall inequality has increased dramatically, while the contribution of farm activities has declined. By 1996, almost 90 percent of overall inequality can be explained by the uneven regional distribution of non-farm activities. This result confirms the previous findings that changing patterns of rural inequality have been closely related to the shift in the structure of the rural economy from agricultural production to non-farm activities (Hare 1994; Rozelle 1994). The questions that next arise are: Why has the non-farm economy diverged so rapidly? And, are there policy options to reverse this trend?

The Role of Public Inputs in Inequality

Having analyzed the changing inequality patterns from regional and income source perspectives, the instruments crucial in determining the effectiveness in inequality-reduction strategies can now be identified. The government has many instruments available for implementing growth and distributive policies. Amongst these instruments, which ones influence regional inequality the most, controlling for the effects of others? This is one of the main issues to be analyzed.

Here, we consider only three public inputs, largely due to data availability: education, roads, and rural communication. Because of lack of systematic data on public inputs at a county level, we can only use the public input data from the Agricultural Census in 1996. For a detailed description of the data source, see Fan et al. (2000). The education variable is defined as the average years of schooling for each county. The road variable refers to road length per rural resident. Rural communication is the number of telephones per rural resident.

In order to quantify the importance of public inputs, two production functions for the agricultural and nonagricultural sectors were estimated.

For agricultural production, we regress the logarithm of farm labor productivity on the logarithm of conventional inputs (arable land, fertilizer, machinery, and irrigation) as well as on the logarithm of public inputs, including education, roads, and rural communication. Because data on capital for non-farm activities at the county level were not available, per capita consumption of electricity is used as a proxy for capital in the regression of non-farm labor productivity. In this regression analysis, the same three public inputs as in the regression for agricultural production are included, apart from the electricity variable.

Table 3.5 reports the estimated coefficients for the above two regressions. The adjusted R^2s for the agricultural and nonagricultural production are high at 0.961 and 0.873, respectively, implying good fitness. All the variables are in logarithms, thus the coefficients in the table are equivalent to the elasticities of a particular input with respect to a particular output. The coefficients for all the variables are positive and statistically significant at the 5 percent level. In the equation for agricultural production, the coefficient for land is high at 0.528, suggesting the land being a major determinant of labor productivity in farm activities. Among the three public inputs, schooling and rural communication have the largest and second largest elasticities. In the equation for non-farm productivity, the coefficients for schooling and phones also rank first and second in terms of magnitude. In particular, the coefficient for the variable of schooling is larger than 1, implying a significant role of education in the growth of non-farm labor productivity.

With the estimated coefficients in table 3.5, we can apply the same factor decomposition methodology used in the above section in order to analyze the contributions of public inputs to inequalities. Figure 3.2 graphs the contributions of the three major public inputs to inequalities in farm and non-farm labor productivity. One striking result is that the contribution shares of the three public inputs are much higher for non-farm labor productivity than for farm labor productivity. The uneven regional distribution of school-

Table 3.5. Estimation Results for Labor Productivity

	Farm	Non-farm
Intercept	2.153 (0.161)	8.323 (0.274)
Land	0.528 (0.029)	
Machinery	0.216 (0.015)	
Fertilizer	0.055 (0.026)	
Irrigation	0.203 (0.023)	
TVE electricity	0.082 (0.015)	
Road	0.072 (0.016)	0.120 (0.035)
Phone	0.108 (0.014)	0.476 (0.034)
Schooling	0.179 (0.033)	1.285 (0.074)
R^2-Adjusted	0.961	0.873

Note: The figures in parentheses refer to the standard deviation.

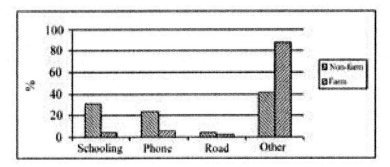

Figure 3.2. Contributions of Public Inputs to Regional Inequality

ing contributed more than 30 percent to non-farm inequality but only 4 percent to farm inequality. The contribution of rural communication to non-farm inequality was 23.5 percent, while its contribution to farm inequality was only 5.5 percent. In total, the contributions of the three public inputs accounted for 12 percent and 58 percent, respectively, of total inequalities in farm and non-farm sectors. The results reveal that improving education levels in less developed areas is the dominant strategy to foster growth of the non-farm sector and reduce non-farm inequality. Because non-farm activities play a dominant role in explaining overall inequality, across-the-board development of education will undoubtedly contribute to the improvement of the overall regional distribution. The key policy implication of this result is that the current nine-year compulsory education system should be further strengthened and strictly enacted, especially in the less developed areas. However, the challenge is how to finance public goods, in particular in the less developed regions, under the current system of fiscal decentralization.

Since fiscal decentralization was introduced, the ability to invest in public infrastructure and services to a large degree depends upon the sources of local revenue. Regions initially endowed with a broader non-farm tax base do not need to charge heavily on the existing and new firms to finance public goods provision, thereby providing a healthy investment environment for the non-farm sector to grow. However, regions with agriculture as the main economic activity have little room to collect additional revenues from the agricultural sector and have an increased tax burden on the existing non-farm sector, thereby greatly inhibiting its growth.

CONCLUSION

In this chapter, a systemic examination is presented of the changing patterns of China's rural regional inequality using a long-period county-level data set. On the basis of the available agricultural census and other data, we further investigated the importance of different kinds of instruments that the government can use for reducing regional inequality. Some findings have emerged from this analysis.

During the study period (1986–1996), China has experienced an upward trend in rural regional inequality. In terms of levels, within-province and within-zone inequality are high, explaining a large portion of the overall inequalities. In terms of trends, the coast-inland gap has widened for both farm and non-farm activities. Thus, the government should pay attention to both interregional and intraregional inequality, when blueprinting growth and distributive policies. A single region-based policy may not be a panacea for reducing regional inequality.

From a structural point of view, the increasing regional inequality can largely be explained by the increasing uneven distribution of non-farm activities, especially by the massive labor shift out of the traditional agricultural sector. By 1996, the distribution of non-farm activities contributed about 90 percent to the overall regional inequality. Understanding the driving forces behind the development of non-farm activities is therefore the key to design rural development policies.

Our decomposition analysis shows that schooling and rural communication have important effects on the growth and distribution of the non-farm sector. However, under the current fiscal arrangements between local and central governments, local governments are primarily responsible for financing the above two activities. Lack of local revenues might be a major reason for the poor regions to underinvest in public goods, which further worsens the existing regional disparity. How to improve the local-central fiscal arrangements and to increase the level of public goods provision in less developed regions of China demands further research.

NOTE

1. The funding from ACIAR for the authors in data collection, data compiling, and paper preparation is acknowledged here.

BIBLIOGRAPHY

Fan, S., X. Zhang, and L. Zhang. "Infrastructure and Regional Economic Development: New Evidence from the Agricultural Census." Paper presented at the International Seminar on the Chinese Census Results, Beijing, September 19–22, 2000.

Fleisher, B. M. and J. Chen. "The Coast-Noncoast Income Gap, Productivity, and Regional Economic Policy in China." *Journal of Comparative Economics* 25, no. 2 (1997): 220–36.

Hare, D. "Rural Nonagricultural Activities and Their Impact on the Distribution of Income: Evidence from Harm Households in Southern China." *China Economic Review* 5, no. 1 (1994): 59–82.

Jian, T., J. Sachs, and A. Warner. "Trends in Regional Inequality in China." *National Bureau of Economic Research Working Paper*, no. 5412. Cambridge, MA: NBER, 1996.

Kanbur, R. and X. Zhang. "Which Regional Inequality? The Evolution of Rural-Urban and Inland-Coastal Inequality in China, 1983–1995." *Journal of Comparative Economics* 27, no. 4 (1999): 686–701.

Lyons, T. P. "Interprovincial Disparities in China: Output and Consumption, 1952–1987." *Economic Development and Cultural Change* 39, no. 3 (1991): 471–506.

Lyons, T. P. *The Economic Geography of Fujian: A Sourcebook*. vol. 2. Ithaca, NY: Cornell East Asia Series, Cornell University East Asia Program, 1997.

Martin, P. "Public Policies, Regional Inequalities and Growth." *Journal of Public Economics*, 73 (1999): 85–105.

Rozelle, S. "Rural Industrialization and Increasing Inequality: Emerging Patterns in China's Reforming Economy." *Journal of Comparative Economics* 19, no. 3 (1994): 362–91.

Shorrocks, A. F. "The Class of Additively Decomposable Inequality Measures." *Econometrica* 48, no. 3 (1980): 613–25.

Shorrocks, A. F. "Inequality Decomposition by Factor Components." *Econometrica* 50, no. 1 (1982): 193–211.

Shorrocks, A. F. "Inequality Decomposition by Population Subgroups." *Econometrica* 52, no. 6 (1984): 1369–85.

Tsui, K. "China's Regional Inequality, 1952–1985." *Journal of Comparative Economics* 15, no. 1 (1991): 1–21.

Yang, D. "Urban-Biased Policies and Rising Income Inequality in China." *American Economic Review* (Paper and Proceedings) 89, no. 2 (1999): 306–10.

Zhang, P. "Zhongguo moncun quyujian jumin de shouru fenpei [International Income Distribution in Rural China." *Jingji Yanjiu* [Economic Research] 2 (1992): 62–69.

Zhang, X. and R. Kanbur. "What Difference Do Polarisation Measures Make? An Application to China." *Journal of Development Studies* 37, no. 3 (February 2001): 85–98.

Zhang, X. and S. Fan. "Public Investment and Regional Inequality in Rural China." *Agricultural Economics* 30, no. 2 (2004): 89–100.

4

The Contribution of Diversification to the Growth and Sustainability of Chinese Agriculture

Funing Zhong

According to the customs records dating back to 1863, China was a net grain importer for close to a century. Its position in the world grain market was reversed during the 1950s, as farmers' production incentives were significantly enhanced by the redistributive land reform. However, there were also misallocations of resources and substantial disincentives as a result of the government procurement programs and central planning, coupled with collectivization and establishment of Rural People's Communes, which depressed further growth of Chinese agriculture. Following the famine of the late 1950s and early 1960s that turned China again into a net grain importer, the government pushed the expansion of grain production at the cost of almost all other sectors in the rural economy. This included demarcating a large section of arable land, previously used for other production, for growing grain with an increased application of chemical fertilizers and pesticides. Simultaneously, several other types of land, such as forests and grassland, marginal land in dry pastoral areas and steep hill areas, as well as marshland and, even, lakes, were converted into grain fields.

These extreme measures resulted in deforestation, salinization, desertification, falling ground water levels, and other serious effects on the environment, thus damaging China's potential capacity for agricultural development. Moreover, there was a decline in the production incentives because of the resultant increase in production costs and decrease in the net income from agriculture, especially from grain production. Thus the two decades preceding the introduction of reforms toward the end of the 1970s, were characterized by very slow growth in Chinese agriculture. Despite all efforts taken by the government to enhance growth during that period, the country's grain production stagnated.

The reforms in agriculture that followed have been well documented. In fact, much of its success is attributed to the introduction and widespread adoption of the Household Responsibility System (HRS). However, the contribution of diversification to the growth of agriculture and the rural economy at large is very much underestimated. An important contributing factor to agricultural and rural economic growth has been the improved resource allocation efficiency through diversification or sectoral structural adjustment—something that might equal the impact of improved incentives through the HRS. Moreover, this structural adjustment and the related market-oriented reforms may have reduced the stress on the environment and hence enhanced the sustainability of China's agricultural development.

This chapter aims to analyze, in section two, the contribution of diversification to Chinese agriculture and the rural economy during the past two decades, and to discuss its implications for future sustainable development, in the final section.

DIVERSIFICATION AND AGRICULTURAL GROWTH

Right from the very beginning of the reforms in 1978 China has witnessed great success in boosting agricultural production. Most observers attribute the fast growth in Chinese agriculture to the improved production incentives facilitated by the introduction of the HRS (Lin 1992; Carter and Zhong 1988; Carter et al. 1996). Roughly the same extent of growth however, has been suggested by some other studies (e.g., Zhong and Zhu 2000), to have occurred because of the diversification or sectoral structural adjustment. Although the HRS and other incentive measures may have stimulated agricultural production with increases in yield (i.e., productivity gains in each line of production), it is diversification that is likely to have provided greater room for improved resource allocation. It might also be responsible for a long-run sustainable growth in agriculture and the improvement of farmers' income.

Chinese agriculture has for the most been dominated by the production of grains. Before the onset of the reforms, more than 80 percent of the total output was accounted for by crops, and 85 percent of this was area sown with grains (largely rice). Not only was it the least profitable sector, farmers even had to incur losses from time to time. As a result, some of the resources were shifted to the production of other crops or to other sub-sectors of the agricultural economy when farmers were given certain powers in decision-making due to market-oriented reforms.

Table 4.1 shows that although output in the crop sector increased by more than 170 percent in real terms during the period 1978–2001, indicating a remarkable 4.4 percent annual growth, the other sectors increased even faster.

Table 4.1. Shares of Gross Output Value of Farming, Forestry, Animal Husbandry, Fishery, and Related Indices (1978–2001)

Year	Shares of Gross Output Value of Farming, Forestry, Animal Husbandry, and Fishery (%)					Indices of Gross Output Value of Farming, Forestry, Animal Husbandry, and Fishery (1978=100)				
	Total	Farming	Forestry	Animal Husbandry	Fishery	Total	Farming	Forestry	Animal Husbandry	Fishery
1978	100.0	80.0	3.4	15.0	1.6	100.0	100.0	100.0	100.0	100.0
1980	100.0	75.6	4.2	18.4	1.7	109.0	106.7	113.8	122.6	104.0
1985	100.0	69.2	5.2	22.1	3.5	160.7	145.3	176.1	203.0	185.2
1990	100.0	64.7	4.3	25.7	5.4	202.6	170.4	179.2	281.4	348.9
1991	100.0	63.1	4.5	26.5	5.9	210.1	171.9	193.6	306.2	373.2
1992	100.0	61.5	4.7	27.1	6.8	223.5	179.1	208.5	331.1	430.4
1993	100.0	60.1	4.5	27.4	8.0	241.0	188.4	225.1	369.1	572.1
1994	100.0	58.2	3.9	29.7	8.2	261.7	194.5	245.2	430.8	686.5
1995	100.0	58.4	3.5	29.7	8.4	290.2	209.8	257.4	494.5	819.7
1996	100.0	60.6	3.5	26.9	9.0	317.5	226.2	272.1	550.9	934.5
1997	100.0	58.2	3.4	28.7	9.6	338.8	236.4	281.1	606.5	1,041.9
1998	100.0	58.0	3.5	28.6	9.9	359.1	247.9	289.2	651.4	1,133.6
1999	100.0	57.5	3.6	28.5	10.3	376.0	258.6	298.5	681.4	1,215.3
2000	100.0	55.7	3.8	29.7	10.9	389.5	262.2	314.6	724.3	1,294.3
2001	100.0	55.2	3.6	30.4	10.8	405.9	271.7	312.4	769.9	1,344.7

Source: China Statistical Yearbook, various editions.
Note: Indices are expressed in real terms.

Animal production increased by 669.9 percent during the same period, while fishery grew by an amazing 1,244.7 percent. The two sectors had annual growth rates of 9.3 percent and 12.0 percent, respectively. Timber and fruits do not form part of "forestry production" as defined in Chinese agricultural statistics, because the former is regarded as part of industrial production and the latter as part of the crop sector. In spite of this, the growth of the forestry sector still surpasses that of the crop sector. Since animal and fishery productions increased much faster than the crop sector production, the annual growth rate of the whole agricultural sector was 6.2 percent, approximately 50 percent higher than that of the crop sector alone. Apparently, the crop sector had lost a large share of its resources to other subsectors. Such structural changes have brought about a higher growth in agriculture as a whole.

Following the rapid growth of other sectors during the reform era, the structure of Chinese agriculture changed dramatically. While the share of animal production and fisheries showed an increase from 15.0 and 1.6 percent to 30.4 and 10.8 percent, respectively, the share of crop agriculture declined from 80 percent in 1978 to just over 55 percent in 2001.

This structural adjustment of the agricultural sector is likely to be equally significant within each of the subsectors. For example, during the above time period, the yield per unit of sown area increased by roughly 2 percent per year, but the output value of the crop sector increased by 4.4 percent in real terms, although the total sown area remained virtually unchanged. By the use of real output value, the effect of price change across the years can be corrected. This leaves the change in the crop mix, resulting from the shift of grain to cash crops, and from low quality and price to high quality and value, to attribute the differences between the growth rates to. Such structural changes within subsectors are likely to be significant in animal production and fisheries. This is because the products in these two subsectors are more market-oriented and profitable so as to cater to the changing demands of consumers. Elsewhere, we have shown that sectoral structural adjustment has contributed more than 40 percent to the total growth in the crop sector during the reform era (Zhong and Zhu 2003). This level of adjustment was possible only through increases in input applications, improved technology, and innovations in institutional arrangements. The same study also suggests that the reallocation of resources among cropping, forestry, animal husbandry, and fisheries has contributed 43 percent to the growth of agriculture as a whole, whereas the growth in each subsector has contributed the other share of 57 percent. If the within-sector reallocation of resources contributed to the growth of the crop sector, the same could also happen in all the other subsectors. Overall structural adjustment contributed 66 percent to the total growth of agriculture, out of which two-thirds has been caused by the reallocation of resources among subsectors and the rest, one-third, by within-sector adjustments.

Diversification in agriculture has significantly reduced the negative effects on the environment of the "grain first" policy. In the early 1980s, a nationwide debate was conducted on the comparison between yields per unit of arable land and the economic returns of the double- and/or triple-harvest cropping system in southern *Jiangsu* and similar areas, which formed the first challenge to the current farming policy and practice. This resulted in a declined intensity in the multicrop farming system, in favor of farmers' income. A shift of resources from grain to other more profitable output was encouraged because of the success in reducing farming intensity, which in turn was a strong signal that economic returns and farmers income ought to play a more important role in making production decisions. Furthermore, the reduction in farming intensity reduced the stress on the environment as well. Marginal arable land in hilly and pastoral areas was changed back to forests and grassland. The change in land usage might have been introduced with the intention of increasing farmers' income through more efficient resource allocation under suitable natural conditions. In turn, it is the effect of change on improving the environment that has been getting more and more recognition. Recently, large-scale programs of "turning arable land back to grassland" and "turning arable land back to forests" have been started in North, Northwest, and Southwest China, which may have a great impact on the environment of such fragile areas.

Contribution of Diversification to Farmers' Income

With the growth of total agricultural output in the reform era, there was a corresponding increase in the Chinese farmers' income. An increase of up to seventeen times in current values, or over four times in real terms (see table 4.2) was visible with the growth of per capita net annual income from 133.6 *yuan* in 1978 to 2,366.4 *yuan* in 2001. This was a very significant achievement compared to the 134 percent growth in nominal terms from 57 *yuan* in 1952 to 133.6 *yuan* in 1978, before the reform era.

Table 4.2. Per Capita Net Annual Income of Rural Households

Year	Value (yuan)	Index (1978 = 100)
1978	133.6	100.0
1980	191.3	139.0
1985	397.6	268.9
1990	686.3	311.2
1995	1,577.7	383.7
2000	2,253.4	483.5
2001	2,366.4	503.8

Source: China Statistical Yearbook (2002).
Note: Values are in current prices; indices are in real terms.

Similar to the growth of the total agricultural production, the growth of farmers' income is attributable to both productivity and diversification gains. The increase in productivity requires an increased application of modern inputs. Its effect on farmers' income will generally be less than that on total output. Therefore, we can argue that diversification may have made a greater contribution to farmers' income than to production.

A detailed picture of rural incomes by source and by types of income can be obtained from table 4.3, with data available since 1985. Income from primary agricultural production showed an increase of 290.6 percent in nominal terms. Though very remarkable in itself, this rise was far less impressive than the growth of incomes from the secondary and tertiary sectors, which increased by more than seventeen times and twelve times, respectively—during the same time period (1985–2001). The contributory steps taken by diversification into other, noncrop sector activities, and employment in secondary and tertiary industries, clearly explain why the increase in average rural income was substantially higher than in farming.

In household business, net income from farming rose by 327.3 percent, whereas that from animal husbandry and fisheries increased by, respectively, 307.9 and 704.2 percent, during the period 1985–2001. However, the comparison is somewhat misleading as the government raised the grain procurement price by more than 100 percent between the years 1993 and 1996. Additionally, the procurement prices for cotton and edible oils were also increased significantly. Furthermore, if we divide the period into three subperiods, the income from animal husbandry increased faster than that from farming in two of the three periods: from 1985 to 1990 (86.3 vs. 70.5 percent) and from 1995 to 2001 (65.8 vs. 8.0 percent). Evidently, income from farming has been more or less constant since the mid-1990s, while the other subsectors account for the near-full increase in income from rural household business during that period.

The changes in the structure of the total net income led to a decline in the share of income from agriculture, or primary production, from a high of 75 percent in 1985 to a mere 49.2 percent in 2001. In comparison, incomes from secondary and tertiary production increased from 7.4 and 10.0 percent in 1985 to 22.5 and 22.6 percent in 2001, respectively (table 4.4). Taking the incomes generated by household business in farming, forestry, animal husbandry, and fisheries together unmistakably shows a decline in the share from farming from 83.5 percent in 1995 to 76.7 percent in 2001, whereas the income from forestry, animal husbandry, and fisheries increased from 16.5 to 23.3 percent.

Had the increase in net incomes from all other sources shown the same pace as the income from farming, per capita net income would have been 1,699.04 *yuan* in 2001, almost 30 percent less than that actually earned by rural households. Thus it is very clear that diversification accounts for a

Table 4.3. Composition of Per Capita Annual Net Income of Rural Households (in yuan)

Item	1985	1990	1995	2000	2001
Net Income	397.60	686.31	1,577.74	2,253.42	2,366.40
By Source					
Wages Income	72.15	138.80	353.70	702.30	771.90
Net Income from Household Business	295.98	518.55	1,125.79	1,427.27	1,459.63
Farming	202.10	344.59	799.44	833.93	863.62
Forestry	6.16	7.53	13.52	22.44	22.10
Animal Husbandry	51.96	96.81	127.81	207.35	211.96
Fishery	3.59	7.11	15.69	26.95	28.87
Industry	2.18	9.15	13.63	52.67	54.57
Construction	7.41	12.18	34.53	46.73	45.43
Transport, Post, and Telecommunication Services	8.47	13.45	27.76	63.63	63.17
Wholesale and Retail Trade and Catering Services	6.13	12.69	34.26	78.54	78.84
Social Services	3.27	6.55	17.18	28.09	30.12
Culture, Education, and Health Care				6.86	7.56
Others	4.73	8.49	41.97	60.08	53.38
Transfer Income and Property Income	29.47	28.96	98.25	123.85	134.87
Grouped by Type of Income					
Productive Income	367.69	657.35	1,479.49	2,129.58	2,231.58
Primary	298.28	510.86	996.51	1,125.34	1,165.17
Secondary	29.47	70.68	287.24	488.89	532.61
Tertiary	39.95	75.81	195.74	515.35	533.80
Nonproductive Income	29.91	28.96	98.25	123.84	134.82

Source: China Statistical Yearbook (2002).
Note: Values are in current prices.

significant part of the farm income. Possibly, more than half of the net income of farmers today has resulted from the diversification or reallocation of resources during the last twenty years or so.

The diversification-induced rapid income growth is likely to have contributed toward a reduction of stress on the environment, and a more sustainable development in Chinese agriculture. The driving force behind the degradation of the environment was the widespread poverty in rural China that forced farmers to overuse their land, which again resulted in a vicious circle of deteriorating environment and further reduction in incomes, in many cases. It was this "poverty trap" that diversification helped break through. The increasing production costs is yet another issue in Chinese agriculture, which is facing tough foreign competition, following the country's accession to the World Trade Organization (WTO) in 2001. Furthermore,

Table 4.4. Composition of Per Capita Annual Net Income of Rural Households (in %)

Item	1985	1990	1995	2000	2001
Net Income	100.00	100.00	100.00	100.00	100.00
By Source					
Wage Income	18.1	20.2	22.4	31.2	32.6
Net Income from	74.4	75.6	71.4	63.3	61.7
Household Business					
Farming	50.8	50.2	50.7	37.0	36.5
Forestry	1.5	1.1	0.9	1.0	0.9
Animal Husbandry	13.1	14.1	8.1	9.2	9.0
Fishery	0.9	1.0	1.0	1.2	1.2
Industry	0.5	1.3	0.9	2.3	2.3
Construction	1.9	1.8	2.2	2.1	1.9
Transport, Post, and	2.1	2.0	1.8	2.8	2.7
Telecommunication					
Services					
Wholesale and	1.5	1.8	2.2	3.5	3.3
Retail Trade and					
Catering Services					
Social Services	0.8	1.0	1.1	1.2	1.3
Culture, Education,	0.0	0.0	0.0	0.3	0.3
and Health Care					
Others	1.2	1.2	2.7	2.7	2.3
Transfer Income and	7.4	4.2	6.2	5.5	5.7
Property Income					
Grouped by Type of Income					
Productive Income	92.5	95.8	93.8	94.5	94.3
Primary	75.0	74.4	63.2	49.9	49.2
Secondary	7.4	10.3	18.2	21.7	22.5
Tertiary	10.0	11.0	12.4	22.9	22.6
Nonproductive Income	7.5	4.2	6.2	5.5	5.7

Source: Data from *China Statistical Yearbook* (2002).
Note: Values are in current prices.

diversification could endow Chinese agriculture with improvements in viability and sustainability in financial terms, as it is capable of reallocating agricultural production, based on regional comparative advantages, and hence is expected to reduce average production costs.

FUTURE PERSPECTIVES FOR SUSTAINABLE DEVELOPMENT IN CHINESE AGRICULTURE

The improvement in resource allocation through rural diversification is likely to have resulted from various factors, such as:

- The shift of resources to more profitable sectors which was made possible by the relaxation of controls on agricultural production and marketing.

- Reduction and/or removal of barriers to interregional marketing, thereby making possible specialization in production, economics of scale and cost reduction.
- Better utilization of rural surplus labor as a result of the reduced control on factor markets, particularly the labor market.
- Changes in consumer demand, especially for farm products, which provided farmers with additional opportunities to shift their resources to high-value products.

By and large, the past two decades have been characterized by market-oriented reforms, first permitting and then stimulating structural adjustment in Chinese agriculture and diversification in the rural economy. Along with some fundamental changes in political, economic, and social institutions, it has made possible better utilization of regional comparative advantages, and stimulated profit seeking. Agricultural development in China could benefit from such changes in two ways: (1) reduction in stress on the environment and (2) reduction of production costs and increase in farmers' net income.

Chinese agriculture is being subject to more and more pressure following its WTO membership, particularly in matters pertaining to bulk commodity production. However, the membership could also provide Chinese agriculture and rural economy with greater opportunities to further adjust resource allocation through its access to bigger markets and, thus, generate higher incomes. In addition, the WTO commitments may help further deepen the market-oriented reforms and eliminate most, or even all, of the remaining barriers to interregional trade. A speedy infrastructural improvement could be possible with the faster growth of the economy after WTO accession, which would in turn facilitate further specialization, diversification, and interregional trade. Domestic resource costs for most major farm products have significant regional differences. Therefore, if accompanied by interregional trade, specialization in agricultural production may improve not just resource allocation efficiency and farmers' income but also competitiveness of Chinese farm products in the world market. Diversification has a more important role to play in the sustainable development of Chinese agriculture and rural economy since the accession to the WTO.

Materializing potential gains in rural diversification/structural adjustment, can be possible only by means of several important policy measures such as promoting the market-oriented reform and increasing and improving the allocation of public investment in infrastructure, research and extension, and information and communication. However, it is the creation of non-farm job opportunities on a large scale that is of paramount importance. This would require prioritizing the labor-intensive industries and coordinating various policy measures, such as technology innovation, institutional reform, and public spending, to ensure further growth in this crucial area of the economy, which most often links the farm economy with industrial and trade sectors, and provides peasant farmers with crucial additional incomes.

BIBLIOGRAPHY

Lin, J. Y. "Rural Reform and Agricultural Growth in China." *American Economic Review* 82 (March 1992): 34–51.

Carter, C. A. and F. Zhong. *China's Grain Production and Trade*. Boulder, CO and London: Westview Press, 1988.

Carter, C. A., F. Zhong, and F. Cai. *China's Ongoing Agricultural Reform*. San Francisco: 1990 Institute, 1996.

National Bureau of Statistics of China. *China Statistical Yearbook*, various issues. Beijing: China Statistics Press.

Zhong, F., Z. Xu, and L. Fu. "Regional Comparative Advantage in China's Grain Production: Implications for Policy Reform." *China's Agriculture in the International Trading System*. Paris: OECD, 2001.

Zhong, F. and J. Zhu. "The Role of Structural Adjustment in Chinese Agricultural Growth (Jiegou Tiaozheng Zai Woguo Nongye Zengzhang Zhong De Zuoyong)." *Chinese Rural Economy* (*Zhongguo Nongcun Jingji*) 7, 2000.

Zhong, F. and J. Zhu. "The Contribution of Diversification to China's Rural Development: Implications of Reform for the Growth of the Rural Economy." Paper presented at the workshop, "The Dragon and the Elephant: A Comparative Study of Economic and Agricultural Reforms in China and India," New Delhi, March 25–26, 2003.

5

Growth and Regional Inequality in Asia's "New Dragons"

China and Vietnam Compared

Max Spoor

China and Vietnam, until the late 1970s, were poor agrarian societies with a considerable degree of equity, and "inward looking" economies, with trade links that were mostly with the Council for Mutual Economic Assistance (CMEA)[1] (such as with Vietnam), or focused on domestic trade, investment resources and a large home market (such as in the case of China). They have emerged in the past two decades as Asia's "new dragons," with stable and high growth rates, and an impressive development trajectory. These two Asian Transition Economies (ATEs) have been able to develop rapidly, diminish poverty drastically, and integrate into the regional and world economy at an ever-increasing speed as active exporters and importers.

However, growth has also been accompanied with increasing sectoral and spatial inequality: urban-rural, as well as coastal-inland (China) and delta-highland (Vietnam). It is widely accepted that increased inequality of income (let alone unequal human development in a more broad sense) can have a detrimental effect on sustainable growth, while also reducing the "poverty elasticity of growth." In fact, in China and Vietnam, income inequality has started to become a severe problem, which needs to be addressed by the policy makers. China initiated its "Great West Development Strategy" in the year 2000, in order to redirect public investments (and attract private ones) to address the development problems of the Western inland provinces. Meanwhile, the Vietnamese government is also concerned about rapidly increasing inequality. In both cases there is a rising awareness, not only of the negative relation between rising inequality, economic growth and poverty reduction, but also of the possible social effects such as dissatisfaction, social disturbances, upheavals, and conflicts. Furthermore, in both countries, rural poverty is still widespread, the rural poor being

93

mostly agricultural producers or (in Vietnam) landless workers who have no additional income from the rural non-farm economy.

Both countries show substantial similarities in their growth paths, and have also become increasingly more unequal, not only in terms of income or expenditure, measured by the "money metric approach," but also in access to and the quality of (public) social services, and lately in assets such as land. Unequal growth can be partly explained by analyzing three factors: the importance of heavy industry in the pre-reform development strategy, the degree of decentralization and the increased openness in trade (as suggested by Kanbur and Zhang 2005). However, geographical characteristics of these two economies are quite different. While Vietnam has a longer coastline and a relatively small inland area, China is much bigger with most of its land being inland.

The purpose of this chapter is to analyze the similarities and differences in growth patterns, poverty reduction and rising inequalities in China and Vietnam, and to draw some lessons for policy making in the near future in both countries. The chapter will proceed in the following manner: The second section will briefly highlight and compare the largely similar "pre-reform" development strategies of China and Vietnam. Nevertheless, there is one particular difference for the case of Vietnam. North Vietnam had followed the socialist model since 1954. Only in 1976, right after the end of the war, the reunified Vietnam made a (rather unsuccessful) move toward collectivization of agriculture in the South. In the third section, the follow-up transition from a command (or centrally planned) economy toward a market economy is analyzed. After the late 1970s and early 1980s, both Vietnam and China embarked on reforms that started in the agricultural sector, and moved their economies into an export-led (and strongly export-promotion) development strategy, which brought about a very successful "transition."

In the fourth section, a comparative analysis will be presented on the growth strategy and macroeconomic performance of both economies, particularly in view of their integration with the Asian (regional) economy, and the world market at large. Both countries, especially China, have been able to attract Foreign Direct Investment (FDI) in large volumes (after they exhibited good growth performance based on domestic investment resources). Not only have they generated consistent high economic growth rates over the past two decades, they have also achieved a greater growth rate of exports (and imports), which shows that they have been able to shift from an inward-looking to an outward-looking economic development strategy. In the fifth section, the positive performance in poverty reduction is analyzed along with an emphasis on the increasing rural-urban as well as regional inequality. The membership of the World Trade Organization (WTO) (China in 2001 and Vietnam expected in 2006) is also seen as a factor that can increase spatial and sectoral inequality.

The conclusion drawn in the final sixth section is clear: both new dragons have done impressively well in terms of growth and poverty reduction, but are at a point where inequality, in particular spatial inequality, can become an obstacle to growth and a cause of social instability, which could have very negative consequences for FDI and macroeconomic growth perspectives. The importance of inequality (left behind on the development agenda since the "growth with equity" debate of the 1970s) is by now widely recognized, although opinions differ on what its linkage is to growth and on how to reduce inequality. On the political agenda of the governments of the two Asian dragons it is a crucial issue to be addressed. This is because inequality and a lack of (domestic) regional development might endanger their power-base in spite of a spectacular macroeconomic record. Finally, it is argued here that the agricultural sector that was the foundation of successful economic reforms and growth, and it should once again become the focus of attention of the policy makers. Moreover, rural public investment will require expansion, in order to decrease inter- and intraregional inequality.

THE TALE OF THE "PRE-REFORM" SOCIALIST TRANSITION

China and Vietnam[2] were very poor agrarian societies at the outset of their first major societal transition, provoked by the revolutions led by the Chinese and Vietnamese Communist Parties in 1949 and 1945, respectively. Primarily led by peasant-based organizations, both revolutions changed the previously dominant feudal agrarian structures and promised "land to the tiller." Land scarcity was important in both countries, with Vietnam being in the worst position, only comparable with that of Bangladesh, whereas the labor-land ratio of China could be compared with that of Indonesia.

China and Vietnam used comparable strategies in their socialist development models (which are now called "pre-reform"), though with several specific differences. Firstly, a radical redistributive land reform crushed the power of the landlord class and made land accessible to the tiller. Since land remained in the hands of the State, there was neither private ownership of land nor the issue of private land titles. Secondly, this land reform was followed rather quickly by a full-scale collectivization of agriculture, bringing in the peasant families into collective (and in some cases state) farms. In China the collective farms were later to be absorbed by the large People's communes. However, this did not happen in Vietnam.

Thirdly, the economy was converted into a command economy, while planning was introduced in the industrial and domestic trade sectors, and production targets were annually set for the collectives and communes.

Fourthly, in both China and post–1954 North Vietnam, heavy industry was the crux of the development model. This translated itself into the absorption of most public investment, and also led to experiments such as the "Great Leap Forward" (1959), with the infamous "backyard steel furnaces." However, in China rural industrialization became an important vehicle for economic growth and employment creation, though not in the case of North Vietnam. Finally, decentralization was also an important part of the development model. Both China and Vietnam witnessed greater autonomy of regional and local administrations, but due to different reasons. In China it was because of the sheer size of the economy and in Vietnam, mainly forced by the war that the United States waged against North Vietnam. In Vietnam, the war also resulted in the movement of productive assets from the cities to the countryside to escape destruction.

Both economies were relatively isolated from the rest of the Asian region, and the world economy at large. While China could depend on its own large domestic market, North Vietnam depended on Soviet (also on Chinese) aid, while it was boycotted from most international trade and was only involved in the CMEA trade block (until 1995, when the USA lifted the embargo[3]). The war had severely affected North Vietnam, and when the reunification was realized in 1976, Vietnam still suffered serious setbacks in its level of development. After the Second World War, Vietnam was only slightly poorer than Thailand, South Korea, and Indonesia, and considerably richer than China (see table 5.1).

However, this picture rapidly changed in the 1960s and 1970s, when North Vietnam (and the South as well, but in a different manner) suffered as a result of the devastating war. Thailand and South Korea were growing rapidly since 1960, while Indonesia was stepping up since the early 1970s. In 1980, Vietnam had a per capita income of only 18.4, 29.7, 40.5 and 71.0 percent in comparison with that of South Korea, Thailand, Indonesia, and China, respectively. In the 1990s, when market reforms (and the overall transition toward a market economy) in China and Vietnam were already in full swing, Vietnam's per capita income was even further behind: 11.9, 22.4, 41.3, and 56.0 percent in 1995. Since then it has started to recover some of the lost territory, in comparison with other countries in the region.

Table 5.1.   Income Per Capita Compared between Vietnam and Asian Countries (in %)

	1950	1960	1970	1975	1980	1990	1995	1998
Vietnam's per capita income relative to:								
China	149.9	118.7	93.9	81.2	71.0	56.0	52.9	52.8
Indonesia	78.3	78.4	61.6	47.2	40.5	41.3	42.1	54.6
South Korea	85.5	72.3	37.6	22.5	18.4	11.9	11.8	13.8
Thailand	80.5	74.1	43.4	36.2	29.7	22.4	21.1	27.0

Source: Klump and Bonschab (2004: 6).

THE SUCCESS OF CHINA AND VIETNAM AS
"ASIAN TRANSITION ECONOMIES"

The two most successful transition economies are, without any doubt, China and Vietnam, exemplifying that the ATEs have fared better with their continuity in institutions and gradualist approach than a large part of the European transition economies, in particular the heirs of the Soviet Union (see Spoor 2004). In the regional Asian context they have also emerged as the new dragons, in particular the spectacularly expanding Chinese economy that is currently heading Asian development (and recovery after the 1997 financial crisis, which did not affect China and Vietnam). Again, there are important parallels that one can highlight in the market-reforms and their outcomes in China and Vietnam, although there are a number of differences, which in part are related to the enormous difference in the sizes of their economies and population (1.3 billion and 80 million in 2002).[4]

Firstly, reforms started in the rural areas and were focused on recovering and improving the stagnating agricultural production and incomes. By lifting restrictive institutional constraints (through the introduction of the Household Responsibility System in China and the Three Contracts System in Vietnam), farmers got more freedom to define their product-mix and sell a larger part of their surplus at market prices. Furthermore, the domestic terms of trade that had been unfavorable for agriculture for a long period were substantially improved. In the early 1980s this led to an enormous boost in China's agricultural incomes, and hence savings that were utilized as development finance for other sectors such as industry. In Vietnam, which had suffered a food crisis after the 1979–1980 agricultural season, domestic food self-sufficiency was attained fairly soon. In fact after further reforms initiated through the *Doi Moi* policies of 1986, the country became one of the world's main rice exporters (Spoor 2003).

Secondly, in the 1980s both countries moved toward an investment-led development strategy that focused on expanding manufactured exports, rather than remaining dependent on the agricultural and rural sectors alone. In China this meant a strong emphasis on the development of the industrial and exporting potential of the Eastern Coastal provinces amongst others by opening up special economic zones, and benefiting from nearby Hong Kong and Taiwan. The latter, since the early 1990s, provided large-scale investments and transfer of highly sought after technology and management knowledge. Vietnam did so on a much smaller scale, with assembly industries and agro-based industries as focal sectors. In both cases the SOEs (State Owned Enterprises) remained important. They were not privatized, but were exposed to tougher hard-budget constraints and financial autonomy.[5]

Thirdly, China and to a lesser extend Vietnam, followed a "gradual-ist path" of transition, in which two-tier price systems remained in use, while compulsory procurement was phased out slowly. No full-scale privatization was implemented in the strategic sectors of the economy, whereas SOEs remained important. In China, the Township and Village Enterprises (TVEs), largely public and partly private, became an impor-tant motor of economic growth and employment creation, absorbing enormous inflows of migrating labor out of the agricultural sector. In Vietnam, intensification of agriculture was chosen as the primary strategy of growth and development, since there was much less labor migration to urban areas to be absorbed by the emerging assembly and light industry sectors. Both countries followed a transition strategy that was guided by the state, instead of completely relying on the emerging market forces as was done in the former Soviet Union (Fan et al. 2002; Cornia and Popov 2001; Spoor 2003).

As seen in table 5.2, growth performance has been exemplary in both China and Vietnam. While starting at very low levels of income/capita, since 1981 China's GDP has been growing at a five-year average of between 7.9 and 12.0 percent, with an overall mean (for 1981–2004) of 9.6 percent. In the case of Vietnam, GDP growth rates have been at a five-year average of between 4.8–8.2 percent (with an overall mean of 6.9 percent) over the same period.

Neither of the two has been overly affected by the Asian financial crisis in 1997, except for the growth of their intraregional exports. The fact that there was hardly any contagion was partly a lucky coincidence, as in both cases (with Vietnam more so than China) the financial system was still weakly developed in relation to the size of the economy. The financial sectors were and still are largely state-bank dominated and hence not so much integrated with the international financial system in Asia. It was also the outcome of prudent policy interventions of both governments, who did not follow the capital account liberalization advice provided by the International Mon-etary Fund (IMF). This helped in not being drawn into a crisis that spread within a matter of months from a single commercial bank breakdown in Thailand into a major regional economic collapse. For example, the crisis of 1997 caused a decade's setback for Indonesia with regard to its develop-ment indicators, increasing poverty levels dramatically (Stiglitz 2002).

Table 5.2.   China and Vietnam: GDP Growth (1981–2004)

	1981–1985	1986–1990	1991–1995	1996–2000	2001–2004
Vietnam	7.2	4.8	8.2	7.0	7.3
China	10.8	7.9	12.0	8.3	8.7

Source: World Economic Outlook Database, International Monetary Fund (2005).

FROM AN "INWARD-" TO AN "OUTWARD-" LOOKING ECONOMIC STRATEGY

The difference in growth performance of China and Vietnam, compared to the other major developing Asian economies (Indonesia, South Korea, Malaysia, Philippines, and Thailand), is indeed striking, as seen in table 5.3. While these five "Asian tigers" were hit hard by the Asian financial crisis (with an average contraction of –7.7 percent of GDP in 1998, in particular Indonesia and Thailand being hit most severely), the new dragons stayed afloat, steadily displaying high growth rates. Moreover, when the rest of the "emerging" East and Southeast Asia was struck by a growth slump in 2001, in particular affecting Malaysia and the Philippines, China and Vietnam's growth performance remained constant (7.5 and 6.9 percent versus an average 2.4 percent for the average of the other five Asian tigers).

Throughout the entire period of 1981–2004 (and still continuing), China and Vietnam have shown economic growth at sustained and high rates. Vietnam commenced its growth path in 1981, after showing negative growth in 1980, while China was already on its way up after the introduction of Deng Xiaoping's program of the "four modernizations" (Agriculture, Industry, Defense, and Science and Technology).

During the era of "command" or "centrally planned" economy, China and Vietnam were scarcely involved in trade and were certainly not integrated into the emerging regional Asian economy where countries such as South Korea, Singapore, Malaysia, Japan, Taiwan, and later Thailand were rapidly developing. These strongly promoted exports (Wade 1990; Amsden 1989) and followed a successful blend of import substitution and export orientation. China and Vietnam remained distinct from these "East Asian Miracle" countries (World Bank 1994), mainly because of domestic and external political factors that defined their strategies. China's transition to a market economy began with the 1978 four modernizations policy,

Table 5.3. Comparing Growth of China and Vietnam in the Asian Region (1996–2004)

	1996	1997	1998	1999	2000	2001	2002	2003	2004
China	9.6	8.8	7.8	7.1	8.0	7.5	8.3	9.3	9.5
Vietnam	9.3	8.2	5.8	4.8	6.8	6.9	7.1	7.3	7.7
Indonesia	8.0	4.5	–13.1	0.8	4.9	3.8	4.4	4.9	5.1
South Korea	7.0	4.7	–6.9	9.5	8.5	3.8	7.0	3.1	4.6
Malaysia	10.0	7.3	–7.4	6.1	8.9	0.3	4.1	5.3	7.1
Philippines	5.8	5.2	–0.6	3.4	4.4	1.8	4.3	4.7	6.1
Thailand	5.9	–1.4	–10.5	4.4	4.8	2.2	5.3	6.9	6.1
Average*	7.3	4.1	–7.7	4.8	6.3	2.4	5.0	5.0	5.8

Source: World Economic Outlook Database, International Monetary Fund (2005).
Note: *Average of growth rates of Indonesia, South Korea, Malaysia, Philippines, and Thailand.

right after the new leadership consolidated its power, by crushing the post-Maoist "Gang of Four." Vietnam started on its path of transition shortly after a renewed collectivization in the South (1976–1980) failed dramatically, and the Northern agricultural system nearly came to a halt.[6]

Two phenomena are striking to note in the comparison of China and Vietnam with the rest of the regional Asian economy. First, after the initial growth spurts of the 1980s, caused by high rates of growth in agricultural and the subsequent growth in rural incomes, the motor of growth shifted to the urban industrial sectors and (in China) TVEs, primarily driven by expanding manufacturing exports. This is much clearer in the case of China, and less clear for Vietnam, as in the latter country exports in at least part of the 1990s were still dominated by rice and coffee. In both cases there was also a shift (partly related to changing domestic demand and to increasing imports) from land-intensive toward labor-intensive crops. Export growth in the past decade is also higher on average compared with the surrounding Asian tigers, although Vietnam was somewhat affected by the 1997 crisis (showing only a small increase in exports in the following year), as seen in table 5.4.

Initially, growth was clearly based on endogenous resources and domestic investment. After the two nations sustained long-term growth, a growing volume of FDI was attracted since the early 1990s. It is clear that FDI was not the cause of growth, but rather was attracted by it (figure 5.1). Over the years, FDI has started to play an important role in modernizing the economy further, mainly in China. Foreign companies represent an important part of export production, often in the form of joint ventures. Nevertheless, FDI still forms a relatively limited share of overall investment, and in terms of share of GDP it is even decreasing. China is now receiving more than $50 billion USD per year in FDI, which makes it the second largest recipient in the world after the USA. Vietnam, at a much smaller scale, has an annual inflow of FDI of around $1.3–1.4 billion USD (which at a per capita basis is around half of the Chinese level).

A deliberate policy for prohibiting the flow of FDI into Western inland provinces had been followed in China to stimulate growth in the Eastern coastal areas, which had comparative advantages for exports (Riskin et al.

Table 5.4.   China, Vietnam, and the Asian Region: Export Growth Rates (1996–2002)

	1996	1997	1998	1999	2000	2001	2002
China	—	—	—	6.1	27.9	6.8	22.4
Vietnam	41.2	24.6	2.4	23.2	25.2	6.5	7.4
Indonesia	9.7	7.3	–8.6	–0.4	27.7	–9.3	3.7
South Korea	3.7	5	–2.8	8.6	19.9	–12.7	8
Malaysia	5.9	0.4	–6.8	15.3	16.1	–10.4	6
Philippines	17.8	22.8	16.9	18.8	8.7	–15.6	9.5
Thailand	–1.9	3.8	6.8	7.4	19.5	–6.9	5.7

Source: Weeks et al. (2004).

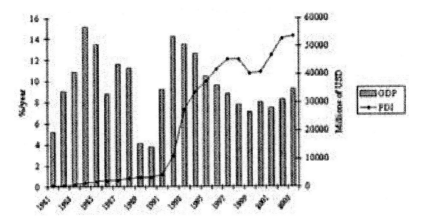

Source: World Economic Outlook Database, International Monetary Fund (2005); UNCTAD Investment Database, Trade and Development Report (2004).

Figure 5.1.   GDP Growth and FDI in China (1981–2003)

2004). This was possibly a very good decision to stimulate growth, but it was also a very "east coast"-biased policy, in contradiction (as we will see below) to balanced growth and sustained poverty reduction. Maybe there was a belief in the old-fashioned "trickle down" theory of growth toward poverty reduction, but it is now obvious that in spite of successes in this field the Chinese model is highly unequal in sectoral as well as in spatial terms (Kanbur and Zhang 2005; Riskin et al. 2004)

In Vietnam, the concentration of FDI was mostly in the industrial centers such as Hanoi, Haiphong (in the North), and even more in particular in Ho Chi Minh City (in the South). Figure 5.2 shows that just like in the case of China—FDI, in Vietnam, followed sustained economic growth, rather than initiating it.

POVERTY REDUCTION AND REGIONAL INEQUALITY

The growth performance of China and Vietnam, the rapid integration into the regional and world markets, and the substantial volumes of FDI influx, had a flipside too. Specifically, during the past decade, they also show a growing regional or spatial inequality and a diminishing poverty reduction elasticity of growth. Poverty reduction, as such, has been impressive, whatever poverty line is used. Particularly the number of rural poor in China has diminished drastically. According to Chinese official estimates, the number has gone down from a massive 260 million in 1978 to just 30 million in 2000 (Riskin et al. 2004: 16). However, using the international comparable poverty line of 1 USD (PPP)/day, the number is estimated to

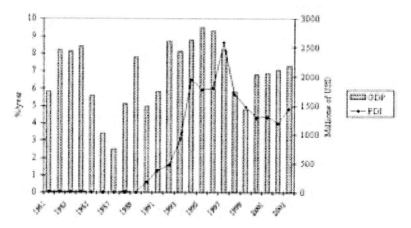

Source: World Economic Outlook Database, International Monetary Fund (2005); UNCTAD Investment Database, Trade and Development Report (2004).

Figure 5.2.   GDP Growth and FDI in Vietnam (1981–2003)

be substantially higher (Fan et al. 2002).[7] In the early 1980s and during the mid-1990s, rural incomes rose and rural poverty diminished with a strong increase of the Domestic Terms of Trade (ToT_D) in favor of agriculture. A reverse movement has begun since then. As Riskin et al. (2004: 2) asserts:

> National and regional policies in China have indeed been conflicting with poverty reduction in many aspects. In particular, many policies, including income distribution, fiscal support, the social welfare system and restrictions in population movement have been "urban-biased."

Apart from the stagnating income development in rural areas, basic health services and education have also suffered in terms of access and quality. Fiscal decentralization provoked a near perverse effect for poorer communities, forcing them to pay for public goods provisioning on the basis of a weak (and by far insufficient) tax base. This caused large differences in the budgetary expenditure per capita between urban and rural areas and coastal and inland provinces. The location and severity of poverty in rural areas have been more pronounced in the inland provinces, notably in the Southwest (including Tibet), the Northwest and part of the Central areas (Risken et al. 2004; Fan et al. 2002). As Riskin et al. (2004: 2) noted:

> The slow development of the central and western regions has become a serious obstacle in reducing poverty in these provinces where much of China's poor population is concentrated.

In Vietnam, poverty has been and still is concentrated in rural areas (as in China). But during the past decade, a significant reduction in poverty headcount has been recorded, namely from 58 (1993) to 37 (1998) to 29 percent

Table 5.5. Poverty Reduction in Vietnam (1993–2002)

| | Headcount Ratio | | |
	2002	2003	2004
Red River Delta	63	29	22
North Central Coast	75	48	44
Southeast	37	12	11
Northern Uplands	82	64	44
Central Highlands	70	53	52
South Central Coast	47	35	25
Mekong River Delta	47	37	23

Source: Weeks et al. (2004)

(2002) (using a somewhat different poverty line than the 1 USD (PPP)/capita). However, there are significant regional differences (table 5.5).

The Northern Uplands, the North Central Coast and the Central Highlands especially lag behind. Furthermore, in the same period rural poverty reduced from 66.4 to 45.5, and further to 35.6 percent, while urban poverty reduced much more rapidly from 25.1, 9.2 to 6.6 percent (Weeks et al. 2004). Akin to China, Vietnam's poverty-stricken areas are those where most of the ethnic minorities are residing. Moreover, a recent study by International Food Policy Research Institute (IFPRI) (Minot et al. 2003) revealed that rural poverty in Vietnam is correlated with degraded land resourses and increased distance from towns.

Finally, income inequality, one of the focal points of this chapter, has increased substantially in both countries (table 5.6). The methodology of calculating the Gini coefficient (in terms of expenditure or income data) is not always comparable, but the indicators show a disturbing trend. China has changed from an egalitarian society into a strongly unequal country, similar to many of the former Soviet transition economies. Vietnam showed initially a slightly more moderate income inequality, which has now also risen to above 0.40 (measured by the Gini coefficient). It is also interesting to note that three-quarters of regional inequality in Vietnam is *within*-province, rather than *between*-province.[8]

In addition, regional inequality in China and Vietnam has increased substantially since the early 1990s (Tarp Jensen and Tarp 2005; Kanbur and

Table 5.6. Income Inequality in China and Vietnam (1992–2002)

| | Gini Coefficient | | |
	1992	1995	2001
China	0.376	0.415	0.447
Vietnam	0.34	0.35	0.42

Source: Weeks et al. (2004)

Zhang 2005; Wan and Zhou 2004). In China there is a striking contradiction between the Eastern Coast provinces and the North- and Southwestern, and some of the Central provinces. While the former has the advantage of relatively good social services and high levels of public and private investment, the latter lacks the same. In Vietnam, it is the Southeast (with Ho Chi Minh City in its center) and the Red River Delta (with Hanoi and Haiphong), in comparison with the Northern Uplands, the North Central Coast and the Central Highlands that defines spatial income inequality.

CONCLUSION:
GROWTH, SPATIAL INEQUALITY, AND RURAL POVERTY

Without any doubt, China and Vietnam have shown amazing growth performance over more than two decades, since the early 1980s. They have emerged as two Asian Dragons, as it were, from the clay of the rice fields in these very poor agrarian societies. In comparison with the Asian Miracle countries these twin ATEs have done very well indeed. Besides, with their sustained high growth rates and relatively low initial inequality, they have been able to reduce the poverty headcount in a spectacular manner, especially in the rural regions. However, income inequality has been growing rapidly since the 1990s, which is argued to become an obstacle to further growth and also to further poverty reduction, especially in rural areas. Inequality is not only sectorial (urban-rural) but has been more and more recognized to be spatially determined (Kanbur and Zhang 2005).

The sectors in which growth is generated, principally those involved in the rapidly growing exports and those receiving FDI, have a very unequal spatial distribution. For China the newly acquired membership of the WTO, though advantageous, has been increasing inequality. Firstly, recent studies show that the poor agricultural producers and workers are the losers of WTO accession (Anderson et al. 2004). Secondly, from the mid-1980s until recently Chinese policies have been detrimental for agricultural producers. These policies have been paving the way for extensive obligations in the aggregate measurement of support (AMS) that the Chinese government accepted to abide by while entering the WTO. The new minimal tariffs for agricultural goods also lowered domestic prices. For Vietnam, the negotiations to become a member of the WTO will most likely be concluded in 2006. While this will be largely positive for Vietnam's trade position, it might well be detrimental for poor agricultural producers, and in general it will further deteriorate spatial and sectorial inequality (Weeks et al. 2004).

In the case of China, the government and the Party recognized the seriousness of growing inequality and responded with the introduction of a large-scale public investment program in 2000 that focused on the western inland provinces. However, what is even more needed is the establishment of dynamic links between the inland (largely agricultural) economies and the fast-growing Eastern coast provinces. In Vietnam, spatial inequality is reflected in the important (income) gap between North and South and between the growth poles of the two Deltas, represented by the three largest cities (Hanoi, Ho Chi Minh City, and Haiphong) and the surrounding highlands. Geographically, Vietnam is in a more advantageous position for reducing spatial inequality because of its long coastline, and the direct benefit to many provinces from FDI and trade openness. Nevertheless, in both countries concerted public policy interventions are needed to undo the growing regional income inequality, that hinders sustainable economic growth and human development. These interventions need to be primarily "rural biased" in order to be effective.

NOTES

1. Council for Mutual Economic Assistance (CMEA), which was the Soviet-dominated Common Market, with Vietnam, Cuba, and Mongolia as associated members. The CMEA was disbanded in 1992, after the collapse of the Soviet Union.

2. Until 1975, North Vietnam is meant here.

3. It is interesting that the first visit of a Vietnamese prime minister since the end of the war to the USA that took place in June 2005, coincided with the ten-year anniversary of the lifting of the trade embargo.

4. See United Nations Human Development Report Data, Website www.undp.org.

5. There is widely accepted criticism on the still existing soft loans that the state-run financial system in China and Vietnam still provides to low-performing SOEs, although both governments are trying to put a hold to these, knowing that this weakness is a ticking time-bomb for the financial system (which actually caused the 1997 Asian Financial Crisis).

6. In 1980 I was visiting Hanoi, and there was a severe food shortage. It was also—cynically enough—the year after the devastating border war between China and Vietnam, in which China "punished" its southern neighbor for having toppled the (friendly to China) regime of Pol Pot in Cambodia. The year 1980 was a conjuncture of critical moments, which led to a reconsideration of the economic strategy, and the beginning of economic reforms. In Vietnam it would still take years of internal debates before by 1986 the *Doi Moi* policy was pronounced, which laid the basis of the successful transition to a market economy.

7. The national poverty line used by the Chinese government is substantially lower than the internationally comparable 1 USD (PPP)/capita poverty line, used by the World Bank.

8. Compare, for the case of China, the decomposition analysis that is given in chapter 3 in this volume.

BIBLIOGRAPHY

Amsden, A. *Asia's New Giant: South Korea and Late Industrialization*. New York and Oxford: Oxford University Press, 1989.

Anderson, K., J. Huang, and E. Ianchovichina. "The Impacts of WTO Accession on Chinese Agriculture and Rural Poverty." Pp. 101–16 in *China and the WTO: Accession, Policy Reforms, and Poverty Reduction Strategies*, edited by D. Bhattasali, S. Li, and W. Martin. Washington, D.C.: World Bank, 2004.

Cornia, A. G. and V. Popov. *Transition and Institutions: The Experience of the Gradual and Late Reformers*. UNU/WIDER Studies in Development Economics. Oxford and New York: Oxford University Press, 2001.

Fan, S., L. Zhang, and X. Zhang. "Growth, Inequality, and Poverty in Rural China." *Research Report*, no. 125. Washington, D.C.: IFPRI, 2002.

Jones, D., C. Li, and A. L. Owen. "Growth and Regional Inequality in China during the Reform Era." *William Davidson Working Paper*, no. 561. Ann Arbor, MI: William Davidson Institue, University of Michigan Business School, 2003.

Heltberg, R. "Spatial Inequality in Vietnam: A Regression-based Decomposition." *World Bank Working Paper*, Washington, D.C.: World Bank, 2003.

IMF. *World Economic Outlook Database*. Washington, D.C.: IMF, 2005.

Kanbur, R. and X. Zhang. "Fifty Years of Regional Inequality in China: A Journey through Central Planning, Reform and Openness." *Review of Development Economics* 9, no. 1 (2005): 87–106.

Klump, R. and T. Bonschab. "Operationalising Pro-Poor Growth: A Country Case Study on Vietnam." Paper written for the OPPG program, October 2004.

Liu, A. Y. C. "Markets, Inequality and Poverty in Vietnam." *Asian Economic Journal* 15, no. 2 (2001): 217–35.

Minot, N., B. Baulch, and M. Epperecht. *Poverty and Inequality in Vietnam: Spatial Patterns and Geographic Determinants*. Washington, D.C.: IFPRI, 2003.

Riskin, C. (ed.). *The Macroeconomics of Poverty Reduction in China*. Regional Bureau for Asia and the Pacific and Bureau of Development Policy. New York: UNDP, 2004.

Stiglitz, J. *Globalization and Its Discontents*. London: Penguin Books, 2002.

Spoor, M. (ed.). *Transition, Institutions, and the Rural Sector*. Lanham, MD, London, and Oxford: Rowman and Littlefield, Lexington Books, 2003.

Spoor, M. "Transition and Development in Asian Perspective." Paper presented at the Conference "Transicion y Desarrollo en Asia." Barcelona: CUIMPB, 8–9 July 2004.

Tarp Jensen, H. and F. Tarp. "Trade Liberalization and Spatial Inequality: A Methodological Innovation in a Vietnamese Perspective." *Review of Development Economics* 9, no. 1 (2005): 69–86.

Wade, R. *Governing the Market: Economic Theory and the Role of Government in East Asian Industrialization*. Princeton, N.J.: Princeton University Press, 1990.

Wan, G. and Z. Zhou. "Income Inequality in Rural China: Regression-base Decomposition Using Household Data." *WIDER Research Paper*, no. 2004/51. Helsinki: World Institute of Development Economics Research, 2004.

Weeks, J., N. Thang, R. Roy, and J. Lim. *Seeking Equity within Growth: The Macroeconomics of Poverty Reduction. The Case Study of Vietnam*. Report. New York: UNDP, 2004.

World Bank. *The East Asian Miracle*. Washington, D.C.: World Bank, 1994.

World Bank. *From Plan to Market*. World Development Report. Washington, D.C.: World Bank, 1996.

World Bank. *Knowledge for Development*. World Development Report. Washington, D.C.: World Bank, 1998.

6

Land Policy in Vietnam's Agrarian Transition

A Case Study of *Namđinh* Province

Nguyen Do Anh Tuan

Vietnam witnessed an economic boom during the 1990s with the economy growing at more than 7 percent per year. The major resultant implications for poverty alleviation were substantiated in the Vietnam's Living Standard Survey (VLSS) of 1993–1998—the poverty rate had declined from 55 to 37 percent (GSO 2000). To all appearances, it seemed the fruits of economic growth were still equally distributed, as the Gini coefficient for expenditure exhibited only a slight increase from 0.33 to 0.35 during this period. In contrast with other developing countries, this is in fact a low-to-moderate representation of inequality (Haughton et al. 2001: 15). However, it rose strongly later on. All the same, with 90 percent of the poor living in rural areas, the structure of inequality has raised the hackles of both policy makers as well as international donors such as the United Nations Development Programme (UNDP), World Bank, and Asian Development Bank (ADB).

Mostly attributed to the widening urban-rural gap, rising inequality means that growth in the economy did not trickle down sufficiently to the agricultural sector and the rural poor. This had in turn led to various disruptions and unrest in rural areas at the end of the 1990s. This raises an even more serious threat to social cohesion, as those who were left behind by the economic boom would increasingly feel marginalized and estranged. The primary concern here is that since economic and industrial growth failed to substantially affect and improve the rural poor, Vietnam must have to look for other alternatives wherein the rural poor benefits from a growth spurt in the agricultural sector (Saith 1990).

Agricultural growth, productivity, and commercialization can be enhanced only with the firm foundation and efficient usage of land. A truly effective way to do this would be by putting land under the management of farm households. From 1988, the cooperative system has been dismantled followed by the reallocation of land to peasant households—resulting in a substantial increase in land productivity. Unfortunately, controversies over the structure of land use and farm size have followed too. Agricultural commercialization has developed the agricultural sector as well as improved living standards in peasant households, but further land consolidation and labor specialization are necessary. There is apprehension however that land concentration would lead to (renewed) rural class differentiation that would contradict the ideology of the ruling party. Additionally, the well-known inverse relationship between farm size and productivity proves that the existing equal distribution of land has contributed to the achievement of both objectives: efficiency and equity.

This chapter tries to contribute to this debate with the help of an empirical investigation of *Namđinh* Province, which is typical for the Red River Delta area of Vietnam. Such an investigation intends to clarify the relationships between land concentration and income inequality and between farm size and agricultural productivity that resulted in the course of Vietnam's agrarian transition.

RESEARCH AREA AND DATA COLLECTION

The crux of this study is the analysis of land policies and agricultural growth in the specific province of *Namđinh*. Issues that need primary attention here are various. Primarily, land policies are politically sensitive—it is often political rather than economic considerations that rule. A prime example that bears witness was the debate on the removal of land ceilings between the Vietnamese Communist Party (VCP) and the National Assembly. Increasing rural unrest brought corruption and signs of "rural despotism" (Akram-Lodhi 2001a), and their close relation to land disputes to the forefront in this debate. However, an effective land policy promoting agricultural growth and commercialization, with the backing of political consensus and with the support of empirical evidence will go a long way in building a possible relation between land consolidation and inequality in Vietnam's rural sector.

The second significance of policy implications will largely reflect the characteristics of the case study. The *Namđinh* Province was selected for two reasons. The first of these is that it carries the three typical characteristics of Vietnam's agriculture at present: (i) a high density of population and limited arable land availability; (ii) transition from centrally planning to market system in the entire economy, and from collective to smallholding peasant household system in agriculture; and (iii) transformation from subsistence to commodity production in agriculture.[1] The second reason is that its location in the Red River Delta signifies that the *Namđinh* Province's economy is primarily dominated by the agricultural sector, generating around 40 percent of GDP, while more than 85 percent of the population still live in rural areas. Consequently the development of the agricultural sector has a decisive impact on the entire province's economic performance and living standards of most of its inhabitants. Adding to this the limited availability of land, a further increase in agricultural productivity and peasant incomes would require the application of land-saving technology and diversification toward crop production (or animal husbandry) that provides a higher per unit value. This, in turn, demands the endorsement of commodity exchange between agricultural and nonagricultural sectors, through improved working of markets.

With favorable natural conditions compared to other provinces in Vietnam, *Namđinh* has the potential to develop commodity production in agriculture. *Namđinh* has many factors in its favor—high rice yields, a good rural transportation system, and educated people. The efficient agricultural extension system of the province operates quite successfully in transferring new technologies to peasant households as well as in adapting new seeds that are appropriate to the climatic conditions of the province. Yet, in spite of all its resource endowments, *Namđinh* is still one of Vietnam's poorest provinces and the degree of commercialization of agricultural output is lower than the average.

Some clarification is required for the data used in this chapter. The secondary data were mostly obtained from the General Statistical Office (GSO) of Vietnam, the *Namđinh* Statistical Office (NSO), and *Namđinh* Office of Agriculture and Rural Development (NOARD). In addition, discussions with policy makers of *Namđinh* gave a good overview on the specific problems and the institutional environment of the province. Primary data was collected through a survey conducted in June–July 2000. Income levels of peasant households were taken as proxies for the levels of agricultural commercialization in this survey. As a result, the three districts *Haihau*, *Yyen*, and *Trucninh* were selected on the basis of descending order of income level. The typical characteristics of *Namđinh* Province's agricultural sector, such as soil, crop structure, irrigation system, employment structure, and market access are reflected in these districts. Twenty peasants were interviewed in each commune. Within these commune groups, the standard followed for selection was six households with high levels of income, eight with medium levels, and six with low levels. This ex-ante assessment of income levels was based on the discussions with policy makers and staffs at provincial, district, and commune levels. Nevertheless, households within each income group in a commune were randomly selected.

The survey was conducted for a research project "Analysis on Challenges for the Development of Agricultural Commercialization in *Namđinh* Province," sponsored by the Vietnam-Netherlands Research Program (VNRP).[2]

Aspects such as agricultural production, land allocation, technology, markets and prices, investment, and credit were covered in the questionnaire. The chairmen or accountants of cooperatives in nine communes assisted in the interviews. They were in charge of handing out questionnaires to peasant households in the sample and explaining the questions. Members of the research group came to collect and check the records a month after the peasants filled in the questionnaire.[3] If questionnaires were incomplete or inconsistent, they were returned to the households for correction. After data cleaning (excluding inconsistent records) was done, the survey ended up with 155 records of peasant households in nine communes of *Haihau*, *Yyen*, and *Trucninh* districts in the *Namdinh* Province.

BACKGROUND ON LAND USE IN VIETNAM

Land Supply and Land Use Structure in Vietnam

A very densely populated country, Vietnam has only 0.15 ha of land per capita. When compared to other neighboring countries in Southeast Asia, the land availability is too low to generate sufficient income by marketing surplus from the agricultural sector. Though the State has invested considerably in land reclamation and irrigation systems, land expansion is limited.[4] In addition the growth in agricultural population is accelerating such that the increase in land supply is not enough to maintain the current level of land per capita. The cultivated area increased 1.5 percent annually during the 1990s, whereas agricultural population rose at 2.0 percent per year on average. As a result, land per capita in the agricultural sector was slightly reduced (Jamal and Jansen 1988: 10; GSO 1999b: 12, 28).

With this limited availability of land, the acceleration of agricultural labor productivity, income and marketed surplus depends largely on the pattern of land use. Besides crop intensification, higher agricultural income also requires increase in land yield, both in kind and in cash. There is no doubt that agricultural diversification toward crops with higher unit value and husbandry plays an increasingly important role to raise the marketed surplus with the sustainability of the ecological system reaching a critical threshold.[5] However, in practice, agricultural production is mostly subsistence-oriented; with changes in the crop structure occurring at a slow pace.[6] Around 65 percent of agricultural output is generated from food production and over 70 percent of cultivated land is used for this purpose. In addition, land fragmentation is also a hurdle to the application of new technology in agricultural production and the transformation in crop structure (GOIA 1998). Land consolidation takes place gradually. Only 10 percent of households participate in land sales and purchases—and of these merely 2.5 percent work in the agricultural sector.[7] The majority of land sales and purchases are concentrated on forestland

and the areas of new settlements, such as the Northern Mountains, the Central Highlands and the Eastern South. In the Red River Delta, the sales and purchases of land are even more rare (GSO 2000: 215, 220).

Stagnation of structural change in the economy is a major obstacle in promoting land consolidation in Vietnam. With over 7 percent growth per year, the 1990s maintained a high level of economic growth. The industry and service sectors even showed growth as high as 9 percent per annum, while agricultural growth was merely 4 percent per annum. As a result, the agricultural share in total GDP declined from nearly 40 percent to 25 percent. Nevertheless, over three-quarters of the population still lives in rural areas, and a majority of the new labor force is absorbed by the agricultural sector (World Bank 2000: 140–45). In reality, the agricultural sector has become the sink of underemployment (Jamal and Jansen 1988), and high population pressure has put constraints on further development of land and labor markets in the agricultural sector. Such slow structural changes in turn may be explained by the overconcentration of industries in some big urban centers like Hanoi and Ho Chi Minh City (UN 1994: 4).

Hence, apart from the limited availability of land there are three other major challenges in the promotion of efficient use of land and agricultural commercialization in Vietnam: The first is the slow transition of crop structure, the second is the fragmented land and impeded land concentration, and the third is the stagnation of structural change due to the concentration of industries in some big cities only.

Land Policy and Land Allocation

With the promulgation of Resolution 10 in April 1988, Vietnam's agriculture has changed fundamentally. The peasant economy dismantled and replaced the collective system. For annual crops peasant households were allocated land for ten to fifteen years, and for perennial crops the allocated duration was longer. Peasant households became autonomous units in agricultural production. The households owned the entire surplus and were free to sell it in the market, after paying a fixed amount of agricultural tax. In practice, the land was distributed in two or three rounds.[8] In the first round, about 70 percent of the former cooperative land was equally distributed among peasant households to till for meeting their basic consumption needs. In the second round, approximately 30 percent of cooperative land was given to households that were able to farm more efficiently.[9] In the third round, the land was rented subsequent to a bidding process.[10] The land distributed in the second and third rounds were completely for commodity production. This land allocation practice created "a relatively equitable land-holding peasantry based on a modified Chayanovian (household demographic composition) principle" (Watts 1988: 483).

These changes provided strong incentives to the peasants, particularly in increasing food production. Since 1989, food export was the most important achievement, despite the fact that Vietnam had been an importer of food up to 1988 (0.5–1.0 million tons). These changes, however, largely limited themselves to subsistence crops, which in turn resulted in the sluggish development of commercial crops. From the perspective of land policy, there were four limitations to agricultural commercialization. Firstly, the duration of land use rights was not long enough to encourage investment by households in agricultural production. Secondly, land transfer was not allowed, thus discouraging land consolidation, specialization, and commodity production in agriculture. Thirdly, land use rights were not used as collateral, thus preventing households from taking bank loans for agricultural investment. Fourthly, local governments still played a dominant role in deciding crop patterns for specific types of land. Due to the bias toward food security, most of the land was used for food production. There was also not enough encouragement for agricultural diversification and commercialization.

The 1993 Land Law was typically meant to deal with the above limitations. The new law and corresponding regulations brought about three fundamental changes in land use. First, it established a framework within which farming families were provided stable and long-term land use rights, namely twenty years for land planted with annual crops such as rice, and fifty years for perennial crops and forestry. At the end of the rental period when reallocation issues arose, the users were given priority, thus ensuring stability. Second, land use rights could be transferred, mortgaged, rented, exchanged, or inherited, although some discretion has been left to the local authorities. Peasants have to comply with several obligations, such as crop specification, the responsibility for the protection of the land, and the payment of taxes on transfers and other revenues stemming from land use. Third, peasants were given Land Use Certificates (LUC), which ensured rights to the land, thus encouraging agricultural investment. In addition, documentation of land rights was expected to reduce uncertainty and information asymmetry between borrowers and lenders by making land a credible form of collateral.

Nevertheless, in order to obtain the right balance between efficiency, equity, and sustainability in allocating land and still ensure food security, several constraints or safeguards were stipulated in the Law. First, land ceilings were imposed on different types of agricultural and forest land.[11] Second, land in excess of the level held prior to October 1993 was treated differently, as excess land only received a land use right for half the allocation period of land held below the Land Law limits. If the household wanted to use the land, even after the completion of this period, the State could lease the land back to the household for a limited time period. Third, a three-year limit was imposed on the leasing of annual cropland between individuals. Fourth, crop specification still remained.

By lifting the legal limitation that restricted the farm size, Resolution 6 of 1998 was intended to intensify the commoditization of land. However, it was so controversial that the Politburo of the VCP was unable to reach an agreement on it, and the National Assembly at first refused to pass the changes in the land law suggested by it. If implemented forcefully, it is expected that this resolution may formalize and further stimulate the existing operation of the informal land market. It intends to remove all legal restrictions on farm labor hiring, thus creating more dynamism in the rural labor market.

LAND USE IN *NAMĐINH* PROVINCE

Land Availability and Land Use Structure

In *Namđinh* Province, land is a scarce resource. Table 6.1 shows that land per capita in *Namđinh* is among the lowest in the country.

In the 1995–1998 period the rural population grew by 2.3 percent, whereas the agricultural land decreased by 0.3 percent per annum on average. Thus, land per capita shrunk from 605 m^2 to 588 m^2. It is expected that 2,000 hectares of land will be reclaimed, though this will be offset by the decline in the existing area due to construction (NOARD 2000: 17). With the increasing population pressure on land, growth of commodity production in agriculture, and income growth in general will depend largely on labor migration into nonagricultural sectors and improved efficiency of land use.

During the 1990s, cropping accounted for about 80 percent of the total agricultural output, with animal husbandry and fishery making up the remaining 20 percent (table 6.2). The cropping share is slightly lower than that of the whole country, largely due to the relatively high agricultural labor surplus in the province. Table 6.3 shows that within the crop sector most land is used for food purposes, especially for rice production.

Table 6.1. Agricultural Land (1995–1998)

(ha)	1995		1998	
	Land/Household	Land/Person	Land/Household	Land/Person
Vietnam	0.70	0.15	0.72	0.15
Mekong River Delta	1.22	0.23	1.16	0.22
Red River Delta	0.26	0.06	0.25	0.06
Namđinh	0.24	0.06	0.23	0.06
	(2,381 m^2)	(605 m^2)	(2338 m^2)	(588 m^2)

Source: Calculation by author based on data from GSO (1999b); NSO (1999).

Table 6.2.   Agricultural Activity Structure[a] (1990–1998) (% of total output)

	Namđinh			Vietnam		
	Cultivation	Husbandry[b]	Service	Cultivation	Husbandry[b]	Service
1990	79.0	20.1	0.8	80.2	16.6	3.1
1995	79.3	19.7	1.0	80.4	16.6	3.0
1998	77.7	21.3	0.9	80.4	16.9	2.7

Source: Calculation by author based on data from GSO (1999b); NSO (1999).
Notes:
a : 1994 constant prices.
b : Animal husbandry and fishing.

Namđinh has made significant investment in irrigation systems.[12] The crop intensity of the province is quite high, and even higher than that of provinces in the Mekong River Delta. In the late 1990s Namđinh yielded the largest quantity of rice (5.8 tons/ha per year) in the country, due to the application of new technologies (GSO 1999b: 77–78). However, Namđinh does not perform better than other provinces in yields of crops with higher unit value (except peanuts) (table 6.4).

The Practice of Land Allocation

By the year 2000, Namđinh had granted land certificates to almost all peasant households.[13] The original land allocation of 1988 was based on an equal distribution of each category of land use amongst all households. Even land for bidding in the third round of allocation was relatively equally distributed, as there was strong population pressure and a rural labor surplus. Landholdings per household are small, due to the limited availability of land as well as a low level of land concentration. In the survey, there is only one household cultivating more than 0.5 ha. Though the yields recorded are at relatively high levels, the small land size prevents households from participating in commodity production. It is, however, difficult to obtain more land. Bidding for the cooperative land is an option, but the opportunity to use this type of land is quite limited. Furthermore, there was no evidence of land transfer among households in the survey. Taking this

Table 6.3.   Land Use Structure (1998)

	Crop Intensity	Land Use Structure (%)		
	(crops/year)	Food	Rice	Non-Food
Vietnam	1.44	73.0	62.9	27.0
Mekong River Delta	2.11	86.7	83.8	13.3
Red River Delta	1.55	82.7	73.1	17.3
Namđinh	2.23	87.7	81.8	12.3

Source: Calculation by author based on data from GSO (1999b); NSO (1999).

Table 6.4. Yields of Selected Crops (1998)

(tons/ha)	Rice	Grain[a]	Peanut[a]	Soy Bean[a]	Banana	Orange
Vietnam	2.70	2.30	1.43	1.11	13.68	5.62
Mekong River Delta	4.07	2.55	1.96	2.05	11.53	6.95
Red River Delta	5.13	2.80	1.51	1.35	18.27	4.43
Namđinh	5.75	2.63	2.32	1.24	17.50	3.83

Source: GSO (1999b).
Note:
a : In rice equivalents.

into account, *Namđinh* peasant households have tried to consolidate the existing plots of land. In the late 1990s, land in *Namđinh* was divided into a smaller number of plots, with larger areas per plot and also a closer contiguity between plots, when compared to other provinces in the Red River Delta with the same average farm size (table 6.5).

Land allocation was based on the rational use of the irrigation system and maintenance of edges among plots. Therefore, land was often first allocated to groups of relatives (around 5–7 households), after which it was further distributed within the group. Such land distribution made it easier for households to cooperate, preventing land fragmentation that generates difficulties for water supplies and increases production cost.

LAND ACCUMULATION AND INCOME INEQUALITY

Land concentration in Vietnam is of deep concern to the ruling Communist Party as it may be the cause of severe income inequality. Such a process would be started with the commercialization of the peasant economy, where peasants are forced to earn cash and pay taxes on these or other goods and services such as health care and education. Another possibility is that they might be forced to grow cash crops within the context of a given settlement area. Participation in market exchange makes the peasants vulnerable to risks and seasonal uncertainties, while landlords might benefit

Table 6.5. Land Fragmentation

	Land/ Household (m²)	Number of Plots	Largest Plot (m²)	Distance from House (m)	Smallest Plot (m²)	Distance from House (m)
Vietnam	10,140	5	4,830	1,200	2,250	900
Mekong River Delta	18,260	2	10,000	1,700	5,290	1,000
Red River Delta	2,370	8	600	1,200	150	700
Namđinh	2,370	5	1,077	1,016	169	220

Source: IFPRI (1996); Household survey in *Haihau, Yyen,* and *Trucninh* districts (in *Namđinh* Province), conducted in June–July 2000.

with their large landholdings and larger capital, having access to markets and new technologies (Bernstein 1992).

In consequence, peasant households are less competitive than landlords, when shifting resources from maximizing use-value to maximizing exchange-value and profit. Hence, a sudden mishap may lead peasant households to fall into debt, and therefore to mortgage or even sell their land to meet the debt servicing. Thus peasants end up as a part of the unemployed labor force in urban areas or become tenants of the landlords. Landlords on the other hand become richer as they increase the price of inputs sold to peasants (linked to credit), raise the land rent (as land is in high demand), or reduce the price of output (by contracting it in advance). The vicious cycle is continued further with more and more peasants being forced to sell their land.

Two implications arise from the above framework:

1. Inequality of landholdings and income must be higher within each region than between regions.
2. With less availability of land in a region, inequality of landholdings and income in that region will be higher.

However, this study proves that this is not the case with *Namđinh* Province in the late 1990s. Table 6.6 shows that the land is fairly equally distributed. Both the Gini and Theil T coefficients of land inequality are very low. For the province as a whole, the values of these inequality measures classified by individuals are lower than those classified by households. The percentage of Theil T for the entire province that can be attributed to inequality within regions declines from 67.2 to 51.0 as the classification changes from households to individuals. This finding suggests that the landholding of each household is determined by the household size, and that the minor differences between households in sizes of landholdings are largely explained by differences in household size and in overall land availability among districts and communes.

The percentage of Theil T that can be attributed to inequality within regions is high, casting doubts on the equality of land distribution in each region. However, looking at land distribution in each district, one notes that the two implications of the above-mentioned analytical framework are not verified. In the *Haihau* district, inequality of landholdings is the highest in the province, but the percentages of inequality within the region are the lowest. In addition, the land inequality within all three communes is lower than that of the district as a whole. Furthermore, both inequality measures are lowest in Haiminh where the land/household and land/person ratios are also the lowest, relative to other communes in the district. As a result, land shortage is not the cause of land concentration and the emergence of landless peasants.

Table 6.6. Land Distribution in *Namđinh* (per household and per capita)

	Gini		Theil T		Land Size (m²)	
	Household	Person	Household	Person	Household	Person
Region						
Namđinh	0.220	0.198	0.067	0.054	2,329	475
	(67.2)	(51.0)				
District						
Haihau	0.212	0.203	0.075	0.073	2,196	465
	(51.0)	(30.6)				
Commune						
Haiminh	0.087	0.046	0.016	0.003	1,231	205
Haitan	0.181	0.111	0.068	0.024	2,084	571
Haigiang	0.115	0.123	0.021	0.024	2,709	542
District						
Yyen	0.079	0.170	0.061	0.023	2,565	507
	(51.0)	(30.6)				
Commune						
Yenninh	0.218	0.260	0.031	0.017	1,991	364
Yenduong	0.206	0.094	0.066	0.016	3,052	603
Yentri	0.198	0.097	0.064	0.017	2,803	567
District						
Trucninh	0.168	0.199	0.046	0.062	2,176	450
	(82.0)	(71.4)				
Commune						
Trungdong	0.215	0.208	0.073	0.074	1,852	348
Liemhai	0.128	0.150	0.027	0.035	2,204	511
Tructuan	0.097	0.132	0.016	0.030	2,549	524

Source: Household survey in *Haihau, Yyen,* and *Trucninh* districts (in *Namđinh* Province), conducted in June–July 2000.
Note: The number in the parentheses below the Theil T shows the percentage of the Theil T resulting from inequality within regions.

Yyen district has the lowest land inequality in the province. However, table 6.6 shows that 93 percent of the Theil T for households originates from inequality within communes. This can be explained from the difference in size of households within communes like *Yenduong* and *Yentri,* rather than the land inequality within communes. It is worth noting that the Gini coefficients in *Yenninh* commune are high compared to other communes in the survey, while land/household and land/person ratios are low. This is mainly attributed to the fact that there are three non-farm households in the survey, who fully participated in handicraft activity. Ignoring these households, while calculating the Theil T, the land distribution in *Yenninh* commune becomes even more equal than the other two communes of the district.

In *Trucninh* district, the land distribution is also relatively equal, as the Gini and Theil T coefficients of the district are lower than those of the entire province. Land inequality within communes like *Liemhai* and *Tructuan* is small, as the values of both inequality measures for these communes are lower than those for the entire district. Only the case of *Trungdong* commune, with high inequality relative to the district and other communes in the entire province, should be considered with caution. A careful look at *Trungdong* commune reveals that it is likely to have two different types of land distribution (table 6.7). The reason is that the survey in *Trungdong* commune was conducted in two villages (*Trunglao* and *Dongthuong*) with considerable differences. Table 6.7 shows that the Theil T coefficients of those two villages are lower than those of the commune. In addition, the land/household and land/person ratios in *Dongthuong* are respectively 1.6 and 1.8 times higher than those in *Trunglao*. It means that a considerable share of land inequality in *Trungdong* commune stems from land inequality *between* villages rather than *within* villages. Furthermore, the land/household and land/person ratios of *Trunglao* village are the lowest among communes in the survey. The Theil T coefficients of this village are also very low compared to *Dongthuong* village and other communes in the province. It implies once again that land scarcity does not necessarily lead to land concentration.

In brief, these detailed observations bring about three tentative conclusions. First, most land inequality comes from differences in landholding between regions, rather than within regions. Secondly, sizes of landholdings depend primarily on the household size. Thirdly, there is no evidence of a negative relationship between land availability and land inequality within regions. Instead, land concentration may take place in regions where there is strong development of non-farm activities.

Table 6.8 shows that there exists an ambiguous correlation between land and income distribution, as is evident from two factors. First, the Gini and Theil T coefficients of income inequality are considerably higher than those of land inequality. Second, the relationship of both coefficients with income per capita resembles the well-known inverse U-shaped curve (Kuznets 1954; Ahluwalia 1976).

Table 6.7. Land Distribution in Trungdong Commune—Trucninh District (per household and per capita)

	Theil T		Land Size (m²)	
	Household	Person	Household	Person
Commune				
Trungdong	0.073 (65)	0.074 (41)	1,852	348
Village				
Trunglao	0.045	0.006	1,456	260
Dongthuong	0.049	0.046	2,292	458

Source: Household survey in *Haihau, Yyen,* and *Trucninh* districts (in *Namđinh* Province), conducted in June–July 2000.
Note: The number in the parentheses below the Theil T shows the percentage of the Theil T resulting from inequality within regions.

Table 6.8. **Distribution of Land and Income in *Namđinh* (per capita)**

	Per Capita Income			Land/person		
	Gini	Theil T	Income (1,000 đ)	Gini	Theil T	m²
Region						
Namđinh	0.372	0.218 (78.4)	2,219	0.198	0.054 (51.0)	475
District						
Haihau	0.316	0.168 (83.7)	2,880	0.203	0.073 (30.6)	465
Trucninh	0.291	0.136 (84.0)	1,762	0.199	0.062 (71.4)	450
Yyen	0.408	0.281 (69.9)	2,072	0.170	0.023 (92.8)	507

Source: Household survey in *Haihau, Yyen,* and *Trucninh* districts (in *Namđinh* Province), conducted in June–July 2000.
Note: The number in the parentheses below the Theil T shows the percentage of the Theil T resulting from inequality within regions.

In comparison to other provinces, the income inequality is relatively high because of differences in crop and employment structures. First, it is evident from table 6.9 that there is an inverse relationship between income per capita and the share of crop agriculture in total income. Second, even in the *Haihau* district that has the most fertile soil in the province, the richest commune is *Haitan*—with the primary share of total income coming from husbandry. In two other districts that have less fertile land, the richest communes are *Trungdong* and *Yenninh*, whose income had a big chunk of non-farm activities.

The question that arises here is whether the rich tend to accumulate land as a direct result of the increasing income inequality. Apart from obstacles imposed by the aforementioned land policies, rich households are further hampered in their quest to accumulate land by the low expected returns from agricultural investment. In the late 1990s, land rental markets were still embryonic. This was due to many reasons—first of which was that land transfer and land lease was very rare among households. In the survey, there are only four cases of households that rented-out land, while thirty-nine households rented-in land after bidding on land of the commune. Up to 75 percent of households reported that they would like to rent-in more land because of labor surplus and to ensure food consumption.

The second factor is that cultivation of annual food crops in land allocated in the first and second rounds is mainly used for the purpose of self-consumption. Table 6.10 shows that cash revenue from food production is lower than cash expenditure on food (except in the *Haihau* district). In other words, food producers are often short for food.

Table 6.9. Composition of Income Sources of Peasant Households in Namđinh

	Total	Annual Crops	Perennial Crops	Husbandry	Non-Farm	Wages and Subsidies	Income per Capita (1,000 đ)	Gini Co-effi-cient
			Composition of Income (%)					
Region								
Namđinh	100	38.9	2.2	21.5	28.3	9.1	2,219	0.372
District								
Haihau	100	37.0	2.2	24.6	25.0	11.2	2,880	0.316
Commune								
Haiminh	100	15.7	1.9	26.2	53.2	3.1	2,084	0.243
Haitan	100	36.5	1.6	32.8	16.2	13.0	3,834	0.307
Haigiang	100	48.1	2.9	15.8	19.7	13.6	2,668	0.277
District								
Yyen	100	38.6	0.3	13.5	38.4	9.1	2,072	0.408
Commune								
Yenninh	100	27.1	0.0	8.6	52.0	12.3	2,624	0.399
Yenduong	100	53.5	0.2	20.4	14.2	11.7	1,514	0.385
Yentri	100	36.4	0.7	12.2	46.7	4.0	2,006	0.391
District								
Trucninh	100	41.1	4.0	26.3	21.6	7.0	1,762	0.291
Commune								
Trungdong	100	31.2	0.5	24.6	37.0	6.7	2,241	0.252
Liemhai	100	46.3	7.7	27.2	8.8	10.0	1,575	0.311
Tructuan	100	47.1	3.8	27.5	18.2	3.5	1,311	0.200

Source: Household survey in *Haihau, Yyen,* and *Trucninh* districts (in *Namđinh* Province), conducted in June–July, 2000.

The third factor is that households are unwilling to rent-out land. Out of these households, 80 percent reported that they wanted to ensure subsistence consumption. Out of the households that wanted to rent-in more land, only 40 percent belonged to the high-income group (Groups 3 and 4 classified by the adjusted criteria of the VLSS in 1998), and only 30 percent are willing to invest in food production.

FARM SIZE AND PRODUCTIVITY

When an inverse relationship between farm size and land productivity exists, a higher inequality of land will lead to a decline in land productivity, hence reducing efficiency of agricultural production (Griffin 1979; Ghatak and Ingersent 1984; Ellis 1988). The first and foremost issue that plagues small farmers is low opportunity cost of labor combined with high oppor-

Table 6.10. Composition of Cash Income of Households in *Namđinh*

	Total	Annual Crops	Perennial Crops	Husbandry	Non-Farm	Wages and Subsidies
Region						
Namđinh	100	−12.1	3.5	49.7	44.8	14.1
District						
Haihau	100	3.3	2.3	45.7	33.1	15.5
Yyen	100	−17.9	0.5	38.0	64.3	15.2
Trucninh	100	−21.0	7.3	63.9	37.9	11.8

Source: Household survey in *Haihau, Yyen,* and *Trucninh* districts (in *Namđinh* Province), conducted in June–July, 2000.

tunity costs of land; therefore they use land intensively and generate high yields. In contrast, in large farms the opportunity cost of land is relatively low and land is used extensively, hence yields are low.

The second factor is that large farmers may take advantage of their better access to credit and input supplies. In addition, the more farm size increases, the more difficult the management and supervision of labor becomes (high monitoring costs). Therefore, such market imperfections often lead large farmers to substitute capital for labor. In many developing countries, land and capital are scarce while labor is abundant. Such a pattern of resource allocation leads to social inefficiency. Therefore, an equal distribution of smallholdings is highly appealing.

However, productivity is not related to farm size in the case of *Namđinh* Province for two reasons. First, most households still have considerable surplus of labor. Second, the production of subsistence crops being dominant in agriculture, peasant households are unwilling to purchase capital goods or hire labor. As a result, there is no clear evidence of an inverse relationship between farm size and land productivity in *Namđinh* as can be seen from table 6.11.

Nevertheless, there are significant differences in land productivity within *Namđinh* Province due to differences in soil fertility, crop intensity, and ir-

Table 6.11. Rice Yields, Farm Size, and Land Fragmentation

	Rice Yield per Crop (kg/sao)	Farm Size (sao/household)	Number of Plots
Region			
Namđinh	195.7	6.52	5.38
District			
Haihau	213.2	6.10	2.86
Yyen	173.8	7.08	8.54
Trucninh	196.1	6.04	5.30

Source: Household survey in *Haihau, Yyen,* and *Trucninh* districts (in *Namđinh* Province), conducted in June–July, 2000.
Note: Land is measured in *sao*: 1 *sao* = 360 m^2.

rigation. Differences in yields may also result from differences in the degree of land fragmentation. As can be seen from table 6.11, yield is negatively correlated to the number of plots per households and positively related to area per plot.

CONCLUSION

The degree of land concentration within the *Namdinh* Province is not related to income inequality or to the level of production efficiency. A major factor explaining the low rice productivity is land fragmentation. There is no doubt that increased land concentration and development of the rural land market is a necessary condition for the further commercialization of agriculture. By encouraging skillful farmers to make further investment in agriculture and the transfer of labor surplus into the nonagricultural sector, the demand for agricultural commodities will increase.

This process of land concentration requires the removal of market bottlenecks and a reduction of transaction costs. For low-income households, land transfers and land leases require access to an efficient and accessible rural financial system for setting land rents (on the basis of the discounted return from land) and for providing credit to those households who intend to accumulate land. In addition, the administrative procedures for land transfers have to be simplified to reduce the relatively high transaction costs.

Developing rural non-farm activities is the most effective way to absorb the labor surplus from agriculture as it fosters land concentration among the remaining peasant households. Households can further be encouraged to make long-term investments in land if the duration of land use rights is extended to fifty years for annual crops, and if the three-year's limit on land leases is removed.

It can also be concluded that land ceilings are neither an efficient nor an effective instrument. In the Red River Delta, farms are still very small. However, in the Mekong Delta, where there is large-scale informal land concentration, land users simply use relatives to escape the formal ceilings, or shift to crops that allow longer land leases. Such land ceilings impede agricultural specialization and reduce production efficiency.

NOTES

1. Typical characteristics of Vietnamese agriculture are reviewed in Nguyen (1998).

2. This project focuses on three challenges to agricultural commercialization: land use, technology transfer, and access to market and terms of trade for agricultural products. The present chapter discusses a specific issue, regarding land use, inequality, and efficiency.

3. The questionnaire includes questions that cross verify in order to ensure the reliability of data filled in by households.

4. Up to 1997, there had been only 1.4 million hectares of unused arable land (Chung 1997: 58; GSO 1999b: 28).

5. There is evidence of overuse of inputs such as fertilizers and pesticides. During the 1990s, only 4 percent of agricultural growth was explained by the increase in total factor productivity (Jamal and Jansen 1998: 11–12; World Bank 2000: 46).

6. Marketed surplus accounts for less than 50 percent of total agricultural output (GSO 2000: 224–25).

7. Around 80 percent of agricultural land sale is purchased by urban households (GSO 2000: 220). It suggests that land is used as an asset against risk; hence its productivity is not expected to be high.

8. Due to the difference in history, the processes and patterns of land allocations differed greatly between the two major deltas in the North and South (Chu et al. 1992; Porter 1993; Nguyen 1995). In the North, after Resolution 10 in 1988, land was allocated on the basis of need (usually indicated by the size of households) and also the ability to farm the land (i.e., the number of household members who can work on the land). This method of land allocation was based on the policy of an equal distribution of each category of land use among all households; that is, all households in a commune are allocated an equal share of the high yield and inferior yielding lands. In practice, depending on the size of the commune and diversity of fertility of land, a household may have received many small plots (each household having on average, four to fifteen plots with total area of 0.23 ha) dispersed widely over the commune (IFPRI 1996: 13). Such land fragmentation imposed negative effects on production efficiency, land productivity, impeding marketed surplus and commodity exchange of agriculture.

The South being oriented toward the establishment of commercial small farms, land was returned largely to previous owners or holders. Compared with the Red River Delta, land distribution in the Mekong River Delta was more unequal but middle peasantry is still dominant in agricultural production. Furthermore, in the Mekong River Delta land was less fragmented, relative to the Red River Delta.

9. However, in practice the plots for distribution in the second round were often those reserved for the newly returned, such as discharged soldiers. In the long run, land of this category would become land distributed in the first round.

10. Often set aside for bidding, were unused lands or water surfaces, areas difficult to cultivate and would require special investment.

11. Agriculture land for annual crops, in the two Southern regions is not supposed to exceed hectares per household. In other provinces and cities under central authority, the limit was two hectares. For perennial crops, agricultural land for each household must not exceed ten hectares for the flat or delta land and thirty hectares for midrange and mountainous land. The limit for the allocation of cleared land, bare hills, wasteland, and reclaimed coastal land to households and individuals is stipulated by the People's Committee of the provinces and cities.

12. More than 80 percent of land is irrigated in the province (NOARD 2000: 19).

13. This achievement is superior to the general index of the entire country recorded at 75 percent only (NOARD 2000: 38; Nguyen 1998: 35).

BIBLIOGRAPHY

Ahluwalia, M. S. "Inequality, Poverty and Development." *Journal of Development Economics* 6 (1976): 307–42.

Akram-Lodhi, A. H. "Landlords Are Taking Back the Land: The Agrarian Transition." *Working Paper*, no. 353. The Hague: Institute of Social Studies, 2001a.

Akram-Lodhi, A. H. "Vietnam's Agriculture: Is There an Inverse Relationship?" *Working Paper*, no. 348. The Hague: Institute of Social Studies, 2001b.

Bernstein, H. "Notes on Capital and Peasantry." Pp. 160–77 in *Rural Development: Theories of Peasant Economy and Agrarian Change* edited by J. Harris. London: Routledge, 1992.

CECARDE. *Nong nghiep nong thon trong giai doan cong nghiep hoa, hien dai hoa* (Agricultural and Rural Sector during Industrialization and Modernization). Hanoi: National Politics Publishing House, 1997.

Chu, V. L., et al. *Hop tac hoa nong nghiep Viet Nam: lich su—van de—trien vong* (Vietnam's Agricultural Collectivization: History—Problem—Perspective). Hanoi: Truth Publishing House, 1992.

Chung, C. H. "Agricultural Growth within a Strategy for Sustainable Rural Development in Vietnam: A World Bank Perspective." Proceedings of the Conference on Strategies for Agriculture and Rural Development in Vietnam, Hanoi, 21–22 March 1997.

Dao, T. T. *Kinh te ho nong dan* (Peasant household economy). Hanoi: National Politics Publishing House, 1997.

Dao, T. T., et al. (eds.). *He thong nong nghiep luu vuc song Hong* (Agrarian System in the Red River Basin). Hanoi: Agricultural Publishing House, 1998.

Deininger, K. and G. Feder. "Land Institutions and Land Markets." *Working Paper*, no. 2014. Washington, D.C.: World Bank, 1999.

Do, T. P. "Ngheo doi o nong thon va giai phap." (Rural Poverty and Solutions). *Journal of New Rural Sector* 4 (1998).

Dollar, D., P. Glewwe, and J. Litvack. (eds.). *Household Welfare and Vietnam's Transition*. World Bank Regional and Sectoral Studies. Washington, D.C.: World Bank, 1998.

Ellis, F. *Peasant Economics. Farm Households and Agrarian Development.* Cambridge, MA: Cambridge University Press, 1988.

Faruqee, R. and K. Carey. "Land Markets in South Asia: What Have We Learned?" *Working Paper*, no. 1754. Washington, D.C.: World Bank, 1997.

General Office of Land Administration (GOLA). *Nghien cuu nhung van de chu yeu ve kinh te xa hoi co quan he truc tiep den viec to chuc thuc hien va hoan chinh phap luat dat dai hien nay* (Study on Socioeconomic Issues Relating to the Enforcement of the Present Land Law). Final Report, Mimeo. Hanoi, 1997.

General Office of Land Administration (GOLA). *Hoi nghi chuyen de ve chuyen doi ruong dat nong nghiep, khac phuc tinh trang phan tan, manh mun trong san xuat* (Conference on Agricultural Land Exchange to Avoid Land Fragmentation). Report, Mimeo. Hanoi, 1998.

General Statistical Office (GSO). *Ket Qua Dieu Tra Kinh Te—Xa Hoi Ho Gia Dinh 1994–1997* (Results from Household Surveys 1994–1997). Hanoi: Statistics Publishing House, 1999a.

General Statistical Office (GSO). *So Lieu Thong Ke Nong—Lam Nghiep—Thuy San Viet Nam 1990–1998 Va Du Bao Nam 2000* (Statistical Data on Agriculture-Forestry-Fishery in Vietnam during 1990–1998 and Forecasting for 2000), Hanoi: Statistics Publishing House, 1999b.

General Statistical Office (GSO). *Dieu Tra Muc Song Dan Cu Viet Nam 1997–1998* (Vietnam Living Standard Survey, 1997–1998), Hanoi: Statistics Publishing House, 2000.

Ghatak, S. and K. Ingersent. *Agriculture and Economic Development.* Brighton: Wheatsheaf Books, 1984.

Griffin, K. *The Political Economy of Agrarian Change: An Essay on the Green Revolution.* London: Macmillan, 1979.

Haughton, D., J. Haughton, and N. Phong. (eds.). *Living Standards during an Economic Boom: The Case of Vietnam.* Hanoi: Statistical Publishing House with UNDP, 2001.

Hayami, Y. "Strategies for the Reform of Land Property Relations. Agricultural Policy Analysis for Transition to a Market-Oriented Economy in Vietnam." Hanoi: FAO, 1994.

IFPRI. *Rice Market Monitoring and Policy Option Study.* Washington, D.C.: International Food Policy Research Institute, 1996.

Jamal, V. and K. Jansen. "Agrarian Transition in Vietnam." *Working Paper*, no. 128. Geneva: International Labor Office, 1988.

Johnston, B. F. and P. Kilby. *Agriculture and Structural Transformation.* New York: Oxford University Press, 1975.

Kuznets, S. "Economic Growth and Income Inequality." *American Economic Review* 45 (1954): 1–28.

Netting, R. McC. *Smallholders, Householders: Farm Families and the Ecology of Intensive, Sustainable Agriculture.* Stanford, CA: Stanford University Press, 1993.

Minot, N. and F. Goletti. *Rice Market Liberalization and Poverty in Vietnam.* Research Report no. 114. Washington, D.C: International Food Policy Research Institute, 2000.

Namđinh Office of Agriculture and Rural Development (NOARD). *Quy hoach phat trien nong lam nghiep thoi ky 2001–2010 tinh Namđinh* (Planning for the Development of Agriculture and Forestry during the Period of 2001–2010 in *Namđinh* Province). Report, Mimeo. *Namđinh*, 2000.

Namđinh Office of Land Administration (NOLA). *He thong so lieu dat dai tinh Nam Dinh* (Land Archives of *Namđinh* Province). Draft Report, Mimeo. *Namđinh*, 2000.

Namđinh Office of Planning and Investment (NOPI). *Ke hoach phat trien kinh te—xa hoi nam 2000 tinh Nam Dinh* (Socio-Economic Development Plan for 2000 of *Namđinh* Province). Report, Mimeo. *Namđinh*, 1999.

Namđinh Statistical Office (NSO). *Statistical Yearbook 1998.* *Namđinh*, 1999.

Ngo, V. L. "Reform and Rural Development: Impact on Class, Sectoral, and Regional Inequalities." Pp. 165–207 in *Reinventing Vietnamese Socialism: Doi Moi in Comparative Perspective*, edited by W. S. Turley and M. Selden. Boulder, CO: Westview Press, 1993.

Nguyen, N. H. *Tinh hinh dat dai trong nong nghiep va nong thon Viet Nam* (Situation of Land Use in Vietnam's Agriculture). Report, Mimeo. Hanoi: General Office of Land Administration (GOLA), 1997.

Nguyen, N. H. and N. D. Pham. *Ve viec sua doi, bo sung luat dat dai gan voi chien luoc giao dat nong nghiep, dat lam nghiep* (On Amendment and Supplement to Land Law for a Strategy on Agricultural and Forestry Land Allocation). Report, Mimeo. Hanoi: General Office of Land Administration (GOLA), 1997.

Nguyen, S. C. *Agriculture of Vietnam 1945–1995*. Hanoi: Statistical Publishing House, 1995.

Nguyen, T. K. *An Analysis of Peasant Agricultural Production in the 1988–1997 Period: the Challenges to Further Development of Commodity Production*. MA Thesis, Mimeo. Hanoi: ISS/NEU, 1998.

Porter, G. *Vietnam: The Politics of Bureaucratic Socialism*. Ithaca, NY: Cornell University Press, 1993.

Saith, A. "Development Strategies and the Rural Poor." *Journal of Peasant Studies* 17, no. 2 (1990): 171–244.

Saith, A. *The Rural Non-Farm Economy: Processes and Policies*. Geneva: International Labor Office, 1992.

Timmer, P. "Food Policy and Economic Reform in Vietnam" in *The Challenge Reform in Indochina*, edited by B. Ljunggen. Cambridge, MA: Harvard Institute of International Development, 1994.

United Nations. *An Agriculture-Led Strategy for the Economic Transformation of Vietnam: Project and Policy*. Hanoi: UNDP, 1994.

United Nations. *Poverty Elimination in Vietnam*. Hanoi: UNDP, 1995.

Watts, M. J. "Recombinant Capitalism: State, Decollectivization and the Agrarian Question in Vietnam." Pp. 450–505 in *Theorizing Transition: The Political Economy of Post-Communist Transformations*, edited by J. Pickles and A. Smith. London: Routledge, 1988.

World Bank. *Vietnam—Economic Report on Industrialization and Industrial Policy*. Hanoi: World Bank, 1995.

World Bank. *Vietnam 2010—Entering the 21st Century: Pillars of Development*. Hanoi: World Bank, 2000.

7

The Effects of Economic Policy Reforms on the Economic Environment of Farm Households in China

An Empirical Analysis for *Jiangxi* Province

Zhigang Chen, Nico Heerink, and Peixin Zhu[1]

The economic reforms implemented since the 1980s in China have had a major impact on the economic environment of farm households. Prices received by farmers for their commodities, and those paid for inputs used in producing these commodities, government quota requirements with respect to strategic products, as well as the availability of alternative market channels have all changed fundamentally during the last two decades. The impact of these reforms, however, differs considerably between the various regions of China. In particular, the introduction of the Governor's grain bag responsibility system in the mid-1990s assigned a greater autonomy in agricultural policy decision-making to provincial governments. Besides, differences between regions in agro-climatic conditions, access to markets, state of the infrastructure, and income level imply that market liberalization policies have rather diverging impacts on the rural economy in different parts of the country.

Available studies on the impact of past and expected future economic policies on the agricultural sector in China tend to concentrate on the country as a whole (see, e.g., Huang 1997; Lin 1997; Huang and Chen 1999; Yu and Frandsen 2002; OECD 2005; Gale et al. 2005). An important exception is the study by Diao et al. (2003) that analyzed the regional impact of China's accession into the World Trade Organization (WTO), finding that farmers in China's least developed regions benefit little or even suffer from WTO accession because (traditional) agriculture is still an important source of their livelihood. In general, however, insights into regional and local differences in the effects of economic reforms are notoriously lacking. A better understanding of these, and the factors causing them, may play an important role in designing improved national and regional policies aimed at rural poverty reduction and food security.

The purpose of this chapter is to undertake an empirical analysis of the impact of China's economic policy reforms implemented since the beginning of the 1980s on the economic environment of farm households in *Jiangxi* Province. Although it is located in the Southeast of China, not far away from the country's centers of economic growth, *Jiangxi* is one of China's poorer provinces. In 2004, its per capita GDP equalled 78 percent of the national per capita GDP (NBS 2005: tables 3.1 and 3.11). Its economy is dominated by agricultural production, with rice as the major crop. These features make *Jiangxi* an interesting case for examining differences between national and regional impact of economic policies on the agricultural economy. The focus of this analysis will be on the impact of the reforms on the prices of the main agricultural products produced by farmers in *Jiangxi* Province, that is, rice and pork, and on the prices of the main inputs used in agricultural production, that is, chemical fertilizers.

METHOD USED

Time-series data from 1981 to 2003 on agricultural input and output prices in *Jiangxi* Province and in China as a whole, and data on relevant explanatory variables for the same period have been collected from statistical yearbooks, such as the *China Statistical Yearbook, China Price Yearbook*, and the *Jiangxi Statistical Yearbook*. Data on world market prices and the official exchange rate over the same period were obtained from the International Monetary Fund (IMF).

There are two main reasons for choosing 1981 as the starting year for the analysis. First, the macroeconomic data used come mostly from official statistical publications. Before 1981, these data are scarce and inaccurate. Second, the focus of the analysis is on price changes facing farm households. Before the start of the household responsibility system at the end of the 1970s, agricultural production levels and prices were decided and planned by governments at various levels. Although data for 2004 were also available at the time of the analysis, we decided not to include these data because that year marked the start of another major reform in policy making (see Gale et al. 2005; and chapters 1 and 8 in this volume). The method that we apply cannot be used to examine the impact of such a recent reform.

The chapter will—after this introduction—consist of two further sections. In the second section the data analysis is discussed and the results are presented. Regression analysis is used to examine the impact of policy reforms on prices of the major agricultural products and inputs in *Jiangxi* Province.

This is done in two steps. First, the impact of world market prices on national prices is examined for tradable inputs and outputs. Dummy variables are used to examine differences between the various stages of economic reforms, as distinguished in chapter 1. The results indicate the extent to which world market prices affected national prices during different reform periods, taking into account the adjustments that were made to the exchange rate.

As a second step, the relation between national prices and regional prices within *Jiangxi* Province is examined. The same stages in economic reforms are distinguished through the use of dummy variables. The results show the extent to which input and output prices were determined by national policies or by internal supply and demand factors and policies within *Jiangxi* Province during different reform periods. A frequent problem in time-series analysis of agricultural prices is that of serial correlation of the residuals. The Durbin-Watson test is used to test for serial correlation.[2] If the test result is positive, a first-order auto-regression term AR(1) and, if necessary, a second-order auto-regression term AR(2) is added to the regression equation to correct for it. In the third section, a short concluding note is presented, reflecting on the results.

THE REGRESSION MODEL USED AND THE RESULTS OBTAINED

In this section, we present the regression results for the prices of grain, pork, and chemical fertilizers, respectively.

Grain

Grain is the main agricultural output in *Jiangxi* Province. It consists largely of rice, which dominates arable cropping throughout the province. In 2004, rice constituted 95 percent of grain output in the province (NBS 2005: table 13–17).

Figure 7.1 shows the changes over time of the retail grain price index in *Jiangxi* Province and in China as a whole. The two series have been corrected for inflation by the consumer price indices (CPI) for *Jiangxi* Province and China, respectively. The figure shows that both prices remained roughly constant during the 1980s, increasing rapidly between 1990 and 1995, and then declining slightly until 2003. Between 1990 and 1995, the grain price in *Jiangxi* Province increased more rapidly than the grain price for China as a whole. The figure also shows the development of the world market price of rice, converted into *yuan* by using the official exchange rate and deflated by the CPI for China, since 1981. There was no upward trend in the world market price in the first half of the 1990s. Since the middle of the 1990s,

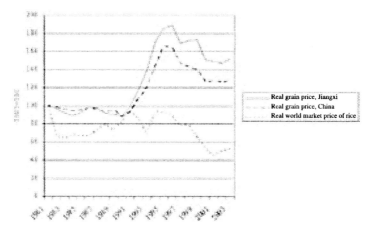

Figure 7.1.   Grain Prices (1981–2003)

however, the trend in the world market price of rice is very similar to that of grain prices in China and in *Jiangxi* Province.

The following two equations are used to explain grain prices in China and in *Jiangxi* Province, respectively:

$$\text{PGRAINC} = c_0 + c_1\text{DUM2} + c_2\text{DUM3} + c_3\text{PRICEW} + c_4\text{DUM2} \times \text{TPGRAIN} + c_5\text{DUM2} \times \text{YPC} + c_6\text{DUM3} \times \text{TPGRAIN} + c_7\text{DUM3} \times \text{YPC} + u_1 \quad (1)$$

$$\text{PGRAINJ} = d_0 + d_1\text{DUM2} + d_2\text{DUM3} + d_3\text{PGRAINC} + d_4\text{DUM2} \times \text{TPGRAINJ} + d_5\text{DUM2} \times \text{YPCJ} + d_6\text{DUM3} \times \text{TPGRAINJ} + d_7\text{DUM3} \times \text{YPCJ} + u_2 \quad (2)$$

where:

PGRAINC:	Real market price of grain in China (1981 = 100)
DUM2:	Dummy variable reflecting the second stage of grain market reform (= 1 for 1985–1993)
DUM3:	Dummy variable reflecting the third stage of grain market reform (= 1 for 1994–2003)
PRICEW:	Real world market price of rice (Thai white rice) in *yuan* (1981 prices; official exchange rate) per metric ton.
TPGRAIN:	Total production of grain in China (in 10,000 tons)
YPC:	Real GDP per capita in China (1981 = 100)
PGRAINJ:	Real market price of grain in *Jiangxi* Province (1981 = 100)
TPGRAINJ:	Total production of grain in *Jiangxi* Province (in 10,000 tons)
YPCJ:	Real GDP per capita in *Jiangxi* Province (1981 = 100)
u_1, u_2:	Error terms with standard properties
$c_0, ..., c_7, d_0, ..., d_7$:	Unknown coefficients.

Since rice constitutes the largest share of grain produced in China, the world market price for rice is used as the world market price indicator.[3] It is expected to be positively related to the domestic price. Total grain production is expected to have a price-depressing effect, while per capita income is expected to boost grain prices. These two variables are multiplied by dummy variables reflecting the second and third reform period, respectively, to estimate whether the impact of domestic factors differed between these two reform periods.

The regression results are reported in the column titled "full equation" in table 7.1. The column titled "final equation" reports the results when only the significant variable(s) and the constant are included in the regression equation. As expected, the world market price of rice is positively related to the domestic grain price. Domestic demand and supply factors, however, also played a significant role during the second reform period. These domestic factors were apparently responsible for the rapid increase in the real market price of grain during that period (see figure 7.1). During the third period of grain market reform, supply and demand factors no longer significantly affected the real prices, indicating that the grain market in China has become more integrated into the world market.

The regression results for the grain price in *Jiangxi* Province are shown in table 7.2. As expected, the national price is a major determinant of the grain price in *Jiangxi* Province (see also figure 7.1). But local supply and demand factors also played an important role during the second stage of the grain market reform. Apparently, these local factors were responsible for the above-average increase in the real market price of grain during the period 1985–1994, which is evident from figure 7.1. During the third period of grain market reform, supply and demand factors within *Jiangxi* Province no longer significantly affected real prices. Despite the greater provincial

Table 7.1. Regression Results for Grain Price, China

Explanatory Variable	Expected Impact	Full Equation		Final Equation	
Constant	+	83.0***	(6.18)	62.9***	(6.19)
DUM2	—	127	(1.58)	163*	(1.89)
DUM3	—	11.6	(0.23)	46.0***	(9.10)
PRICEW	+	0.02	(1.09)	0.06***	(3.69)
DUM2 × TPGRAIN	–	–0.004**	(–1.79)	–0.005**	(–2.07)
DUM2 × YPC	+	0.02***	(2.74)	0.03***	(2.76)
DUM3 × TPGRAIN	–	0.001	(1.33)	—	—
DUM3 × YPC	+	–0.004	(–1.65)	—	—
R^2		0.93		0.91	
R^2-adjusted		0.90		0.88	
No. of observations		23		23	
Durbin-Watson		1.76		1.74	

Notes: t-statistics are in parenthesis.
* Significant at 10% level, ** Significant at 5% level, *** Significant at 1% level.

Table 7.2.   Regression Results for Grain Price, *Jiangxi* Province

Explanatory Variable	Expected Impact	Full Equation		Final Equation	
Constant	—	1.21	(0.10)	−7.34	(−0.78)
DUM2	—	59.3	(1.57)		
DUM3	—	16.5	(0.65)	27.4***	(4.86)
PGRAINC	+	0.96***	(8.01)	1.03***	(11.1)
DUM2 × TPGRAINJ	–	−0.05**	(−1.84)	−0.01**	(−1.79)
DUM2 × YPCJ	+	0.02***	(4.44)	0.02***	(3.73)
DUM3 × TPGRAINJ	–	0.01	(0.61)	—	—
DUM3 × YPCJ	+	−0.00	(−0.28)	—	—
AR(1)	—	—	—	0.35*	(1.47)
R^2		0.99		0.99	
R^2-adjusted		0.99		0.99	
No. of observations		23		23	
Durbin-Watson		1.66		1.70	

Notes: t-statistics are in parenthesis.
* Significant at 10% level, ** Significant at 5% level, *** Significant at 1% level.

autonomy in balancing grain supply and demand and stabilizing prices during this period, the grain market had become more integrated into the national market, and hence into the world market.

These results are consistent with the findings of Park et al. (2002) and Huang et al. (2004). Park et al. (2002) find that agricultural markets had become fairly integrated during the early 1990s, but some large, poorer areas were not completely integrated into national markets at that time. Huang et al. (2004) found that agricultural markets had continued to develop in the late 1990s and even farmers in China's poor, remote villages are nowadays linked to China's regional markets (and hence to the world market).

The findings of those two studies and ours suggest that government interventions in the grain market have become ineffective. According to Park et al. (2002: 74–75) this may be explained from the fact that agricultural markets have become much larger and difficult to control, and that new incentives to public grain managers has reduced their willingness to implement state grain policies

Pork

The main livestock product in *Jiangxi* Province is pork; in 2004, it constituted 68 percent of the total meat output in the province (NBS 2005: tables 13–21). As discussed in chapter 1, pork prices and markets in China were gradually liberalized during the period 1985–1993, but contrary to grains and oil-bearing crops, pork was not included in the regional food security policies in the second half of the 1990s.

Figure 7.2 shows the developments of real pork prices since 1981 in *Jiangxi* Province and in China as a whole, as well as in the world market. Again,

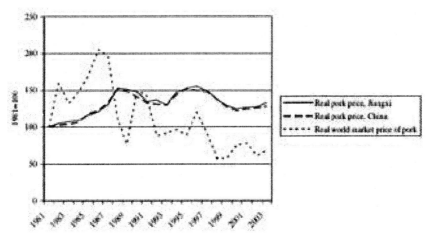

Figure 7.2.   Pork Prices (1981–2003)

regional and national prices follow quite similar patterns, with the former (expressed in an index number) slightly exceeding the latter in some periods. Both prices increased about 50 percent between 1981 and 1988, and have been fluctuating since then. The world market price of pork increased more than 100 percent between 1981 and 1986, but declined rapidly since then to a level that was around 20–40 percent lower in 1998–2003 than the level at the start of our analysis (in 1981).

Similar specifications as those used for grain prices are used to explain the development of pork prices in China and in *Jiangxi* Province:

$$\text{PPORKC} = c_0 + c_1\text{DUM2} + c_2\text{DUM3} + c_3\text{PPORKW} + c_4\text{DUM2} \times \text{TPPORK} + \\ c_5\text{DUM2} \times \text{YPC} + c_6\text{DUM3} \times \text{TPPORK} + c_7\text{DUM3} \times \text{YPC} + u_1 \quad (3)$$

$$\text{PPORKJ} = d_0 + d_1\text{DUM2} + d_2\text{DUM3} + d_3\text{PPORKC} + d_4\text{DUM2} \times \text{TPPORKJ} + \\ d_5\text{DUM2} \times \text{YPCJ} + d_6\text{DUM3} \times \text{TPPORKJ} + d_7\text{DUM3} \times \text{YPCJ} + u_2 (4)$$

where:

PPORKC:	Real market price of pork in China (1981 = 100)
DUM2:	Dummy variable reflecting the second stage of grain market reform (= 1 for 1985–1993)
DUM3:	Dummy variable reflecting the third stage of grain market reform (= 1 for 1994–2003)
PPORKW:	Real world market price of swine meat in 0.01 *yuan* (1981 prices; official exchange rate) per pound
TPPORK:	Total production of pork in China (in 10,000 tons)
YPC:	Real GDP per capita in China (1981 = 100)
PPORKJ:	Real market price of pork in *Jiangxi* Province (1981 = 100)

TPPORKJ:	Total production of pork in *Jiangxi* Province (in 10,000 tons)
YPCJ:	Real GDP per capita in *Jiangxi* Province (1981 = 100)
u_1, u_2:	Error terms with standard properties
c_0, ..., c_7, d_0, ..., d_7:	Unknown coefficients.

A distinction is made again into the periods 1985–1993 and 1994–2003 to reflect the fundamentally different market environment of farmers during these two periods.

The regression results for the pork price in China are shown in table 7.3. The world market price of pork does not have a significant impact on the domestic pork price. This is no surprise, given that the share of pork that is traded internationally is negligible.[4] The results also show, however, that domestic supply and demand factors do not significantly affect the domestic price of pork. Given the gradual liberalization of the pork market during the period of analysis, this result is unexpected. Problems with the reliability of reported statistics on livestock production in China are likely to play a major role here (Zhong 1997; Fuller et al. 2000; Ma et al. 2004).

Table 7.4 summarizes the regression results for the pork price in *Jiangxi* Province. As expected, the national pork price is a major determinant of the regional price of pork. However, besides the national pork price, domestic supply and demand factors within *Jiangxi* Province also significantly affect the price of pork; total pork production and per capita GDP in *Jiangxi* have a significant effect on the pork price in both periods. The estimated coefficients (–0.44 and 0.04 in the second reform stage; –0.05 and 0.001 for the

Table 7.3. Regression Results for Pork Price, China

Explanatory Variable	Expected Impact	Full Equation		Final Equation	
Constant	—	109***	(14.5)	119***	(15.2)
DUM2	—	−69.2*	(−2.12)	10.5*	(1.69)
DUM3	—	39.9**	(2.31)	17.8**	(2.21)
PPORKW	+	−0.02	(−0.65)	—	—
DUM2 × TPPORK	−	0.07	(2.50)	—	—
DUM2 × YPC	+	−0.03	(−1.96)	—	—
DUM3 × TPPORK	−	0.05	(0.85)	—	—
DUM3 × YPC	+	−0.005	(−1.95)	—	—
AR(1)	—	1.23***	(7.16)	1.23***	(7.16)
AR(2)	—	−0.85***	(−4.95)	−0.85***	(−4.95)
R^2		0.92		0.85	
R^2-adjusted		0.85		0.81	
No. of observations		21		21	
Durbin-Watson		2.02		1.95	

Notes: t-statistics are in parenthesis.
* Significant at 10% level, ** Significant at 5% level, *** Significant at 1% level.

Table 7.4. **Regression Results for Pork Price, *Jiangxi* Province**

Explanatory Variable	Expected Impact	Full Equation		Final Equation	
Constant	—	−15.2***	(−2.23)	−14.4**	(−2.56)
DUM2	—	0.48	(0.07)	—	—
DUM3	—	4.54	(0.34)	—	—
PPORKC	+	1.17	(17.9)	1.16***	(21.3)
DUM2 × TPPORKJ	−	−0.46*	(−1.60)	−0.44***	(−4.32)
DUM2 × YPCJ	+	0.04**	(1.79)	0.04***	(4.38)
DUM3 × TPPORKJ	−	−0.07	(−1.13)	−0.05***	(−2.84)
DUM3 × YPCJ	+	0.001*	(1.71)	0.001**	(2.39)
AR(1)	—	−0.41	(−1.47)	−0.43*	(−1.83)
R^2		0.99		0.96	
R^2-adjusted		0.98		0.98	
No. of observations		22		22	
Durbin-Watson		2.11		2.12	

Notes: t-statistics are in parenthesis.
* Significant at 10% level, ** Significant at 5% level, *** Significant at 1% level.

third reform stage) reveal that the impact of local demand and supply factors has weakened considerably during the period 1994–2003 as compared to the period 1985–1993. This finding again confirms that agricultural product markets have become more integrated in recent years.

Fertilizer

Prices of chemical fertilizers have been affected by various policy reforms as well. As discussed in chapter 1, a provincial responsibility system aimed at meeting excess demand for chemical fertilizers and stabilizing prices was implemented in 1995. In addition, measures were taken to limit private enterprise involvement in domestic fertilizer distribution and to control international trade in fertilizers. Comparisons of international fertilizer prices with domestic prices at the retail level indicate that China has protected its domestic fertilizer industries since 1993.

The evolution of real fertilizer prices since 1981 in *Jiangxi* Province, China as a whole and on the world market is shown in figure 7.3. National and regional prices fluctuated slightly until 1994, but much less than the world market price for urea, the most important type of fertilizer imported into China. From 1994 to 2000, fertilizer prices first showed an increase until 1995–1996, and then rapidly declined, with the decline in *Jiangxi* Province being slightly larger than the decline in China as a whole, but much smaller than the urea price decline at the world market.

World market prices for chemical fertilizers are available for urea, superphosphate, and potassium. Imports of urea have dominated fertilizer imports in China until the mid-1990s, constituting 30–55 percent of total imports. Since 1997, however, urea imports have been negligible. The share of phosphate and potash fertilizers rapidly increased to 70–85 percent during the period 1997–2002 (FAOSTAT 2005).[5]

The model specification used for estimating the impact of world market prices on the fertilizer price in China and in *Jiangxi* Province is therefore as follows:

$$
\begin{aligned}
\text{PFERTC} = {} & c_0 + c_1\text{DUMF} + c_2\text{PUREAW} + c_3\text{DUMF} \times \text{PUREAW} + \\
& c_4\text{DUMF} \times \text{PPHW} + c_5\text{DUMF} \times \text{PKW} + c_6\text{DUMF} \times \text{TPGRAIN(t-1)} \\
& + c_7\text{DUMF} \times \text{TPFERT(t-1)} + u_1 \qquad\qquad\qquad\qquad (5)
\end{aligned}
$$

$$
\begin{aligned}
\text{PFERTJ} = {} & d_0 + d_1\text{DUMF} + d_2\text{PFERTC} + d_3\text{DUMF} \times \text{TPGRAINJ(t-1)} + \\
& d_4\text{DUMF} \times \text{TPFERTJ(t-1)} + u_2 \qquad\qquad\qquad\qquad (6)
\end{aligned}
$$

where:

PFERTC:	Real market price of chemical fertilizers in China (1981 = 100)
DUMF:	Dummy variable reflecting the introduction of the provincial governor responsibility system (-1 for 1995-2003)
PUREAW:	Real world market price of urea in *yuan* (1981 prices; official exchange rate) per ton.
PPHW:	Real world market price of superphosphate in *yuan* (1981 prices; official exchange rate) per ton.
PKW:	Real world market price of potassium in *yuan* (1981 prices; official exchange rate) per ton.
TPGRAIN(t − 1):	Total production of grain in China in preceding year (in 10,000 tons)
TPFERT(t − 1):	Total production of pure fertilizers in China in preceding year (in 10,000 tons)
PFERTJ:	Real market price of chemical fertilizer in *Jiangxi* Province (1981 = 100)
TPGRAINJ(t − 1):	Total production of grain in *Jiangxi* Province in preceding year (in 10,000 tons)
TPFERTJ(t − 1):	Total production of pure fertilizers in *Jiangxi* Province in preceding year (in 10,000 tons)
u_1, u_2:	Error terms with standard properties
$c_0, ..., c_7, d_0, ..., d_4$:	Unknown coefficients.

A distinction is made in the analysis between the periods before and after the introduction of the provincial responsibility system, the most relevant reform for the purpose of our analysis. The world market price of urea is expected to affect the domestic price throughout most of the period 1981–2003. World market prices for superphosphate and potassium are expected to play a role in the last period only, with the price of urea playing a less important role in that period. Local demand and supply factors are expected to play a role only after the introduction of the provincial responsibility system. Chemical fertilizer is used in China especially for producing rice. Total grain production is therefore included in the equation as the main demand variable. To avoid causality problems, local demand and supply of fertilizer are lagged one year.

The regression results for the impact of world market prices on the domestic price of fertilizers are shown in Table 7.5. The world market price of urea has a significant impact on the price of chemical fertilizers in China, while the world market price of superphosphate has played a role since the mid-1990s. The result further shows that the total production of fertilizers significantly affects the domestic price since the start of the policy reform. This means that supply factors are taken into account in setting fertilizer prices since provincial governments have become responsible for meeting input demand and stabilizing fertilizer prices.

Table 7.5. Regression Results for Chemical Fertilizer, China

Explanatory Variable	Expected Impact	Full Equation		Final Equation	
Constant	—	97.5***	(15.8)	84.9***	(14.6)
DUMF	—	−7.61	(−1.69)	—	—
PUREAW	+	0.02	(1.25)	0.03***	(2.90)
PPORKC	+	1.17	(17.9)	1.16***	(21.3)
DUMF × PUREAW	−	0.03	(0.96)	—	—
DUMF × PPHW	+	0.09*	(1.60)	0.11***	(4.36)
DUMF × PKW	+	−0.08	(−0.71)	—	—
DUMF × TPGRAIN (t − 1)	+	0.00	(0.23)	—	—
DUMF × TPFEERT (t − 1)	−	−0.01***	(−3.52)	−0.01***	(−3.36)
AR(1)	—	0.39	(1.06)	0.76***	(3.60)
R^2		0.92		0.90	
R^2-adjusted		0.86		0.87	
No. of observations		21		21	
Durbin-Watson		1.89		2.06	

Notes: t-statistics are in parenthesis.
* Significant at 10% level, ** Significant at 5% level, *** Significant at 1% level.

The regression results for the fertilizer price in *Jiangxi* Province, summarized in table 7.6, confirm that local factors have played a role in determining the market price of chemical fertilizer since the mid-1990s. Apparently, these local factors are responsible for the more than average decline in regional fertilizer prices between 1996 and 2000 that is evident from figure 7.3.

Figure 7.4 shows the changes in the ratio of the fertilizer price to the grain price in *Jiangxi* Province and in China over the whole 1981–2003 period. Both ratios have declined significantly during the period 1990–1994, and showed relatively minor changes during the other years. Hence, the profitability of fertilizer application has increased considerably since the beginning of the 1990s. The decline in *Jiangxi* Province was even larger than the decline in China as a whole, indicating that farmers in *Jiangxi* province have

Table 7.6.   Regression Results for Chemical Fertilizer, *Jiangxi* Province

Explanatory Variable	Expected Impact	Full Equation		Final Equation	
Constant	—	2.91	(0.21)	−0.63	(−0.05)
DUMF	—	13.0	(0.61)	—	—
PFERTC	+	1.01***	(6.80)	1.05***	(7.93)
DUMF × TPGRAINJ (t − 1)	+	0.001	(0.05)	0.007*	(1.61)
DUMF × TPFEERTJ (t − 1)	–	−0.41**	(−2.13)	−0.35**	(−2.19)
AR(1)	—	0.65***	(2.96)	0.64***	(3.00)
R^2		0.92		0.90	
R^2-adjusted		0.86		0.87	
No. of observations		21		21	
Durbin-Watson		1.89		2.06	

Notes: t-statistics are in parenthesis.
* Significant at 10% level, ** Significant at 5% level, *** Significant at 1% level.

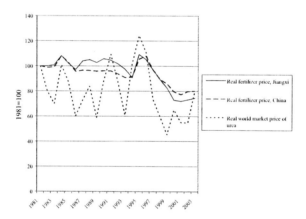

Figure 7.3.   Fertilizer Prices (1981–2003)

profited relatively more from these price developments. At the end of the 1990s, the regional fertilizer-grain price ratio was about 40 percent of its value at the end of the 1980s (in China: 60 percent). As can be seen from figures 7.1 and 7.3, the principle cause of this trend was the rapidly rising grain price in *Jiangxi* Province during the period 1990–1994.

CONCLUSION

The empirical results presented in this paper show that domestic supply and demand factors had a significant impact on grain and pork prices in *Jiangxi* Province during the period 1985–1993. Since 1994, the role of domestic supply and demand factors has diminished substantially for pork prices

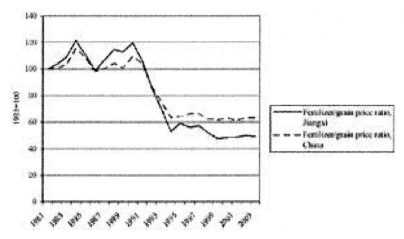

Figure 7.4.   Fertilizer/Grain Ratios (1981–2003)

and even totally disappeared for grain prices. These findings confirm results of earlier studies that found that agricultural product markets have gradually become more integrated in China (Park et al. 2002; Huang et al. 2004). As a result of the opening up of China's agricultural sector to the world market and the inability of the Chinese government to control domestic grain trade, the world market price has become the main determinant of regional and national prices since 1994. The world market prices for pork, on the other hand, do not significantly affect domestic pork prices due to the negligible size of international trade in pork with respect to China.

The reform of the fertilizer market has only been partial; to protect farmers from excessive price fluctuations, the state successfully intervened in fertilizer prices (see figure 7.3). Although world market prices do affect domestic prices of fertilizer, domestic supply of fertilizers play a role as well. Likewise, local supply and demand factors within *Jiangxi* Province affect fertilizer prices in this province since the mid-1990s when provincial governments became responsible for meeting input demand and stabilizing fertilizer prices. These local factors explain the more than proportional decline in fertilizer prices in *Jiangxi* Province between 1996 and 2000.

Farm households in *Jiangxi* Province have benefited relatively more from these developments than farmers elsewhere in China. As can be seen from figures 7.1 and 7.3, they experienced substantially larger grain price increases during the period 1990–1994 and slightly larger price declines in chemical fertilizers during the period 1998–2000. Since the autumn of 2003, grain prices in *Jiangxi* Provinces as well as in other parts of China have increased rapidly after several years of decline, while prices of chemical fertilizers also started to increase. At the same time, the Chinese government started a new policy aimed at raising rural household incomes and promoting grain production (see chapter 1). The impact of these changes in the economic environment of farm households on their production decisions and income levels is the subject of chapter 8.

NOTES

1. The authors would like to thank Zhu Jing for her helpful comments on a draft version of this chapter. Financial support by the Netherlands Ministry of Development Cooperation (DGIS-SAIL program) and the European Union (INCO-DC program) is gratefully acknowledged.

2. The Breusch-Godfrey LM test gives similar results for all presented models.

3. In 2004, rice constituted 44 percent of total cereals production in China (NBS 2005: table 13-17).

4. The export share of pork production in China equaled 0.6 percent in 2004 (NBS 2005: tables 13-21 and 18-8).

5. More recent data on fertilizer imports in China are not provided in the FAO-STAT database.

BIBLIOGRAPHY

Diao, X., S. Fan, and X. Zhang. "China's WTO Accession: Impacts on Regional Agricultural Income—A Multiregion General Equilibrium Analysis." *Journal of Comparative Economics* 31 (2003): 332–51.

FAOSTAT. *Statistical Database.* Rome: Food and Agricultural Organization, 2005. www.fao.org.

Fuller, F., D. Hayes, and D. Smith. "Reconciling Chinese Meat Production and Consumption Data." *Economic Development and Cultural Change* 49 (2000): 23–43.

Gale, F., B. Lohmar, and F. Tuan. *China's New Farm Subsidies.* USDA Outlook WRS-05-01. Washington, D.C.: United States Department of Agriculture (USDA), Economic Research Service (ERS), 2005.

Huang, J. "Agricultural Development, Policy and Food Security in China." Pp. 3–38 in *Proceedings Workshop Wageningen—China,* edited by P. W. J. Uithol and J. J. R. Groot. *Report* 84. Wageningen: AB-DLO, 1997.

Huang, J. and C. Chen. "Effects of Trade Liberalization on Agriculture in China: Institutional and Structural Aspects." *Working paper,* no. 42, Bogor, Indonesia: CGPRT Centre, 1999.

Huang, J., S. Rozelle, and M. Chang. "Tracking Distortions in Agriculture: China and Its Accession to the World Trade Organization." *World Bank Economic Review* 18 (2004): 59–84.

Lin, J. Y. "Institutional Reform and Dynamics of Agricultural Growth in China." *Food Policy* 22 (1997): 201–12.

Ma, H., J. Huang, and S. Rozelle. "Reassessing China's Livestock Statistics: An Analysis of Discrepancies and the Creation of New Data Series." *Economic Development and Cultural Change* 52 (2004): 445–73.

National Bureau of Statistics of China (NBS). *China Statistical Yearbook 2005.* Beijing: China Statistics Press, 2005.

Organization for Economic Cooperation and Development (OECD). *Review of Agricultural Policies—China.* Paris: OECD, 2005.

Park, A., H. Jin, S. Rozelle, and J. Huang. "Market Emergence and Transition: Transaction Costs, Arbitrage, and Autarky in China's Grain Market." *American Journal of Agricultural Economics* 84 (2002): 67–82.

Yu, W. and S. E. Frandsen. *China's WTO Commitments in Agriculture: Does the Impact Depend on OECD Policies?* Paper prepared for the Fifth Annual Conference on Global Economic Analysis, held in Taipei, Taiwan on June 5–7, 2002.

Zhong, F. "Exaggeration and Causes of Meat Production Statistics Overreporting in China." *Chinese Rural Economy* 10 (October 1997): 63–66 (*in Chinese*).

PART II

Farm Household Decision-Making under a Changing Economic and Institutional Environment

8

Farm Household Responses to China's New Rural Income Support Policy

A Village-Level Analysis

Nico Heerink, Marijke Kuiper, and Xiaoping Shi

Since the beginning of 2004, the government of the P. R. China has replaced its centuries-old policy of taxing agriculture by a new policy aimed at subsidizing agriculture and stimulating rural incomes. To this end, agricultural taxes—standing at around 8 percent of agricultural incomes—were drastically reduced. By now they are abolished in most provinces. In addition, farmers growing grain receive a direct income subsidy, new seed varieties and mechanization are subsidized, and large public investments are made in agriculture and rural infrastructure. The main purpose of this policy is to raise rural incomes while at the same time promoting grain production (Gale et al. 2005).

During the first year of its introduction, the policy seems to have met with reasonable success. Per capita real annual net incomes of rural households increased by 6.85 percent in 2004, while the urban-rural income gap slightly decreased from 3.23 in 2003 to 3.20 in 2004 (NBS 2005a: tables 9-3 and 10-2). The population in absolute poverty in rural areas (with an annual per capita net income below 668 *Yuan*) declined from 29.0 to 26.1 million in 2004 (NBS 2005b). Total grain output increased with 9.0 percent to 469.5 million tons in 2004, after steadily declining from 512.3 to 430.7 million tons from 1998 to 2003. This trend reversal was partly caused by a 2.2 percent increase in area sown with grain, and partly by an increase in input use, as is evident from the 5.1 percent rise in national chemical fertilizers' consumption in 2004 (NBS 2005a: tables 13-10, 13-15, and 13-17).

The increase in grain production, however, coincided with rapidly increasing grain prices between October 2003 and April 2004. As a result, average producer prices for grain in 2004 were 28.1 percent higher than in

2003; for rice, the average increase in producer price was even 36.3 percent (NBS 2005a: table 9-9). Moreover, favorable weather conditions in 2004 have contributed to this spur in grain production.

This raises the question to what extent the new rural income support policy has contributed to the rise in rural incomes, and the increase in grain production and chemical input use, and to what extent the rapid grain price increases and other factors were responsible for these achievements. Gale et al. (2005) use simple production costs and profit calculations per *mu* planted with grain to examine this question. They find that the increase in grain prices was the primary factor that increased grain profitability and farm income in 2004; income subsidies and tax reduction played only a minor role. Their analysis, however, does not take into account the extent to which farm households planting decisions and crop management decisions change as a result of price changes and income support measures.

The objective of this chapter is to analyze the impact of direct income payments to grain farmers and agricultural tax reduction on household decision-making with respect to grain production and input use in China. Farm household responses to income policy measures and price changes depend on the available resources within households for earning (on-farm or off-farm) incomes and on the degree to which farm households are integrated into markets. The outcomes may therefore differ considerably between different groups of farm households and between different villages and regions. In this chapter, we use a village-level computable general equilibrium (CGE) model to separate the impact of income support measures from recent price trends in grains and inputs, and to account for differences in household responses to income support measures. The model is applied to two villages in Northeast *Jiangxi* Province, one of China's major rice growing areas. The classification of households within these villages is based on the resources that they have for either earning agricultural incomes or for earning off-farm incomes. Insights obtained from such an analysis are relevant for many other villages and regions in China.

The two selected villages differ fundamentally in their degree of market access. There is convincing empirical evidence that agricultural commodity markets in China have become highly integrated in recent years (Park et al. 2002; Huang et al. 2004). Transaction costs, however, differ considerably between villages with good market access and villages located in remote areas. Moreover, markets for production factors such as agricultural land, labor, and credit face many institutional obstacles in China and remain underdeveloped in rural areas (Bowlus and Sicular 2003; Zhang and Tan 2004). These market imperfections have important implications for farm household choices between grain and high value-added production, and hence for the income gains of different household groups. Model simulations of the impact of price changes and income support measures for two villages with different market access conditions allow us to assess such differences in household responses, income gains, and grain production levels.

The structure of the chapter is as follows. Background information on the two villages and the method of data collection is provided in the second section. The third section describes the structure of the village CGE model. The scenarios that are used for the model simulations and their results are presented in the fourth section. The final section summarizes the major conclusions and draws some policy implications.

DESCRIPTION OF THE RESEARCH SITES

Jiangxi Province is located in the southeast of China, bordering the coastal provinces of *Zhejiang, Fujian,* and *Guangdong* and the inland provinces of *Anhui, Hubei,* and *Hunan.* Although *Jiangxi* Province is not far away from the country's centers of economic growth, *Jiangxi* is one of China's poorest provinces. In 2004, its per capita GDP equaled 8,189 *yuan* (= USD 989) or 78 percent of the national per capita GDP (NBS 2005a: tables 3-1 and 3-11). Agricultural production is an important component of its economy. The primary sector accounted for 20 percent of GDP and 48 percent of total employment in 2004 (NBS 2005a: tables 3-11 and 5-3). *Jiangxi* Province has 3.00 million hectares of cultivated land, with 1.84 million hectares of effectively irrigated land (NBS 2005a: tables 13-5 and 13-8). It is one of China's main grain producing areas, with rice as the major crop. Of the total sown area, 58.5 percent is planted with rice. Rice yields equaled an average of 5,213 kg per ha in 2004, as compared to 6,311 kg per ha for China as a whole (NBS 2005a: tables 13-15 and 13-17). Fertilizer consumption, however, is above the national average. It amounted to 413 kg per ha of cultivated land in 2004, as compared to 357 kg per ha for the whole country (NBS 2005a: tables 13-5 and 13-8).

The two villages selected for the study are *Gangyan* village in *Yanshan* County and *Shangzhu* village in *Guixi* City, both in Northeast *Jiangxi* Province. The selected villages are located in a soil degradation-prone area. The main problem, according the local farmers, researchers, and officials that we interviewed, is natural compaction of the soil. Soil erosion also occurs in some upland areas, but generally not in paddy fields. Natural compaction of the soil is caused by a reduction in the use of organic and green manure, which lowers soil organic carbon and can result in a relatively low bulk density (Lal and Kimble 2001). Increased chemical fertilizer use may aggravate natural soil compaction (Mzuku et al. 2005).

Gangyan village is located in a plain area at about 20 km from the county capital. Almost all arable land (97 percent) in this village is irrigated. *Shangzhu* village is located in a remote and hilly/mountainous area close to the border of *Fujian* Province. Rice is the main crop in both villages, with most irrigated land in *Gangyan* producing two rice crops a year and most irrigated land in *Shangzhu* village producing only one rice crop per year.

An extensive household survey, to collect all data necessary for building village social accounting matrices (SAMs) covering the year 2000, was held in August 2000 and January 2001. A stratified random sample was used for selecting the households, with the hamlets (or "natural villages") within each village forming the strata. In total, 168 households were interviewed in *Gangyan* village and 109 households in *Shangzhu* village, representing around 23 percent of the households in both villages.

Basic information on the two selected villages is presented in table 8.1. Per capita income in the year 2000 was only 1,042 *yuan* (= USD 126) in *Shangzhu* village. In *Gangyan* village, it was 78 percent higher. Rice yields in *Shangzu* village are about 15 percent below those in *Gangyan* village, while fertilizer use is almost 60 percent lower. Off-farm incomes are an important share of household incomes, contributing 45 and 41 percent of household incomes in *Gangyan* village and *Shangzhu* village, respectively. Agricultural tax (and village and township fees) payments make up 5 to 6 percent of total household incomes.

METHODOLOGY

Simulations of the impact of recent price changes and farm income support measures are made with a village-level CGE-model, which allows for

Table 8.1.   Characteristics of the Two Selected Villages

	Gangyan	*Shangzhu*
Households	730	472
Sample size	168	109
Accessibility	Close to city	Remote
Land characteristics	Plain	97% upland
Irrigated land per household (*mu*)	6.06	5.06
Irrigated land/farmland	97%	86%
One-season rice area/total rice area	18.5%	71.6%
Major crops	Rice, vegetables, sugarcane	Rice, bamboo, bamboo shoots
Rice yield (kg/ha)	4,629	3,950
Fertilizer use in rice (kg/ha)	759	481
Income per capita (*yuan*)	1,854	1,042
Off-farm income share (%)	45.4	40.7
Tax payments per household (*yuan*)	492	227
Tax/household income (%)	5.9	5.0

Source: Own household survey (2000).

simultaneous decision-making on production, consumption, and labor supply by farm households. The model applies a macrolevel general equilibrium model structure, but modified in such a way that modeled household behavior is fully compatible with the rural household literature (Singh et al. 1986). Nonseparability of household decisions is built into the village-level equilibrium model using an approach suggested by Löfgren and Robinson (1999). The result is a hybrid village model that accounts for interactions among households within the village, while preserving individual rationality. The position of households in markets as net buying, autarkic, or net selling is made endogenous in the model through the use of mixed complementarity constraints. Both the nonseparability of decision-making and the endogeneity of the household position in markets are departures from existing village models like those in Taylor and Adelman (1996).

Three commodity groups are distinguished in the model:

- Tradables: These are tradable outside the village; their prices being exogenous to the village.
- Village nontradables: These are tradable only within the village; their prices depending on demand and supply within the village.
- Household nontradables: These are not tradable; their (shadow) prices depending on demand and supply of the household.

Data from the household survey and insights obtained during the fieldwork are used for classifying commodities. Traction services, agricultural labor and locally produced consumption goods are classified as village nontradables, while arable land, manure and crop residues, and in *Shangzhu* village forestland and fuelwood, are household nontradables. All the other commodities in the model are tradables. Village prices for agricultural labor and locally produced consumption goods are assumed to be fixed, resulting in demand-driven markets and non-zero profits; traction services is governed by an endogenous village price.

Four household groups are distinguished in each village (see also chapter 10). Household classification is aimed at grouping households that are expected to respond similarly to policy changes within the same group. Household characteristics that are relatively stable and that are expected to strongly affect household responses should be used for this purpose. We expect major differences in responses to rural income support policies for households that mainly depend on agriculture for their incomes and households that earn a large share of their incomes from off-farm employment. The household classification in our analysis is therefore based on the availability of resources for earning agricultural income and/or the availability of resources for earning off-farm income. The ownership of draught power (animals or tractors) was identified as the main resource for earning agricultural incomes in both villages. Explorative data analysis further indicates that the presence of a link outside the province (defined as the presence of a migrated household member or a relative sending remittances) is the most important resource for earning nonagricultural incomes in *Gangyan* village, while in *Shangzhu* village it is the number of educated household members (defined as members with more than four years of schooling).

Table 8.2 shows the resulting classification. The four possible combinations of draught power ownership (yes/no) and presence of a link outside the province (yes/no) define the four groups in *Gangyan* village. In *Shangzhu* village, the first group consists of households with no educated household members; some of these households own draught power, others do not. The households with at least one educated household member are subdivided into three groups. The first group consists of households owning no draught power, the second of households owning draught power and having one or two educated household members, while the last group consists of households owning draught power and having three or more educated members.

Table 8.2.   Classification of Household Groups in the Two Villages

Village		*Household group*			
		1	*2*	*3*	*4*
Gangyan	Owns draught power	No	Yes	No	Yes
	Link outside province	No	No	Yes	Yes
	Number of households	18	23	59	68
Shangzhu	Owns draught power		No	Yes	Yes
	Educated members	None	≤1	1 or 2	≥3
	Number of households	16	14	35	44

Nonseparability of household decisions implies that households face unobservable shadow prices, complicating construction of a SAM on the basis of value flows. Estimates of (partly household-specific) prices derived from the household survey held in 2000 were used to construct a detailed SAM with production activities and production factors disaggregated by the household group. The estimated shadow price for irrigated land indicates a supply-constrained land market in both villages, which confirms observations during the household survey of a highly distorted land market. Household survey data were used for estimating the model elasticities for each of the two villages. The two village SAMs were used for calibrating the model.[1]

Three scenarios are run with the models. They are described in table 8.3. The first scenario assesses the impact of price changes from 2000 until August 2004.[2] Considerable price changes took place over that period in *Jiangxi* Province (and the rest of China). Rice prices initially declined somewhat, but increased rapidly since the autumn of 2003, resulting in a price increase of more than 35 percent over the entire period. Prices of fertilizers, the main variable input in crop production, increased by about 13 percent during the same period, while pork prices declined by almost 3 percent. The inflation rate (as measured by the consumer price index) was almost 5 percent during this period. The first scenario simulates the impact of these price changes on household incomes, production levels, and input use. The second and third scenario add the two main components of the rural income support policy, direct subsidies to grain farmers, and agricultural tax cuts, to the price changes simulated in scenario 1. Farmers in *Jiangxi* Province received a subsidy of 10 *yuan* per *mu* in 2004 for each plot of rice with early rice, late rice, or one-season rice (Gale et al. 2005: table 2). One of the main purposes of these direct subsidies, besides raising rural incomes and reducing income inequality, is to stimulate grain production by rural households. The impact of this direct income support policy is simulated in scenario 2. Agricultural taxes have been cut in all provinces in China since the beginning of 2004 and were abolished in most provinces (including *Jiangxi* Province) by the end of 2005. Scenario 3 simulates the impact of full tax abolition. By comparing the results for these three scenarios, we can separate household responses to the rapid price increases since the autumn of 2003 from the responses to the main new income support measures implemented in 2004 (under constant weather conditions).

Table 8.3. Description of Scenarios Used in Village Model Simulations

		Scenario 1		*Scenario 2*		*Scenario 3*
Price changes	Rice	+36.6%	Rice	+36.6%	Rice	+36.6%
2000–2004	Pork	–2.8%	Pork	–2.8%	Pork	–2.8%
	Fertilizer	+13.0%	Fertilizer	+13.0%	Fertilizer	+13.0%
	CPI	+4.9%	CPI	+4.9%	CPI	+4.9%
Income support policy			10 yuan per mu		Full abolishment of rice land agricultural tax	

SIMULATION RESULTS

In order to disentangle the effects of price changes from those of the rural income support policy, we first present the simulation results for scenario 1 (price changes only) and then compare the results of scenario 2 and 3 with those of scenario 1. For each scenario, we first present the simulation results for household incomes. This will give us insights into the differences in income gains between the major household groups that we distinguish. Next we provide the simulation results for changes in production activities in order to obtain a better understanding of the farm household responses that cause these differences in income gains. Finally, for each scenario we give the simulation results for the use of agricultural inputs that have a direct impact on rural natural resource use (i.e., manure, fertilizer, and pesticides use).

The results for scenario 1 are shown in tables 8.4–8.6. The income gains from the price changes from 2000 to 2004 are about two percentage points smaller in the remote village (*Shangzhu*) than in the village with good market access (*Ganyan*) as can be seen from table 8.4. In both villages, the household groups that possess resources for off-farm employment and have limited agricultural resources (group 3 in *Gangyan* and group 2 in *Shangzhu*), gain substantially less than the other three household groups. Since these are the richest groups in both villages, income inequality was reduced substantially by the price changes that occurred since 2000.

Table 8.5 shows the changes in production activities resulting from the price changes. In *Gangyan*, all four income groups expand their production of two-season rice at the expense of one-season rice, raising pigs, and (to a lesser extent) growing other crops. Due to the intensification of rice cultivation, the demand for traction services increases and as a result the price of this village nontradable goes up by 57 percent. This explains why the two household groups possessing oxen and tractors gain relatively more than the two other household groups. In *Shangzhu*, on the other hand, rice production is strongly dominated by one-season rice, and prevailing agroclimatic conditions impede the expansion of two-season rice. The rapid price increase for rice therefore causes a very significant increase in one-season rice production at the expense of perennials and pigs (and small livestock). The area of irrigated land is constant in the model (see methodology section), so the increase in rice production comes purely from increased input use. The price of traction services increases by 16 percent in this village, which adds to the income gains of the households possessing oxen.

The results for input use are shown in table 8.6. The intensification of rice cultivation in *Gangyan* means that substantially more fertilizers and pesticides

Table 8.4. Income Results for Scenario 1 (Price change scenario)

Village	Household Group	1	2	3	4	Total
Gangyan	Owns draught power	No	Yes	No	Yes	
	Link outside province	No	No	Yes	Yes	
	Income 2000 (*yuan*)	6,204	7,273	9,098	8,997	8,497
	Increase Aug. 2004	27.0%	33.4%	11.8%	24.0%	20.7%
Shangzhu	Owns draught power		No	Yes	Yes	
	Educated members	None	≤1	1 or 2	≥3	
	Income 2000 (*yuan*)	2,861	6,409	5,114	4,969	4,891
	Increase Aug. 2004	21.1%	7.0%	24.6%	17.6%	18.5%

Table 8.5. Production Results for Scenario 1 (Price change scenario)

Village	Household Group	1	2	3	4	Total
Gangyan	Owns draught power	No	Yes	No	Yes	
	Link outside province	No	No	Yes	Yes	
	One-season rice	−54	−72	−61	−73	−68
	Two-season rice	23	39	34	39	36
	Other crops	−7	−9	−11	−11	−11
	Pigs	−51	−79	−46	−55	−53
Shangzhu	Owns draught power		No	Yes	Yes	
	Educated members	None	≤1	1 or 2	≥3	
	One-season rice	58	132	95	90	91
	Two-season rice	4	−59	−53	−45	−42
	Other crops	−3	−4	−2	−22	−10
	Perennials	−69	−16	−83	−50	−61
	Pigs, chicken, ducks	−15	−48	−43	−43	−42

Note: Data in table are percentage changes with respect to the base scenario.

are used by all household groups, while the use of manure is drastically reduced. In *Shangzhu*, the large increase in single rice cultivation causes an even larger increase in fertilizer and pesticides use than in *Gangyan*. And, in contrast to *Gangyan*, the amount of manure used also increases slightly for all household groups. The model results therefore indicate that the rapid grain price increase since the second half of 2003 has caused a large increase in the use of chemical inputs and a more unbalanced use of inputs in both villages.

Scenario 2 simulates the combined impact of the direct income subsidy to grain farmers and the same price changes as in scenario 1. The direct in-

Table 8.6. Input Use Results for Scenario 1 (Price change)

Village	Household Group	1	2	3	4	Total
Gangyan	Owns draught power	No	Yes	No	Yes	
	Link outside province	No	No	Yes	Yes	
	Manure	−51	−22	−46	−27	-31
	Fertilizer	27	41	29	36	34
	Pesticides	26	33	22	31	28
Shangzhu	Owns draught power		No	Yes	Yes	
	Educated members	None	≤1	1 or 2	≥3	
	Manure	21	58	6	11	15
	Fertilizer	64	169	82	86	90
	Pesticides	43	149	79	73	80

Note: Data in table are percentage changes with respect to the base scenario.

come subsidy is paid in *Jiangxi* Province, and in some other provinces, on the basis of grain areas reported for taxation rather than the factual sown area (Guo 2005). It therefore adds to the incomes of rural households, but is not directly related to rice planting decisions.

The simulation results of this scenario are compared with the outcomes of scenario 1 in order to separate the impact of the direct income support from those of the price changes. Table 8.7 shows the impact on income and production activities of the four household groups in each village. The policy has only a modest impact on incomes. The average income increase in *Gangyan* is 1.6 percent and in *Shangzhu* it is 1.3 percent. Again the richest household groups gain least from it, so the income support policy reduces inequality in both villages.

Farm households in *Gangyan* respond to the income increase by raising more pigs (except for the richest group) and switching from two-season rice to one-season rice. Pig production is intensive in the use of external inputs. The income increase means that farmers have more cash available, which they can use for buying such inputs. Moreover, with the increase in wealth, households attach more value to leisure. The shadow price of labor increases 1.7–2.2 percent for household groups 1, 2, and 4, but only 0.9 percent for household group 3 (which has more resources for off-farm employment and fewer resources for agriculture). The higher preference for leisure also induces farm households to switch to a less intensive way of rice

Table 8.7. Income and Production Results for Scenario 2 (Price change and direct income payment scenario)

Village	Household Group	1	2	3	4	Total
Gangyan	Owns draught power	No	Yes	No	Yes	
	Link outside province	No	No	Yes	Yes	
	Household income	2.1	2.1	1.0	1.7	1.6
	One-season rice	2.0	2.2	1.0	1.8	1.6
	Two-season rice	−0.3	−0.2	−0.1	−0.2	−0.2
	Other crops	0.3	0.2	−0.3	0.4	0.1
	Pigs	1.9	2.4	−0.3	4.5	1.5
Shangzhu	Owns draught power		No	Yes	Yes	
	Educated members	None	≤1	1 or 2	≥3	
	Household income	1.6	0.7	1.2	1.4	1.3
	One-season rice	−0.5	−0.1	−0.6	0.1	−0.2
	Two-season rice	0.3	−1.6	1.3	2.0	0.9
	Other crops	0.9	0.6	0.8	0.5	0.7
	Perennials	−3.8	−0.1	1.0	−1.3	−0.7
	Pigs, chicken, ducks	1.3	−1.0	1.1	1.4	0.9

Note: Data in table are percentage changes with respect to scenario 1 (price changes only).

cultivation. As a consequence, the direct income support policy does not reach one of its major goals, namely promoting grain production.

Households in *Shangzhu* also increase pig production (except for the richest group), but their response is smaller and they do not seem to resort to less intensive rice cultivation. This may partly be explained from the smaller increase in the shadow price of labor, which equals 0.8–1.3 percent for household groups 1, 3, and 4, and only 0.1 percent for household group 2 (the household group with fewer agricultural resources).

The third scenario simulates the combined effect of the full tax abolition and the price changes (table 8.8). Results are again compared with those of the first scenario. The results are very similar to those of scenario 2, but the magnitude is much larger. The average income increase in *Gangyan* is 10.7 percent, while in *Shangzhu* it is only 5.0 percent. The absolute income increase caused by tax abolition is much larger in *Gangyan* village (see table 8.1), and the cash available for buying external inputs therefore increases much more. This allows farmers in *Gangyan*, the relatively rich village, to make a much larger switch toward more profitable activities than rice production. It also leads to a much larger increase in the shadow wage in *Ganyan* village (11–17 percent for groups 1, 2, and 4; 7 percent for group 3) than in *Shangzhu* village (3.3–4.3 percent for groups 1, 2, and 4; 0.9 percent for group 2). So, although income inequality within villages declines, the inequality between villages increases.

Table 8.8. **Income and Production Results for Scenario 3 (Price change and tax abolition scenario)**

Village	Household Group	1	2	3	4	Total
Gangyan	Owns draught power	No	Yes	No	Yes	
	Link outside province	No	No	Yes	Yes	
	Household income	16.5	14.8	7.4	11.0	10.7
	One-season rice	15.3	15.6	7.3	11.9	11.1
	Two-season rice	−2.3	−1.8	−0.9	−1.2	−1.3
	Other crops	2.6	1.1	−2.4	2.4	0.7
	Pigs	14.9	17.4	−2.2	29.0	9.9
Shangzhu	Owns draught power		No	Yes	Yes	
	Educated members	None	≤1	1 or 2	≥3	
	Household income	6.8	4.2	3.9	5.9	5.0
	One-season rice	−1.7	0.4	−0.5	−0.4	−0.4
	Two-season rice	1.3	−8.6	16.8	6.3	4.4
	Other crops	4.3	3.5	0.4	3.5	2.3
	Perennials	−16.8	−0.8	3.8	−6.7	−3.8
	Pigs, chicken, ducks	5.5	−6.4	6.5	5.4	3.8

Note: Data in table are percentage changes with respect to scenario 1 (price changes only).

Finally, the consequences of these changes in production patterns for input use are shown in table 8.9. The changes in input use under scenario 2 are very small, and therefore are not discussed here. The changes in input use under scenario 3 are also much smaller than under scenario 1. Household groups 1 and 4 in *Gangyan* village use substantially more manure (provided by pigs). All the other changes in input use are relatively small. The abolition of taxes and the income support policy therefore do not seem to have much effect on soil and environmental quality in these two villages.

Table 8.9.   Input Use Results for Scenario 3 (Price change and tax abolition)

Village	Household Group	1	2	3	4	Total
Gangyan	Owns draught power	No	Yes	No	Yes	
	Link outside province	No	No	Yes	Yes	
	Manure	14.9	1.3	–2.2	8.7	6.5
	Fertilizer	0.7	–0.6	–0.6	–0.2	–0.3
	Pesticides	0.9	0.1	–0.5	0.5	0.2
Shangzhu	Owns draught power		No	Yes	Yes	
	Educated members	None	≤1	1 or 2	≥3	
	Manure	–0.3	0.1	1.7	1.7	1.2
	Fertilizer	–0.8	0.5	4.3	1.7	2.0
	Pesticides	0.3	0.5	2.2	1.2	1.3

Note: Data in table are percentage changes with respect to scenario 1 (price changes only).

CONCLUSION

Since the beginning of 2004, the Chinese government has adopted a new rural income support policy that is more in line with WTO regulations. Its major purpose is to stimulate rural incomes, while at the same time promoting grain production and food self-sufficiency. The two major measures taken in this respect are direct income support payments to grain farmers and abolition of agricultural taxes and fees paid by rural households. The results of the two village models discussed in this chapter show that the policy does not reach its goal of promoting grain production; the large increase in grain production in 2004 was not caused by the income support policy but by the rapid price increases in 2003–2004. The increased incomes resulting from the new policy allow farm households to buy more inputs that can be used in livestock production. Moreover, because leisure is valued higher with increasing incomes, farmers tend to switch to less intensive grain production.

The direct income subsidy is paid to grain farmers in *Jiangxi* Province and several other provinces on the basis of grain areas reported for taxation. If the subsidy payments would be linked to sown areas, as is done in some other parts of China (Guo 2005), payments will be higher for farmers who grow more grain. In that case, the direct income subsidy will have a more positive impact on grain production. In designing a rural income support policy, however, it is important to realize that promoting grain production and stimulating rural incomes are two different goals that may to some extent conflict with each other.

We further find that tax abolition has a much larger impact on incomes and production than the direct income support (at around ten *yuan* per *mu*) of 2004. Tax abolition also tends to widen income inequalities between villages, because the absolute income gain is much larger in relatively rich villages. The switch toward more profitable activities like livestock production is therefore much stronger in these villages. Supplementary policies to stimulate income growth in poorer villages, such as public investments in human capital and rural infrastructure and policies promoting local off-farm employment opportunities, will be needed in order to deal with the growing income gap between villages caused by the agricultural tax abolition.

The simulation results further indicate that the income support policy has a negligible effect on input use, although tax abolitions seem to have a minor positive impact on manure application (and hence improves soil quality) in villages with good market access. It is, however, overshadowed by the rapid grain price increase since the second half of 2003, which has caused a large increase in the use of chemical inputs and a more unbalanced use of inputs.

NOTES

1. A detailed description of the model and its calibration for *Gangyan* village can be found in Kuiper (2005).

2. This scenario is based on trends between 2000 and August 2004 for *Jiangxi* as whole. The authors would like to thank Nie Fengying of the CAAS Scientech Information and Documentation Centre for providing us with the price data for *Jiangxi* Province.

BIBLIOGRAPHY

Bowlus, A. J. and T. Sicular. "Moving Towards Markets? Labor Allocation in Rural China." *Journal of Development Economics* 71 (2003): 561–83.

Gale, F., B. Lohmar, and F. Tuan. "China's New Farm Subsidies." *USDA Outlook* WRS–05–01. Washington, D.C.: United States Department of Agriculture (USDA), Economic Research Service, 2005.

Guo, J. "On the Execution and Problems of China's Agricultural Subsidy Policies, and Some Proposals for Them." Unpublished paper. Beijing: Development Research Council of the State Council, Agricultural Economics Department, 2005 [*in Chinese*].

Huang, J., S. Rozelle, and M. Chang. "Tracking Distortions in Agriculture: China and Its Accession to the World Trade Organization." *World Bank Economic Review* 18 (2004): 59–84.

Kuiper, M. H. *Village Equilibrium Modeling: A Chinese Recipe for Blending General Equilibrium and Household Modeling.* PhD Dissertation. Wageningen, The Netherlands: Wageningen University, 2005.

Lal, R. and J. M. Kimble. "Importance of Soil Bulk Density and Methods of Its Measurement." Pp. 65–86 in *Assessment Methods for Soil Carbon*, edited by J. M. Kimble, R. F. Follett, and B. A. Stewart. Boca Raton, FL: Lewis Publishers, 2001.

Löfgren, H. and S. Robinson. "Nonseparable Farm Household Decisions in a Computable General Equilibrium Model." *American Journal of Agricultural Economics* 81 (1999): 663–70.

Mzuku, M., R. Khosla, R. Reich, D. Inman, F. Smith, and L. MacDonald. "Spatial Variability of Measured Soil Properties Across Site-specific Management Zones." *Soil Science Society of America Journal* 69 (2005): 1572–79.

National Bureau of Statistics of China (NBS). *China Statistical Yearbook 2005.* Beijing: China Statistics Press, 2005a.

National Bureau of Statistics of China (NBS). *Statistical Communiqué of National and Social Development in 2004.* Beijing: NBS, 2005b.

Park, A., H. Jin, S. Rozelle, and J. Huang. "Market Emergence and Transition: Transaction Costs, Arbitrage, and Autarky in China's Grain Market." *American Journal of Agricultural Economics* 84 (2002): 67–82.

Singh, I., L. Squire, and J. Strauss (eds.). *Agricultural Household Models. Extensions, Applications and Policy.* Baltimore: Johns Hopkins University Press, 1986.

Taylor, E. and I. Adelman. *Village Economies: the Design, Estimation and Use of Village-Wide Economic Models.* Cambridge, MA: Cambridge University Press, 1996.

Zhang, X. and K.-Y. Tan. "Blunt to Sharpened Razor: Incremental Reform and Distortions in the Product and Capital Markets in China." *DSGD discussion paper*, no. 13. Washington, D.C.: International Food Policy Research Institute, 2004.

9

Marketing Chains, Transaction Costs, and Resource Management

Efficiency and Trust within Tomato Supply Chains in *Nanjing* City

Ruerd Ruben, Hualiang Lu, and Erno Kuiper

Tomato producers in *Nanjing* City have a choice between different marketing outlets for selling their products. Tomatoes can be sold to consumers through a stall or by traders at the wet market, or sold through the local wholesale market. Delivery conditions and quality demands tend to differ widely amongst these outlets, producing various types of transaction costs and offering farmers different implicit incentives for improving their production system or management regime.

The presence of high transaction costs at local markets implies that the efficiency of exchange transactions can become seriously constrained. The costs for gathering information, searching and mobilizing production factors, monitoring performance, and transporting the produce tend to raise the pure production costs and could therefore reduce the incentives for better input applications or improved crop management (Williamson 1986; Gabre-Madhin 2001). As a consequence, farmers will be less inclined to make the required investment in order to enhance product quality and to guarantee sufficient input applications. A reduction of transaction costs within an open market environment might encourage farmers toward resource intensification, improving productivity and quality. This could also contribute to raising their family income as well as lead to better resource management practices (North 1990). However, during economic transition, traditional relationships between buyers and sellers may easily erode and relational trust or *Guanxi* is bound to disappear.[1]

The objective of this chapter is to understand the simultaneous effects of farm household characteristics, resource endowments and institutional conditions on farmers' choices for different tomato marketing chains, and to assess the marketing chain efficiency by using disaggregated analysis of different categories of transaction costs, and investigate the impact of these transaction costs on product quality management and resource intensification.

Earlier studies on feasible technological and policy strategies for improving resource use and management practices have focused on the application of better production and crop management practices (AVDRC 1998; Li et al. 1998) or the introduction of improved incentives to enhance farmers' willingness to invest in more advanced production methods (Scott 1995; Hueth et al. 1999). Far less attention has been given to the options to improve coordination and relationships of trust amongst agents within the marketing chain, as an alternative strategy to increase welfare and upgrade quality. Reduction of transaction costs throughout the chain could indeed be a helpful measure to control ex-ante and ex-post information problems and related uncertainties that tend to hinder farmers' adoption of improved production methods or management practices (Hobbs 1997; Calabresi 1969). In consequence, when search and negotiation periods can be substantially reduced and price revelation takes place earlier, local farmers will be more inclined to make the necessary investments in order to upgrade their production systems. Tomato marketing chains around the city of *Nanjing* in China offer a particularly challenging setting for the analysis of the importance of transaction costs regarding input use, investment, and management decisions.

The remainder of this chapter is organized as follows. In the next section, recent developments of tomato production and the organization of the tomato marketing chain in *Nanjing* will be discussed. We also analyze farmers' marketing behavior concerning outlet choices. In the third section, a hedonic price model is developed in order to attribute the differences in sales prices and volumes of tomato sales to four specific categories of transaction costs: transportation, information, negotiation, and monitoring. In the next two sections, the impact of these transaction costs are assessed on decisions regarding technology choice (i.e., input use and production-specific investments) and farm management (i.e., production scale and degree of specialization), respectively. We conclude with a discussion and some policy implications.

TOMATO PRODUCTION AND MARKETING IN *NANJING*

As the capital of *Jiangsu* Province, *Nanjing* has a long history for vegetable production, being one the most important agricultural centers at the lower reaches of the Yangtze River. With the implementation of economic reforms and the opening-up policy of China since 1978, the vegetable marketing system in *Nanjing* has been substantially liberalized, and a rapid shift was made from a sellers' market into a buyers' market.

Vegetable production in *Nanjing* experienced rapid growth during recent years. Between 1998 and 2003, the cultivated area under vegetables increased from 53,700 to 131,920 hectares. Vegetable production reached 3,343.4 thousand tons in 2003 and the output value was 3,070 million *yuan*. Acreage, production, and output values recorded an average growth rate of around 16 percent per annum (table 9.1).

After a long period of market development and with the liberalization of the marketing system, *Nanjing* vegetable markets became more competitive (Ahmadi-Esfahani and Stanmore 1997). Currently, the vegetable marketing chain includes large-scale wholesale markets located in net consumption and net production areas, retail markets (mainly wet markets), and local and foreign supermarkets. Traders and transporters play an active role in all these markets.

Baiyunting and *Zijinshan* wholesale markets are the main locations for vegetable wholesale activities in *Nanjing* City. Wet markets, including indoor and open-air markets, are spread throughout the city, representing the most important retail outlets. Through these wet markets, 90 percent of the vegetables are sold. Supermarkets represent a relatively new vegetable outlet for purchasing fresh vegetables. They offer guaranteed sanitary conditions and introduce relatively new and luxury vegetable varieties, as well as prepacked and preprocessed vegetables convenient to consumers. Nevertheless, supermarkets, at the end of 2003, only accounted for 2–3 percent of total vegetable sales in *Nanjing*.

Production technologies for vegetable production are still rather traditional. Due to the small size of the cultivated area (2 *mu* per farm household on average, with 1 ha = 15 *mu*), and the advanced age of most vegetable producers (on average fifty years old[2]), farmers mostly rely on labor-intensive techniques for production, handling, and transport. Input applications are usually low; on average only 62 percent of the vegetable farmers use chemical fertilizers and apply pesticides on their tomato fields. However, although most farmers rely on traditional technologies for to-

Table 9.1. Vegetable Production in Nanjing (1998–2003)

Year	Average (1,000 ha)	Output (1,000 ton)	Value (million yuan)
1998	53.7	1,339	1,240
1999	69.5	1,648	1,647
2000	104.7	2,125	2,076
2001	110.1	2,480	2,318
2002	127.5	3,422	3,100
2003	131.9	3,343	3,070
Growth rate (%)	16.1	16.5	16.3

Source: Nanjing Agricultural and Forestry Bureau (2004).

mato cultivation, some new technologies have been recently introduced. For example, in *Maqun* Township, farmers use small shelters for tomato production to enable early harvesting in order to obtain a higher market price. Small quantities of herbicides are used in tomato production in order to reduce the incidence of wild grasses. Most farmers use some organic manure in producing tomatoes. Nevertheless, farmers experience difficulties in accessing good quality seed, and the use of low quality seeds is considered to be the main factor for crop loss and quality deterioration.

For this study on tomato marketing chains, we collected data from a random sample of eighty-six farm households in *Maqun* Township, located in *Qixia* district of *Nanjing* City. All farmers have been involved in tomato production for more than twenty years. Tomato growers in *Maqun* Township have an average farmland size of 2.6 *mu* (0.17 ha), and cultivate several vegetable varieties in order to harvest during the whole year. This provides regular earnings for their daily needs and insures against different kinds of risks. The farmers rely on three different marketing channels for selling their tomatoes:

1. Direct sales to consumers, through a stall at the wet market (N = 46);
2. Sales transactions with traders that purchase the produce at the stall or deliver to local institutions, called "*Dui*" trade (N = 20); and
3. Delivery to the local wholesale market (N = 20).

The average tomato area is about 0.7 *mu* (0.05 ha), which is around one-quarter of the total farm size. Producers linked to the wholesale market have slightly larger farms but are less specialized in tomato production (with a much higher spread around the mean tomato acreage). Tomato yields are significantly higher for farmers engaged in sales to traders and in direct sales. Tomato prices are significantly different for the three channels, which was to be expected. The price the farmers receive from direct sale is the highest (1.18 *yuan*/kg), whereas prices received from traders or at the wholesale markets are 8 and 35 percent below the direct sales price, respectively (table 9.2).

The wholesale market is located close to the production area and is the most convenient outlet. Usually, women tend to negotiate better prices with traders at the wholesale market, where often more time is required before an acceptable agreement is reached. Direct sale is the more convenient outlet for resource-poor farmers without appropriate means of transportation. Most farmers have only a bicycle available. Especially female tomato sellers face difficulties in transporting tomatoes to the wet market. About 30 percent of the respondents possess a fixed stall at the wet market. Most of them sell their tomatoes directly to consumers, but some also sell part of the produce to traders. The rent is relatively high (5–10 *yuan*/day) and since not all farmers have enough vegetables for daily delivery, only few are able

Table 9.2. Tomato Production and Exchange Characteristics

Marketing Chains	Average Tomato Area (mu)	Average Farm Size (mu)	Distance to Market (km)	Average Yield (kg/mu)	Average Price (yuan/kg)
Direct sale	0.66	2.5	10.8	3,508	1.18
(N=46)	(0.53)	(1.06)	(58.0)	(2,225)	(0.14)
Sale to trader	0.71	2.5	12.0	3,804	1.08
(N=20)	(0.60)	(0.89)	(58.5)	(2,907)	(0.11)
Wholesale	0.78	3.0	2.3	3,057	0.76
(N=20)	(1.31)	(2.17)	(20.5)	(2,515)	(0.97)
Total	0.70	2.6	9.2	3,461	1.06
	(0.78)	(1.37)	(49.4)	(2,458)	(0.15)

Source: Nanjing vegetable marketing chain survey (2002).
Note: Standard deviations are in parentheses.

to maintain a fixed stall. Hence, it is mostly retailers that occupy the stalls at wet markets. Farmers who are engaged in wholesale transactions tend to maintain this channel as an exclusive outlet.

The three market outlets differ in terms of access and trading conditions. The average distance to the market is about 9 km. The wholesale market is located nearby, while the wet market is further away. Different negotiation procedures are typical for each marketing outlet. Farmers who sell tomatoes directly to consumers at the wet market maintain the shortest negotiation time. Consumers buying tomatoes at the wet market care most about quality and less about price. Farmers operating at the wet market are usually better informed about market prices. Traders operating at the wholesale market offer lower prices because they buy larger quantities. The wholesale market is a buyers market where prices tend to be depressed. Consequently, the negotiation time tends to be longer. Another difference in the marketing process refers to the available time space for conducting transactions. Due to the long travel time to reach the wet market and the small transaction volume with each consumer, farmers usually need most time (up to 8 hours), while for sales at the wholesale market they need only 2.5 hours to sell their daily harvest.

TRANSACTION COSTS IN TOMATO CHAINS

The selection of the appropriate marketing outlet for different types of fresh produce depends largely on farm and household characteristics and resource endowments (Lu 2003). Marketing of tomatoes is a complex process that involves various types of transaction costs. Depending on the choice of outlet, costs are made for transporting the produce, visiting different traders to become informed about the ruling price, negotiating on the price and maintaining regular contacts with traders in order to monitor market tendencies. The level and composition of these transaction costs determines the efficiency of the tomato marketing chain.

Following the analytical framework developed by Escobal (1999), we use an indirect procedure for quantifying the observable attributes of transaction costs. Two possible transactions (T^1 and T^2) are considered; the one with lower transaction costs (TC) is assumed to occur (T^*):

$$T^* = T^1, \text{ if } TC^1 < TC^2 \text{ or } = T^2, \text{ if } TC^1 \geq TC^2 \tag{1}$$

Although TC^1 and TC^2 are not directly observable, it is enough to observe vector X, which represents observable attributes that affect transaction costs:

$$TC^1 = \beta_1 X + \varepsilon_1 \quad TC^2 = \beta_2 X + \varepsilon_2 \tag{2}$$

where β_1 and β_2 represent vectors of unknown coefficients, and ε_1 and ε_2 are error terms with standard properties. Empirically, the probability of observing T^1 is equivalent to:

$$\text{prob } (TC^1 < TC^2) = \text{prob } (\varepsilon_1 - \varepsilon_2 < (\beta_2 - \beta_1)X) \tag{3}$$

The existence of transaction costs makes many rural households participate in certain markets. The degree of risk aversion and profit maximization determines the marketing chain choice preference. Different transaction costs between market chains can help to explain their preference. If p is the effective price that determines production and consumption decisions, each household faces the following conditions:

Supply of product	$q = q\,(p, z^q)$	(4)
Demand of product in market j	$c^j = c^j\,(p^j, z^{dj})$	(5)
Idiosyncratic transmission of prices in market j	$p^{sj} = p^{sj}\,(z^{pj})$	(6)
Transaction costs in market j	$TC^j = TC^j\,(z^{ij})$	(7)

where z^q, z^{dj}, z^{pj}, and z^{ij} are exogenous variables that affect supply, demand, sales price, and transaction costs, respectively. Thus, for the retailers of a product in market j, the effective price at the level of each household equals:

$$p^j = p^{sj}\,(z^{pj}) - TC^j\,(z^{ij}) \tag{8}$$

The condition of being a retailer in market j equals:

$$q\,[p^{sj}\,(z^{pj}) - TC^j\,(z^{ij}), zq] - c\,[p^{sj}\,(z^{pj}) - TC^j\,(z^{ij}), z^{dj}] > 0$$
$$\text{or } I\,(z^q, z^{dj}, z^{pj}, z^{ij}) > 0 \tag{9}$$

The model can be estimated using the following probit equation:

$$\text{prob}(\text{Net seller in market } j) = \text{prob}[I(z^q, z^{dj}, z^{pj}, z^{ij}) > 0] \qquad (10)$$

After estimating equation (10), the reduced form of the equation of supply conditional on the selected strategy can be derived:

$$q = q(p, z^q)(\text{prob}[\text{Net seller in market } j]) \qquad (11)$$

The estimation of equation (11) requires estimation in two stages, where the Inverse Mill's Ratio (IMR) is introduced to take into account the endogenous nature of the decision (sell to consumer directly at the wet market or also at other markets). This allows us to correct for a possible selection bias in the sample.[3]

A good that has several characteristics generates a number of hedonic services. Each of these services could raise its own demand and may be associated with a hedonic price. A hedonic price procedure can thus be used to associate transaction costs to the effective price each farmer receives (Rosen 1974). Here, we interpret the price that the farmer receives as a set of quality "premiums" or "discounts" for a series of services that have been generated, or perhaps omitted. Hence, we define the average price as a function of hedonic prices, which is simply the mathematical relationship between the prices received by adding value and the characteristics of the transaction associated with the product. This is:

$$p_j = h(z_{1j}, z_{2j}, z_{3j}, \Lambda\Lambda, z_{kj} | \text{prob}[\text{Net seller in market } j]) \qquad (12)$$

Where p_j is the average price obtained by j-th farm household for the sale of tomatoes, and where $z_{1j}, z_{2j}, z_{3j}, \ldots, z_{kj}$ represents the vector of characteristics associated with the transactions completed by the farm household.

It is clear from the literature of hedonic price functions that $h(\cdot)$ does not strictly represent a "reduced form" of the functions of supply and demand that could be derived from the production or utility functions of the economic agents involved in the transaction (Rosen 1974; Wallace 1996). Rather, $h(\cdot)$ should be seen as a restriction in the process of optimization of sellers and buyers. Rosen (1974), and more recently, Wallace (1996) showed that whereas growing marginal costs exist for some of the characteristics (in this case associated with the generation of different categories of transaction costs) for farmers and/or sellers, the hedonic function could be nonlinear. In this case, the nonlinearity would mean that the relative importance of transaction costs is not the same for all farmers.

The estimation of an equation such as the one proposed here permits us to disaggregate the price received by the farmer into a series of components associated with the attributes of the transactions. A complementary way of interpreting this equation is that the constant estimate represents a price indicator that results from following the "law of one price," with the rest of the equation being the elements that must be discounted from the price due to the differences in the distance of the farmers from the market and other associated transaction costs. Comparing the transaction costs of households with different endowments will allow us to understand the importance of key assets in reducing transaction costs.

We used a composite of various indicators to disentangle the effects of different components of transaction costs on variations in sales prices and volumes sold. The four categories of transaction costs included are:

1. Transportation costs, depending on distance, time, road conditions, and availability of own means of transport
2. Information costs, depending on the number of traders visited before selling and the sources of access to market information
3. Negotiation costs, related to the number of visits made until reaching an agreement on the selling price
4. Monitoring costs, related to the number of years that the farmer is engaged with the trader.

Table 9.3 shows the results of the probit model estimated by using equation (10). This estimation will serve as the basis for estimating both the supply and price equations.

We find a negative relationship between the negotiation time required to reach agreement on the price and the probability that farmers sell their tomatoes directly on the wet market. This means that when farmers want to economize on negotiation costs, they are likely to select the direct marketing channel. In addition, there is a positive relationship between using chemical fertilizer and farmers choosing to sell their tomatoes directly

Table 9.3.   Determinants of Direct Sale (Probit regression results)

Explanatory Variables	Coefficient	Standard Error
Constant	0.7866	1.057
Sex of household head (male = 1)	−0.4625	0.374
Age of household head (year)	0.0174	0.019
Total tomato area (*mu*)	0.1018	0.271
Using chemical fertilizers (yes = 1)	0.6341*	0.368
Average distance to sale point (km)	0.0225	0.024
Negotiation time to reach the price (number of visits)	−1.1528***	0.227
Number of observations	86	
Log likelihood	−35.960	
Pseudo R^2	0.44	

Source: Based on data from *Nanjing* tomato marketing chain survey (2002).
Note: * Significant at 10% level; ** Significant at 5% level; *** Significant at 1% level.

at the wet market. This finding points toward reliance on more input-intensive practices that guarantee a better tomato quality.

Tables 9.4 and 9.5 summarize the estimation results of equations (11) and (12). The supply equation can be interpreted as a reduced form of the model. The results of the price equation as well as the supply equation give a significant IMR, indicating that relevant differences that exist in prices and sales volumes depend on the marketing outlet choice strategy adopted by farmers. Hence, prices received and volumes traded by farmers are significantly different in each of tomato marketing supply chains. The price and sales volume are thus not only dependent on farm household characteristics, but are also influenced by transaction costs.

The sales price is positively influenced by the age of the household head, the number of years the farmer knows the traders (influencing monitoring and negotiation costs), and the average time needed to reach the market (related to transport costs). The use of fertilizers in tomato production leads to a significantly lower (not higher) sales price. But it indirectly leads to higher sales prices by increasing the probability that farmers choose the direct sale marketing channel (tables 9.2 and 9.3). In addition, the possession of a fixed stall at the wet market, the number of years involved in tomato production, and the total tomato volume tend to depress the price.

The traded volume is positively influenced by the education level of the household head, the total area of tomato production, and the number of traders who buy from the same farmer (related to monitoring and negotiation costs). More tomato trade takes place when a long-term relationship

Table 9.4. Determinants of Sales Price (OLS regression results)

Explanatory Variables	Coefficient	Standard Error
Constant	1.304***	0.107
Inverse Mill's ratio	0.160***	0.035
Age of household head (year)	0.004**	0.002
Using fertilizer (yes=1)	–0.075**	0.036
Fixed stall at the wet market (yes=1)	–0.070*	0.038
Number of years producing tomato (year)	–0.005**	0.002
Total tomato production (kg)	–0.00001*	0.000
Access to information from trader (yes=1)	–0.085**	0.042
Number of years knowing the trader (year)	0.053**	0.027
Average time to reach the market (minutes)	0.002***	0.0006
Price below the market average level (yes=1)	–0.372***	0.035
Ratio of marketing effectiveness	–0.004***	0.001
Inverse Mill's ratio × Price below the market average price	–0.149**	0.061
Number of observations	86	
R²-adjusted	0.768	

Source: Based on data from *Nanjing* vegetable marketing chain survey (2002).
Note: Ratio of marketing effectiveness × number of traders with transactions/number of trader contacts.
* Significant at 10% level; ** Significant at 5% level; *** Significant at 1% level.

Table 9.5.　Determinants of Amount Sold (OLS regression results)

Explanatory Variables	Coefficient	Standard Error
Constant	–76.42	41.88
Inverse Mill's ratio	303.11**	129.54
Education level of household head (year)	43.76**	16.93
Total area of tomato (*mu*)	1,587.64***	101.90
Number of traders to whom the farmers sell	71.58**	27.21
Number of years knowing the trader (year)	303.06***	110.64
Average condition of road (good = 1)	713.05***	220.87
Percentage of not sale (%)	–26.28**	10.98
Percentage of waste (%)	–43.39**	17.91
Total volume × Average distance to market	0.02***	0.002
Condition of road × Average distance to market	–63.13***	12.44
Price below average market price × Ratio marketing effectiveness	–17.20***	5.58
Number of observations	86	
R²-adjusted	0.915	

Source: Based on data from *Nanjing* vegetable marketing chain survey (2002).
Note: Ratio of marketing effectiveness × number of traders with transactions/number of traders contacts.
* Significant at 10% level; ** Significant at 5% level; *** Significant at 1% level.

between the farmer and the trader exists (related to monitoring and negotiation costs) and when better road conditions (related to transport cost) are available. Scarcity of market information has a negative effect on the amount sold to the market. The percentage of non-sale (remaining after market closure) and the percentage of product denial (due to inferior quality) also reduce to the total traded volume.

These results indicate that—in addition to farm households' characteristics and resource endowments—the transaction costs involved in tomato marketing supply chains also have significant effects on sale price and traded volume. It is now possible to estimate and disaggregate transaction costs by using the estimations presented in tables 9.4 and 9.5 as a base. Table 9.4 enables us to estimate the tomato price increase each household could have received if that household would not have incurred any transaction costs in their relations with traders. Table 9.5 permits us to assess the effect that a reduction of these costs would have on sales volume.

Table 9.6 shows the discounts in the price farmers received due to the transaction costs incurred. These estimates suggest that prices could have been on average 26 percent higher if farmers would not have incurred any transaction costs. Transactions based on direct sale face the highest price discount. Negotiation costs are higher for direct sales, whereas sales to traders and at the wholesale market incur more information costs.

We can observe that the most important transaction costs are associated with information and negotiation procedures that account for about 20 and 16 percent of the total price discount. Due to the limited develop-

Table 9.6. Discount in Sales Price by Type of Transaction Costs (% of price)

	Information	Negotiation	Monitoring	Transport	Total
By type of marketing chain (%)					
Direct sale	−17.36	−26.44	0.68	9.07	−35.64
Wholesale	−22.28	−17.91	1.40	4.44	−34.35
Sales to traders	−22.84	−14.26	3.22	8.98	−24.91
Total	−19.87	−16.43	2.14	7.72	−26.43

Source: Based on data from *Nanjing* vegetable marketing chain survey (2002).
Note: A negative value indicates discounts in the price the farm household receives, while a positive value suggests a price increase. Calculations based on data in table 9.4.

ment of local organizational and informational networks, farmers face major difficulties in receiving adequate services and market information. Hence, farmers can only get the latter through frequent visits to traders or establishing good contacts with neighbors. Maintaining a fixed stall at the wet market reduces information costs, but substantially raises fixed investments. Farmers with a stall have to sell all their tomatoes during a short period and therefore tend to face higher negotiation costs that could reduce their price margin.

Monitoring costs and transport costs appear to have a positive effect on the price received. The positive effect of monitoring costs is surprising, but may be explained by the fact that farmers made an effort to establish long-term relationships with traders in previous periods, resulting in a higher current price that compensates for the previously made costs. Such long-term relations are more common for sales to traders and wholesalers. Prices paid in wet markets through direct sales or in transactions with traders are substantially higher than those at the wholesale market. Therefore, higher transport costs are more than compensated by the better prices that are paid in distant markets, encouraging farmers to travel a long distance in order to get a more rewarding price.

In conclusion, we register that the direct marketing chain faces the highest negotiation costs, while sales to traders and wholesalers incur the largest information costs. Those farmers who sell their tomatoes directly to consumers at the wet market have to negotiate the price with each individual customer. While the negotiation costs involved in each transaction might be small, the total time involved for all transactions is high. Farmers selling their tomatoes to traders at the wet market or the wholesale market have to collect more information about the market situation to be able to make a good deal with the traders. This strongly increases their information costs and results in a larger price discount.

The impact of transaction costs on sales volume is based on the supply function (table 9.7). The total quantity sold could have been nearly 24 percent higher if transaction costs had not been incurred.

Table 9.7.   Discount in Amount Sold by Type of Transaction Costs (% of volume)

	Information	Negotiation	Monitoring	Transport	Total
By type of marketing chains (%)					
Direct sale	−27.05	−2.54	2.31	−3.41	−30.69
Wholesale	−19.75	−5.10	3.06	−1.47	−23.25
Sales to trader	−2.47	−9.82	8.50	−5.19	−8.98
Total	−18.96	−5.17	4.10	−3.44	−23.47

Source: Based on data from *Nanjing* vegetable marketing chain survey (2002).
Note: A negative value indicates discounts in quantity sold, while a positive value expresses an increase in the quantity sold. Calculations based on data in table 9.5.

Information costs are by far the most important in affecting sales volume. This is again related to the scarce development of commercial organizations and market facilities. Negotiation costs and transportation costs represent up to 8.5 percent of the volume discount. Monitoring costs have a positive effect on sales volume. Information costs strongly reduce the sales volume, since the usual high fluctuations at the wet market require farmers to frequently collect information. Negotiation cost and transportation cost are the most important costs affecting the sales volume in the sales to traders. Marketing through traders involves longer distances to reach the wet market, and consequently also more time is required to sell all the produce. Monitoring costs have a positive effect on sales volume, which may be again related to the establishment of trust through more permanent trading relationships.

In summary, direct marketing involves most transaction costs and thus results in high discounts on the sales volume, while sales with traders encounter the lowest transaction costs. Information costs are most important in discounting sales volume within the direct and wholesale channels, while negotiation cost are most relevant for the marketing transactions with traders.

The aggregate effect of transaction costs on the sales value (see table 9.8) reveals that estimated sales value was on average 44 percent lower due to transaction costs, affecting mostly the farmers involved in direct sales. Information costs proved to be the most important category, followed by negotia-

Table 9.8.   Discount in Sales Value by Type of Transaction Costs (% of sales value)

	Information	Negotiation	Monitoring	Transport	Total
By type of Marketing Chains (%)					
Direct sale	−39.72	−28.11	3.00	5.35	−54.92
Wholesale	−37.64	−22.09	4.50	2.91	−49.62
Sales to sraders	−24.75	−22.69	11.99	3.33	−31.66
Total	−35.06	−20.75	6.33	4.02	−43.70

Source: Based on data from *Nanjing* vegetable marketing chain survey (2002).
Note: A negative value indicates discounts in the sales value and a positive value suggests an increase. Calculations based on data in tables 9.6 and 9.7.

tion costs. Information and negotiation costs have a strong negative effect on tomato sales values; the former affect both the price and the volume, whereas the latter are particularly reducing the price. Nearly all transaction costs have the strongest impact on the direct marketing chain; the only exception is for transport costs that tend to be lower in this particular channel.

TRANSACTION COSTS AND TECHNOLOGY CHOICE

Transaction cost analysis can be used to identify how market and institutional reforms could contribute to improved technology choice and better resource and farm management. A reduction in transaction costs could provide important incentives for the intensification of resource use in vegetable production (Ruben et al. 2001). We therefore focus our attention now on the implications of different transaction costs for input use decisions and technology implementation in tomato production in *Nanjing* City.

We analyze first the effect of transaction costs on fertilizer use in tomato production, using a probit model (table 9.9). Results show that transaction costs have a significant impact on fertilizer use in the direct sales channel and wholesale market channel, but not on fertilizer use of farmers involved in sales transactions with traders. Negotiation costs related to the duration of the contacts between farmers and traders are usually higher at the wet market, where transactions take place for smaller quantities and at an irregular frequency. Such negotiation costs have a distinct negative impact on fertilizer use by farmers involved in the direct sales marketing chain. When these negotiation costs are controlled, farmers may become inclined to use more fertilizers in order to reach high tomato yields and better quality produce. At the wholesale market channels, however, farmers face more negotiation pressures as the traders dominate the negotiation processes for large volume transactions. The negotiation costs have a significant positive impact on fertilizer application for farmers involved in the transactions with traders at the wholesale markets. Since farmers have limited capacities to achieve a good price at the wholesale market with weak negotiation power, they are more likely to increase the fertilizer use to produce more vegetables and improve quality.[4]

We further analyzed the determinants of farm households' investment behavior in vegetable production. To that end, we identified the factors influencing the total amount of production-specific investment for *Nanjing* tomatoes (table 9.10).

Results show that older farmers involved in wholesale trade and less-educated farmers involved in sales to traders are less willing to make production-specific investment in their tomato production. Those farmers usually produce tomatoes in a rather traditional way and prefer to use more labor instead of capital investments. Larger households with more family

Table 9.9. Factors Influencing Fertilizer Use (Probit regression results)

Explanatory Variables	Direct Sale	Trader	Wholesale
Constant	0.697	−2.092	−4.400**
	(0.643)	(1.646)	(1.969)
Tomato area (mu)	−0.145	0.649	−0.352
	(0.478)	(0.830)	(0.740)
Information costs	−0.440	1.052	1.133
	(0.435)	(0.943)	(2.670)
Negotiation costs	−1.200***	−0.012	0.313**
	(0.450)	(0.059)	(0.150)
Monitoring costs		−0.077	−0.140
	(0.342)	(1.972)	
Transport costs	0.010	0.027	0.275
	(0.007)	(0.019)	(0.384)
Log-likelihood	−23.570	−9.684	−5.328
Pseudo R^2	0.222	0.252	0.613
Number of observations	46	20	20

Source: Based on data from *Nanjing* vegetable marketing chain survey (2002).
Note: * Significant at 10% level; ** Significant at 5% level; *** Significant at 1% level.
Standard errors are in parentheses.

Table 9.10. Factors Influencing Production-Specific Investment (OLS regression results)

Explanatory Variables	Direct Sale	Trader	Wholesale
Constant	−1,616.8	3,644.9	−899.5
	(1,262.4)	2,220.2	(5,299.9)
Age square of the household head (year)	—	—	−1.7**
	—	—	(0.68)
Education level households head (year)	—	212.0*	—
	—	(100.7)	—
Household size (persons)	645.2**	−550.5	463.3
	(268.3)	315.9	(487.7)
Tomato area (mu)	2,574.6***	3,005.1***	4,294.7***
	(588.5)	(759.1)	748.1
Pesticides use (yes = 1)	—	−279.0	−1,511.22
	—	−1,040.5	(1,779.6)
Information costs	−493.7	−2,810.2	921.2
	(389.6)	(2,462.5)	(2,413.2)
Negotiation costs	28.4	3,482.2***	5,230.1**
	(24.1)	(801.3)	(2,242.1)
Monitoring costs	—	2,111.0***	−1,830.1
	—	(391.9)	(1,205.2)
Transport costs	−187.8	−2,339.8***	−25.2
	783.0	(738.3)	(113.8)
R^2-adjusted	0.41	0.86	0.93

Source: Based on data from *Nanjing* vegetable marketing chain survey (2002).
Note: * Significant at 10% level; ** Significant at 5% level; *** Significant at 1% level.
Standard errors are in parentheses.

members involved in direct sales are more inclined to make production-specific investment in tomato production. This is mainly because the labor availability permits them to maintain a larger production scale. This is also confirmed by the fact that farmers with larger tomato areas generally maintain higher investment levels.

Different transaction cost categories also have an impact on the level of production-specific investment maintained by tomato producers in the *Nanjing* area, particularly for farmers involved in deliveries to traders. Those farmers spent more resources for production-specific investments to improve their tomato yield and the quality of the produce and to reach economies of scale in their vegetable production. Monitoring costs have a particularly strong impact on investment behavior of farmers selling to traders, indicating the importance of establishing long-term relationships as a guarantee for realizing production-specific investments. Information costs do not show a significant influence on investments, indicating that the availability of knowledge on alternative outlets is not a constraint for realizing production-specific investments. Negotiation costs significantly decrease production-specific investments of farmers engaged in wholesale trade, but enhance investments in transactions with traders. This indicates that quality requirement is higher at the wet market compared to the wholesale market. In a similar vein, transport costs reduce production-specific investments by farmers involved in the trader channel due to the risk of quality loss during transport.

TRANSACTION COSTS AND FARM MANAGEMENT

Transaction cost analysis may also be used to understand managerial decisions regarding the scale of production and the degree of specialization. For this purpose, we used total cultivated vegetable area and the share of tomato area in the total cultivated area as indicators of scale and specialization respectively, and included farm household characteristics and transaction cost components as independent variables.

Results show that age, household size, and input use significantly influence the vegetable production scale and diversification in different marketing channels (tables 9.11 and 9.12). Older farmers prefer to maintain small-scale tomato areas when they are involved in the wholesale market outlet or in direct selling. This is probably due to their declining labor ability. With increasing experience in vegetable production, farmers prefer to specialize in tomatoes. Farmers who use fertilizer and sell to traders tend to cultivate a larger area with vegetables. The same holds for farmers who use pesticides and sell their tomatoes at the wholesale market. Positive effects on specialization are found for farmers who do not use fertilizer and sell directly to consumers and for farmers who use pesticides and sell to traders.

Table 9.11.   Factors Influencing the Scale of Vegetable Production (OLS regression results)

Explanatory Variables	Direct Sale	Trader	Wholesale
Constant	3.10***	2.59***	10.0***
	(0.55)	(0.81)	(1.91)
Age square of the household (year)	–0.003**	—	–0.001***
	(0.00)	—	(0.00)
Household size (persons)	—	0.16	0.44***
	—	(0.20)	(0.14)
History of tomato production (year)	0.01	–0.04	—
	(0.02)	(0.03)	—
Fertilizer use (yes = 1)	0.48	1.61**	—
	(0.35)	(0.62)	—
Pesticides use (yes = 1)	—	—	2.35***
	—	—	(0.54)
Information costs	–0.19	1.95**	1.41
	(0.31)	(0.85)	(1.03)
Negotiation costs	–0.001	–0.003	–0.03
	(0.01)	(0.01)	(0.02)
Monitoring costs	—	–0.18	–0.76*
	—	(0.20)	(0.40)
Transport costs	0.28**	0.63**	5.82***
	(0.02)	(0.29)	(1.32)
R^2-adjusted	0.32	0.47	0.92

Source: Based on data from Nanjing vegetable marketing chain survey (2002).
Note: * Significant at 10% level; ** Significant at 5% level; *** Significant at 1% level.
Standard errors are in parentheses.

Transaction costs in the tomato channel generally provide farmers with incentives to decrease the tomato production area and shift toward less-specialized production systems in order to avoid production and marketing risks. Transport costs, however, have a clear positive impact on the production scale, since a full truckload is an important device to save costs for farmers in all market chains. In addition, higher information costs provide incentives to vegetable farmers to increasing their production scale when delivering to traders. Negotiation costs reduce the degree of specialization for farmers involved in direct sale. When more time is required to reach an acceptable price, farmers face difficulties to sell the daily tomato harvest at the wet market and prefer diversification into other crops. Finally, farmers that have no long-term relationship with agents operating in the wholesale market, and thus incur high monitoring costs, are less likely to specialize in tomatoes in order to prevent dependency.

CONCLUSION

The main aim of this study is to examine the simultaneous effects of farm household characteristics, resource endowments and institutional conditions on farmers' choices for different tomato marketing chains, making

Table 9.12. Factors Influencing the Degree of Specialization (OLS regression results)

Explanatory Variables	Direct Sale	Trader	Wholesale
Constant	0.43***	—	(0.11)
	0.16	—	(1.01)
Age of the household head (year)	—	0.02**	(0.01)
	—	0.01	(0.03)
Age square of the household head (year)	—	0.19**	−0.001
	—	(0.001)	(0.00)
Total area of the farmland (*mu*)	−0.04	−0.02	—
	(0.03)	(0.05)	—
History of vegetable production (years)	0.01	0.01*	—
	(0.06)	(0.01)	—
Fertilizer use (yes = 1)	−0.13*	—	0.10
	(0.06)	—	(0.15)
Pesticides use (yes = 1)	—	0.19*	−0.21
	—	(0.10)	(0.12)
Information costs	−0.06	−0.28	−0.05
	(0.06)	(0.22)	(0.10)
Negotiation costs	−0.13**	−0.004	−0.11
	(0.06)	(0.003)	(0.11)
Monitoring costs	—	−0.03	−0.05***
	—	(0.03)	(0.01)
Transport costs	0.001	−0.001	0.01
	(0.001)	(0.001)	(0.01)
R^2-adjusted	0.25	0.66	0.71

Source: Based on data from *Nanjing* vegetable marketing chain survey (2002).
Note: * Significant at 10% level; ** Significant at 5% level; *** Significant at 1% level.
Standard errors are in parentheses.

use of a transaction costs economic approach for a quantitative evaluation of marketing chain efficiency in *Nanjing* City, *Jiangsu* Province, in China. To this end, we selected a random sample of eighty-six tomato-producing farmers living in *Maqun* Township of *Nanjing* City. Probit models were used to identify relevant farm characteristics influencing market outlet choice, while hedonic price techniques were applied to capture the transaction costs in different tomato marketing chains.

The empirical analysis confirms that factors like age, education, tomato production history, market distance, and the availability of a market stall influence prices and volumes sold in different marketing outlets. Results also show that transaction costs in the study area equal almost 27 percent of the sales price, 23 percent of the sales volume, and 44 percent of the sales value. Different tomato marketing chains typically incur different types of transaction costs: The direct marketing chain faces the highest negotiation costs, while sales to traders and wholesalers incur the largest information costs.

Small-scale vegetable production received early stimulation in Chinese agrarian policies. During the "start-to-develop" period of the 1950s, farmers were allowed to sell the produce freely in the local markets. However, during the period of strict economic planning of the 1960s and 1970s, only state-owned companies were entitled to sell vegetables. With the introduction of the household responsibility system, vegetable production substantially increased. In addition, aspects of quality, variety development, greenhouse production, and improved technologies for disease control received major attention. From the late 1980s onward, the implementation of the veg-etable basket project intends to improve the vegetables circulation regime. During the 1990s, vegetable production further expanded due to favorable market prices of horticultural crops compared to cereals. Consumption increased, however, at even stronger rates, and the vegetable market shifted from a seller's toward a buyer's market.

The economic transition in the Chinese vegetable markets increasingly asks for an accompanying adjustment in the sphere of market integration. Information and negotiation costs represent a major share of transaction costs and can only be reduced when improved market information systems are put in place. The lack of price transparency and absence of personal trust can seriously hamper fluid transactions (Sternquist and Chen 2002). When product quality and input use are becoming more important criteria, we may expect that transaction costs further increase, since such information should be made transparent. Reducing these information costs, making use of traditional *Guanxi* (relationship marketing), is greatly facilitated with the ownership of market stalls that offer more permanent relations with trad-ers. High stall fees (compared to the volume of transactions) can, however, be rather prohibitive as entry costs on the wet market. In addition, the alternative of developing vegetable auctions deserves serious attention as a device for reducing negotiation costs. The expected growing importance of supermarkets in the retail of vegetables (now they represent only a minor share of the market) will probably lead to greater attention for product quality, timely delivery, and preservation characteristics to enhance shelf life of vegetable products. It is therefore likely that preferred-supplier arrange-ments will emerge to control the information and negotiation costs.

Policies that try to enhance the incentives for intensification of resource use in tomato production should particularly focus on the reduction of transaction costs as a feasible strategy for optimizing input use, product quality, and crop management. Main attention should thereby also be given to measures that reduce the negotiation costs. For our sample we found that negotiation costs (i.e., the number of visits before reaching an agreement on the price) has a significant negative impact on the probability of fertil-izer use for farmers involved in direct sale at the wet market and positive impacts for sales transactions with traders at the wholesale market. Con-sequently, increasing market competition and improving transparency of market conditions will inevitably provide the incentives required for quality upgrading. Further institutional reforms can therefore be a necessary strategy for enhancing market-driven conditions for improving farmers' resource use management practices.

NOTES

1. The literal meaning of *"Guan"* is "gate," while *"Xi"* means "connection."

2. Most young people prefer engagement in non-farm activities in the urban area.

3. Heckman's two-stage procedure (Heckman 1979) is applied for this estimation to avoid selectivity bias in different marketing chain choices. At the first stage, a probit analysis is used to determine the probability of farm households choosing a certain marketing chain. From that, the Inverse Mill's Ratio (IMR) is calculated, representing the conditional probability of the household choosing this marketing outlet. At the second stage, the IMR is taken as explanatory variable to control for selectivity bias. A significant coefficient for the IMR indicates that the differences registered in received prices depend on the marketing strategy adopted.

4. Results of a probit analysis for pesticides use show that transaction costs do not have significant effects on pesticides use decisions. The most important factor influencing pesticides use is the cultivated tomato area. For larger farms, the existing constraint on labor availability induces farmers to rely on labor-saving production technologies. Only farmers involved in direct trade are less inclined to increase pesticides use when negotiation costs become higher. This may be related to the increasing risk of product denial when traders recognize the pesticides applications and classify the produce in a lower category.

BIBLIOGRAPHY

Ahmadi-Esfahani, F. Z. and R. G. Stanmore. "Demand for Vegetables in a Chinese Wholesale Market." *Agribusiness* 13 (1997): 549–59.

AVDRC. "Improvement and Stabilization of Year-Round Vegetable Supplies." Taiwan: Asian Vegetable Research and Development Center, 1998.

Calabresi, G. "Transaction Costs, Resource Allocation and Liability Rules: A Comment." *Journal of Law and Economics* 11 (1969): 67–73.

Escobal, J. A."Transaction Costs in Peruvian Agriculture: An Initial Approximation to Their Measurement and Impact." *Research Report* GRADE, Lima, Peru, 1999.

Gabre-Madhin, E. Z. "Market Institutions, Transaction Costs, and Social Capital in the Ethiopian Grain Market." IFPRI *Research Report*, no. 124. Washington, DC: International Food Policy Research Institute, 2001.

Heckman, J. "Sample Selection Bias as a Specification Error." *Econometrica* 47, no. 1 (1979): 153–61.

Hobbs, J. "Measuring the Importance of Transaction Costs in Cattle Marketing." *American Journal of Agricultural Economics* 79 (1997): 1083–1110.

Hueth, B., E. Ligon, S. Wolf, and S. Wu. "Incentive Instruments in Fruit and Vegetable Contracts: Input Control, Monitoring, Measuring and Price Risk." *Review
of Agricultural Economics* 21, no. 2 (1999): 374–89.

Li S., L. Gao, S. Zhou, G. Liu, and W. Liu. "Diversification of Vegetable Growing in
the Middle and Lower Reaches of the Yangtze River. Proceedings of 3rd International Symposium of Vegetables Crops." *Acta Horticultura* 467 (1998): 253–55.

Lu, H. *Tomato Marketing Supply Chain Choice, Efficiency and Transaction Costs Analysis:
A Case Study in Nanjing City, Jiangsu Province, P. R. China.* MSc Thesis, Wageningen
University, the Netherlands, 2003.

Nanjing Agricultural and Forestry Bureau. *Database on Vegetable Production.* Nanjing:
Nanjing Agricultural and Forestry Bureau, 2004.

North, D. *Institutions, Institutional Change and Economic Performance.* New York: Cambridge University, 1990.

Rosen, S. "Hedonic Prices and Implicit Markets, Production Differentiation in Pure
Competition." *Journal of Political Economy* 82 (1974): 34–55.

Ruben, R., M. Wesselink, and F. Saenz. "Contract Farming and Sustainable Land Use:
The Case of Small Scale Pepper Farmers in Northern Costa Rica." Paper presented
at AEEA Seminar, Copenhagen, June 2001.

Scott, G. J. *Price, Product and People: Analysis Agricultural-Markets in Developing Countries.* Boulder, CO: Lynne Rienner, 1995.

Sternquist, B. and Z. Chen. "Food Retail Buyer Behavior in People's Republic of China:
A Model From Grounded Theory." Mimeo, Michigan State University, 2002.

Wallace, N. E. "Hedonic–based Price Indexes for Housing: Theory, Estimation,
and Index Construction." *Federal Reserve Bank of San Francisco Economic Review* 3
(1996): 34–48.

Williamson, O. E. *Economic Organization: Firms, Markets and Policy Control.* Brighton,
Sussex: Wheatsheaf Books, 1986.

10

Off-farm Employment, Factor Market Development, and Input Use in Farm Production

A Case Study of a Remote Village in *Jiangxi* Province, China

Xiaoping Shi, Nico Heerink, Stein Holden, and Futian Qu[1]

The massive rural labor flowing into off-farm employment has become a significant phenomenon in the process of China's economic reform. By 2000, almost 200 million people were involved in off-farm employment (Zhang et al. 2002). Non-farm employment reduces surplus labor in rural areas and allows rural households to acquire other sources of income, including non-farm wage income, self-employment, and remittances from migration. Kung and Lee (2001) have shown that the share of non-farm income in four counties in *Hunan* and *Sichuan* provinces was more than 25 percent of per capita income in China in 1993. Other studies found that income growth of most farmers in the late 1980s and 1990s can be attributed to increased off-farm employment, that is, to self-employment and wage labor (Parish et al. 1995; Rozelle 1996). Migration as an increasingly emerging off-farm activity has also played an important role in the increase of household income in recent years (de Brauw et al. 2001).

Zhang et al. (2001) and de Brauw et al. (2002) stress the impact of off-farm employment on the development of the labor market in the process of rural development. Off-farm activities become more and more important in promoting the development of other factor markets, which is a key characteristic of the transition of the planned economy to a market economy in China (Kung 2002). Several recent studies examine the development of the rural labor market and the evolvement of off-farm activities in rural China (de Brauw et al. 2002; Zhang et al. 2002; Kung 2002). Other studies focus on how the income from off-farm employment is distributed and what the consequences are for the income distribution of households in rural areas of China (Kung and Lee 2001).

Participation in off-farm activities changes resource endowments of households, especially labor and capital used for financing off-farm employment is moving out of farm production. Hence, households may need to restructure their farm production by changing factor and variable input use. A recent study on rural China found that migration has a negative direct impact on farm yields, but that remittances compensate the effect of the labor loss. The overall effect of migration on farm yields is slightly negative (Rozelle et al. 1999). Other studies focused on the long-run effects of off-farm activities on farm productivity. De Brauw (2001) examined the impact of migration, especially remittances, on households' farm investment behavior. He found that migration does not seem to affect household investment in on-farm or off-farm production. Wu and Meng (1996; 1997) found that labor transfer from farm to non-farm activities has no significant impact on grain production, although farmers with a high share of non-farm income invest less in grain production, because there are abundant labor resources in farm production.

Little is known, however, about the impact of off-farm employment on factor and variable input use, especially on the choice of household's input use. Farm production is the main linkage between economy and environment in rural China. The use of factors and variable inputs on farmland are important elements affecting farmland production capacity and environmental quality. How to improve or maintain farmland production capacity and the environment in the long run is an issue that attracts much attention (Huang 2000; Yao 2002; SEPA 1999; Niu and Harris 1996; Huang and Rozelle 1995; World Bank 1992; Zhao 1991).

The development of rural factor markets may play an important role in this respect. With the massive flow of rural labor out of farm production, other factor markets have emerged to some extent. Empirical studies by Yao (2000), Lohmar et al. (2001), and Kung (2002) show the important role of the off-farm labor market in inducing the development of the land rental market. This will facilitate household market interactions, which may intensify or reduce the impact of off-farm employment on land production capacity because of the concurrent change in input use.

Off-farm employment may also affect the agricultural production and input use of those households within the same village with no members working off-farm. Increased income and expenditure of households involved in off-farm employment and the emergence of village-level factor markets are responsible for such indirect effects. The overall objective of this chapter is to analyze the impact of off-farm employment on village factor market development, and to examine the effect of off-farm employment on factor and variable input use in farm production and on land production capacity.

The chapter has three more specific objectives. First, we will explore the development of village land rental markets, oxen rental markets, and labor markets along with different types of off-farm activities (local non-farm activities, self-employment, and migration). Second, we will examine the impact of income obtained from off-farm employment on farm production, especially on factor and variable input use, for different household groups within the same village. Third, we will examine the implications for land production capacity.

In relation to the first objective, household groups are distinguished according to different characteristics that are relevant for off-farm employment and farm production in order to examine the involvement of these groups in village factor markets. A modeling approach, using the same household grouping, is then used to achieve the second and third objective. Microeconomic farm household models are useful tools for analyzing farm household behavior, but they do not capture the income linkages and the general equilibrium effects within a village. A village social accounting matrix (SAM) can present a picture of market and income linkages of household groups within a village, and the interactions with the world outside the village. We will use a village SAM multiplier model derived from the village SAM to simulate the impact of changes in off-farm income on the level of production, factor use, variable input use and household incomes for different groups within a village. Based on that, the implications of changing factor use and variable input use for land production capacity and environmental quality are investigated.

The data underlying this study were obtained from a farm survey held in the summer of 2000 and the spring of 2001 in *Shangzhu* village, *Guixi* county in *Jiangxi* Province. The survey was carried out within the framework of a larger research project on "Economic Policy Reforms and Soil Degradation in Southeast China." The survey included questions on income sources and expenditures as well as on inputs and outputs of production activities. The questionnaire was designed in such a way that the information collected can be used for constructing a village SAM. Three villages, reflecting differences in geography and infrastructure, were selected by the project for intensive farm household data collection (Kuiper et al. 2001). *Shangzhu* village was chosen for this study, because it is located in a mountain area and relatively isolated from outside markets. Local household and market linkages are expected to be stronger for a remote village, and the indirect effects of off-farm employment are therefore expected to be larger than for the other two villages.

Our research area falls in an area characterized by serious soil degradation. The yield of main crops, such as rice, rapeseed, and cotton, is much lower in *Jiangxi* Province than elsewhere in China (Li and Lin 1998; Huang 1999). One important reason is that the soil organic matter content in cultivated land is lower in *Jiangxi* than in neighboring provinces, partly because of less planting of green manure and lower use of animal manure (Li and Lin 1998). Farmers in the research area traditionally plant green manure and apply animal manure on the farm, but modern input use is increasingly replacing these traditional techniques. This study will examine the extent to which the change of input use from traditional inputs to modern input use is related to the growth of off-farm employment.

The rest of the chapter is organized as follows. In the next section, we will develop a theoretical framework of effects of off-farm employment on village factor market development and on factor use and variable input use in farm production. The third section describes the household grouping as well as the presence of factor and output markets and the use of natural resources in the selected village. In the following section, the village SAM and the multiplier simulation results will be presented, respectively. The last part of the chapter will discuss the findings and conclusions.

THEORETICAL CONSIDERATIONS

Off-farm employment enhances the differentiation between households, because only some of them have access to off-farm activities. In rural China, there are limited off-farm employment opportunities due to institutional and noninstitutional entry barriers, such as the household registration system (*hukou*) and the discrimination against rural dwellers in urban jobs. Many households are excluded from the possibilities of working off-farm. Possession of certain skills/education or social capital (*guanxi*) is important to gain access to off-farm employment (Zhao 2001; Zhang et al. 2001). Due to the limited opportunities and strong competition, additional resources of households may be required to gain access to off-farm employment. Hence, differences in human capital, social capital, land, or other resource endowments are an important cause of differentials in access to off-farm activities, and differentiation among households (Zhao 2001; de Brauw and Rozelle 2003).

Local factor markets may be stimulated when households participate in off-farm activities. However, large diversities in factor market development may be observed (Kung 2002), because of institutional barriers and high transaction costs that are characteristic of factor (and commodity) markets in many developing countries (de Janvry et al. 1991). If there is no labor surplus, off-farm employment shifts labor out of farm production by reallocating labor time. The resulting labor shortage on the farm may induce farmers to hire labor or rent out land, which will induce the development of land rental markets and/or labor markets. If there are high transaction costs or institutional barriers in land renting activities and relatively low costs or barriers in hiring labor, households with members participating in off-farm employment will be more likely to hire labor in order to compensate for their own labor loss. In contrast, if high transaction costs or institutional barriers are involved in hiring labor, the land rental markets will tend to be more developed.

Prevailing institutions, land tenure systems and (until recently) quota obligations in China hinder household participation in land rental activities (Lohmar et al. 2001). In particular, land use rights are assigned to households by village leaders for a fixed contract period (mostly thirty years at the moment), based on equality among households. Partial redistributions frequently take place, however, to correct for migration or other demographic changes. Land renting activities emerged on a small scale within villages as an alternative to administrative reallocations (Deininger and Jin 2002). In some areas within China, 10–15 percent of the land is leased inside the village (Huang et al. 2000). When the constraints hindering household land rental activities are removed, land rental markets will be further intensified by off-farm employment (Turner et al. 2001; Lohmar et al. 2001; Kung 2002).

Due to labor moving out of farming and seasonality of agricultural production, labor exchange and labor hiring become more important in rural China. Farmers increasingly hire labor for land preparation and harvesting, or rent a small tractor to plow. In mountainous villages, where tractors cannot be used, oxen rental activities and shared oxen ownership are developed. With the increase of off-farm activities, oxen or tractor renting appear as alternative strategies for saving labor.

Off-farm activities usually increase household income, which may be used to expand agricultural production factors (labor or land, depending on institutional barriers) and input use to increase farm productivity. But it may also be invested in nonagricultural activities or used to increase consumption. To some extent, households without off-farm activities and no access to formal credit may be able to obtain loans from households with off-farm employment within the same village.

Household-market linkages are now widespread in rural China (Benjamin and Brandt 2002). Household-market exchanges are very important in shaping farm household responses to policy changes. When external shocks occur, their impact on households will pass through such linkages. In a perfect market, shocks will be contained in the price changes in the market. However, high transaction costs (e.g., caused by missing or asymmetric information), risk and institutional barriers may lead to missing, imperfect, or thin markets. If substantial market imperfections exist between a village and the outside world, this may cause inside-village markets (Hoff et al. 1993; Sadoulet and de Janvry 1995; Taylor and Adelman 1996). Generally, unfavorable physical conditions of villages cause high transaction costs in trading commodities and factors with the outside world, which may make such villages isolated from exchange with the outside world, resulting in more local exchange between households within the village.

Moreover, the existence of internal village markets is also the result of differentiation of households in the village. When households differ in their resource endowments and (as a result) in their production activities, internal trade will be beneficial to those households. Without differentiation among them (and without exchanges with the outside world), households will be self-sufficient. The typology of village economies developed by Holden et al. (1998) clearly illustrates why households in a village generate strong market exchanges, and why village economies will be important to focus on. Specifically, when households are highly differentiated and transaction costs with the outside world are high, village markets will usually arise, with a price formation independent of market signals from outside the village (Holden et al. 1998).

Off-farm activities accelerate the differentiation of households in rural China and facilitate market exchanges among households. Therefore, any expansion or reduction of off-farm employment may have important implications for inside-village activities. The existence of high transaction costs between village households and the outside world will generate general equilibrium effects within the village economy. The expansion of off-farm employment will usually shift labor out of farm production, which will increase the opportunity costs of the household in farm production and the wage rate at the village labor market if there is no labor surplus in the village. At the same time, increasing income from off-farm activities will induce households to enjoy more leisure (and increase consumption), release cash or credit constraints of the households (if any), and stimulate farmers to use more labor-saving inputs on the farm. Village land or labor markets provide a buffer to compensate the labor-loss effect of households involved in off-farm employment. Application of green manure or animal manure needs much labor input and little or no finance, while use of modern chemical fertilizer needs little labor and much finance. Hence, off-farm employment is expected to stimulate the adoption of labor-saving production technologies that may lead to land production capacity decline in the long run.

SOCIOECONOMIC CHARACTERISTICS OF THE VILLAGE

Shangzhu village is only ten kilometers away from a township, but it takes one hour from the village office to the township by bus because the road is sandy. It takes one more hour from the township to *Guixi* City. *Shangzhu* village has sixteen village groups *(cunming xiaozu)*, which are the basis for land distribution, and thirty-two natural hamlets. *Shangzhu* village is located in a mountainous area and some hamlets are quite far from the village office, which is located along the sandy road in a bigger hamlet named *Xiazhu*. The sandy road ends in a neighboring hamlet. Farmers in the remote hamlets need a half-hour to two-hours walk to reach the village office by mountainous tracks. Several years ago there was a mining enterprise in the village (that belonged to the county government) because the soil is very suitable for making porcelain (*ciqi*). However, when the mine was depleted, more investment was required for moving further inside the mountain, and the enterprise went bankrupt. Now, only some farmers still carry out mining and simple processing.

The total population in the year 2000 was 2,028 persons, within a total of 472 households. A household is defined in our research as a group of people living under the same roof and eating food from the same pot. Some family members temporarily migrated to other places and sent income back to their families. We also recorded them as household members. We sampled 109 households, accounting for 23 percent of the total. In some households, all the members migrated outside the village, while in others only the children lived in the village. We did not interview these households.

The village has four types of land, namely irrigated land (with paddy fields), dry land, forestland, and wasteland. All land is contracted to households, except for some pieces of forestland. No village level redistribution of land was implemented in recent years, except for limited adjustments of irrigated land in some village groups. In the early 1990s some pieces of forestland belonging to the village committee were contracted to household groups instead of individual households. However, until 2000 no profits were generated from this forestland, because household groups argued with the village committee about how to harvest trees and share the benefits. All the paddy and dry land are located in the mountains, and most of them are built with terraces. Wasteland is seldom cultivated because the area is very small and steep.

The main crops are rice and vegetables. Perennial crops, especially bamboo and bamboo shoots, cultivated in the forestland, are also important to households. Labor, chemical fertilizer, animal and green manure, seeds, and oxen plowing are the main variable inputs in farm production. Almost no inputs are applied to forestland, except that farmers sometimes leave bamboo leaves in the field. Livestock production consists of oxen, pig, chicken, duck, and fish, with the latter two being less important.

Local agricultural and nonagricultural employment, self-employment, and temporary migration are the main types of off-farm activities in the village. Local agricultural employment includes crop harvesting, rice transplanting and bamboo shoot digging. Nonagricultural employment includes wood (bamboo) carpentry, house building, and teaching. Self-employment includes shopkeeping, small handcraft making and selling, and transportation. We define household members working off-farm and not living together with other household members as migrants. Most migrants from *Shangzhu* village work outside their counties and even their provinces.

Household Classification

Only a limited number of household groups can be distinguished in a village SAM. Because our sample comprises only 109 households, we decided to distinguish not more than four groups for the village SAM. Obviously, different criteria for grouping will generate different groups. The criteria to be used for grouping should be based on the objective(s) of the research. Because the focus of our research is on the impact of off-farm employment on agricultural production decisions, we use the resources that households have for generating off-farm income and the resources for generating agricultural income as the two main grouping criteria.

Several indicators of off-farm employment resources (such as social networks and education level of household members) and for agricultural production resources (oxen ownership) were carefully examined. The social network of households can be an important determinant of access to off-farm employment. Zhao (2001) shows that social networks are crucial for households for participating in off-farm activities, especially migration to faraway places in China (see also Shi et al. 2004). However, the contents of social networks are quite diversified. They consist of social relationships (*guangxi*) of households, kinship networks, and personal contacts of migrants and other institutions. Our data set contains only limited information on social networks, namely information on household members who have already migrated outside the province and information on remittances (sent by relatives). Exploratory regressions on the impact of such social relationships on participation in off-farm employment, however, did not give statistically significant results (Shi et al. 2004).

Another important potential resource for off-farm employment is the education level of household members. Our exploratory analyses (Shi et al. 2004; Kuiper et al. 2002) indicate that the educational level of the labor force is a very important determinant of household participation in off-farm activities. The average educational level (using the number of years of schooling) of the labor force in *Shangzhu* is around four years. Workers with more than four years of schooling have a high probability (at household level) to participate in off-farm activities, particularly in migration. The educational level (with four years as the threshold) was therefore used as a criterion for grouping households.

Oxen ownership is an important resource for earning agricultural income and an important determinant of input use levels in *Shangzhu* village. More than 80 percent of farm households keep oxen or share oxen with other households. Oxen are mainly used for plowing, while oxen manure is an important source of organic fertilizer. Farmers sometimes reapply rice straw on their fields to improve the soil structure, but the rice straw also increases difficulties during rice transplanting, especially for late rice. However, the rice transplanting becomes easier if fields in which straw is applied are plowed more than once. Our exploratory analysis (Kuiper et al. 2002) shows that oxen ownership is an important determinant of fertilizer use for crop production in *Shangzhu*. It was therefore decided to use oxen owner-ship as the second criterion for grouping households.

Using these two criteria, four household groups are distinguished (see table 10.1). The first group, named "Households with no educated per-sons," consists of households having no laborers with more than four years of schooling. The second group is named "Households with no oxen and at least one educated persons." It consists of households that do not own oxen and have one or more laborers with more than four years schooling. The third group is named "Households with oxen, one or two educated persons" and consists of households owning oxen and having one or two laborers with more than four years of schooling. The last group is the group of households having at least three laborers with more than four years of schooling, and holding oxen. It is named "Households with oxen, at least three educated persons."

Table 10.1. Criteria Used for Grouping Households

No. of Persons with More	Oxen Ownership	
Than 4 Years of Schooling	No	Yes
0	Group 1	Group 1
1 – 2	Group 2	Group 3
3 or more	Group 2	Group 4

Characteristics of Household Groups

A comparison of the four household groups reveals large differences. Table 10.2 shows the basic household group characteristics. Because the dry land area is very small, it is not presented in the table. The first two groups have more per capita contracted irrigated land than the other two groups. The second group has more forestland, as compared with the other three groups. Another basic visible difference among the groups is the average household size. The first group has a much smaller household size, while the household size of the fourth group is the largest. Group 4 also has the largest number of laborers, while Group 3 has the largest number of chil-dren (non-laborers).

Table 10.A.1 in the appendix shows the average number of schooling years of labor force members for each household group and for the total sample. The average length of schooling is largest for household Groups 2

Table 10.2. Basic Household Group Characteristics

Household Groups	No. of House holds	Population	Average Household Size	Average No. of Workers	Per Capita Contracted Irrigated Land	Per Capita Contracted Forestland
Group 1	16	46	2.87 (1.41)	2.25 (1.23)	1.37 (0.46)	0.55 (0.57)
Group 2	14	57	4.07 (0.92)	3.29 (0.91)	1.32 (0.94)	1.14 (1.45)
Group 3	35	152	4.34 (1.33)	2.86 (1.26)	1.19 (0.37)	0.58 (0.67)
Group 4	44	222	5.04 (1.14)	4.02 (0.79)	1.17 (0.40)	0.60 (0.60)
All groups	109	477	4.37 (1.41)	3.39 (1.21)	1.23 (0.50)	0.65 (0.78)

Note: Standard deviations are in parenthesis.

and 4, and lowest for Group 1. The results of pairwise t-tests for household size and number of workers for the four groups are presented in appendix table 10.A.2. The results indicate that the mean household sizes and labor force sizes are significantly different from each other for all combinations of household groups, except for household Groups 2 and 3. Table 10.A3 in the appendix makes a similar comparison of the mean values of per capita irrigated land and forestland for the four household groups. It shows that Group 2, the group with no oxen, has significantly higher levels of forestland per capita than the other three groups. In addition, Group 1 has a significantly smaller area of forestland per capita than Group 4, and a significantly larger area of irrigated land per capita than Group 3. All other differences in irrigated and forestland endowments between the four groups are not statistically significant. The results of the t-tests in the appendix provide complementary evidence that the four groups are different in demographic characteristics and land endowments.

Table 10.3 presents the average incomes, subdivided by income source, for the four household groups. The average per capita income in this village equals 1,386 *yuan*, or $0.46 per capita per day (based on the official exchange rate in 2000, 1 USD = 8.30 *yuan*). Group 1 has the lowest total household income, whereas Groups 1 and 4 have the lowest per capita household incomes. Group 2 (no oxen ownership) has the highest total and average household income. On average, households obtain 43 percent of their income from off-farm activities, 51 percent from farm production (paddy, vegetables, perennial crops, and livestock), and 6 percent from other sources (including government transfers, family member remittances and other assistance from relatives). Family remittances are different from migration remittances. Family remittances refer to the money sent by relatives who do not, or do no longer, belong to the household. Group 2 obtains more than 17 percent of its income from other sources, with 75 percent of this type of income consisting of family remittances. The four household groups have very similar patterns of income sources, except for Group 2, which obtains a relatively small share of its income from farm production.

Table 10.3. Average Incomes from Different Sources per Household Group (*yuan*)

Household Groups	Total Income	Per capita Income	Farm Income	Off-Farm Income	Other Sources
Group 1	3,587	1,248	2,133	1,335	119
Group 2	8,055	1,978	3,249	3,384	1,422
Group 3	6,529	1,503	3,404	3,018	108
Group 4	5,960	1,181	3,160	2,437	363
All groups	6,064	1,386	3,099	2,584	381

Table 10.4 shows the composition of off-farm income. Remittances by migrated household members constitute the largest component for all household groups. Its share of off-farm income is 62 percent on average. Households with less-educated persons (Group 1) obtain a relatively large share of their off-farm income (27 percent) from agricultural employment. Households with no oxen (Group 2) rely more on local nonagricultural employment (43 percent of their off-farm income). For households with oxen and one or two educated members (Group 3), remittances from migration are the main source of off-farm income (77 percent). Somewhat surprisingly, however, households with oxen and three or more educated members (Group 4) rely less on migration remittances, but relatively more on local nonagricultural employment.

Table 10.4. Composition of Off-Farm Incomes per Household Group (*yuan*)

Household Groups	Agricultural Wage Employment	Nonagricultural Employment	Self-Employment	Remittance from Migration	Total
Group 1	356	56	173	750	1,335
Group 2	0	1,441	286	1,657	3,384
Group 3	226	134	332	2,326	3,018
Group 4	178	685	273	1,302	2,437
All groups	197	513	279	1,595	2,584

Output Markets

Food produced for personal consumption is a large share of total farm production. Only a few households buy rice for their own consumption. On aggregate, all household groups are net rice sellers. Since land is distributed equally across households, each household grows paddy and vegetables. Some households with several or all members participating in off-farm activities need to buy rice when they are back in the village during holidays or festivals. They rent out their land when they are absent from the village. Although the rent for land is mostly paid in rice, it is sometimes not enough to meet their consumption needs.

The quality of one-season rice is better than that of late rice and particularly early rice. Hence, there is a low percentage of two-season rice being sold: on average 13 percent of one-season rice and 7 percent of two-season rice is sold (table 10.5). Vegetables are almost entirely for personal consumption, while perennial crops are largely being marketed. Overall, the share of livestock products being commercialized ranges from about one-third to about two-thirds in value terms. Group 2 (no oxen) sells no two-season rice but sells the largest share of its livestock products.

Table 10.5. Percentage of Main Crops Being Sold

Household Groups	One-Season Rice	Two-Season Rice	Vegetables	Perennial Crops	Livestock Products
Group 1	11	7.4	0.0	93	34
Group 2	14	0.0	0.0	91	63
Group 3	13	8.4	0.8	84	51
Group 4	14	10.7	2.6	94	47
All groups	13	6.6	0.9	90	49

Agricultural Labor Market

The agricultural labor market shows more diversity (see table 10.6). There are two types of outside-household labor used in agricultural production, mainly in rice and perennial crop production. These are exchange and hired labor. Exchange labor is used only in rice production, while hired labor is used in both rice and perennial crop production. More than 30 percent of the households use exchange labor in one-season rice production, while 14 percent use hired labor. The large shares of exchange and hired labor used by households show that seasonal agricultural labor markets in the village do exist. Contrary to the other three groups, Group 2 uses little exchange labor and relatively more hired labor. All household groups except Group 2 are net agricultural labor sellers. Group 2 employs 11 percent of the total village agricultural labor; the remainder is employed outside the village.

Table 10.6. Percentage of Households Using Exchange and Hired Labor in Rice and Perennial Production

Household Groups	Exchange Labor		Hired Labor		Perennial Crops
	One-Season Rice	Two-Season Rice	One-Season Rice	Two-Season Rice	
Group 1	38	13	6	0	0
Group 2	7	0	36	0	14
Group 3	31	9	11	3	0
Group 4	36	9	11	0	7
All groups	31	8	14	1	3

Land Rental Market

The land rental market is more developed than the agricultural labor market; the percentage of households in the village participating in land rental activities equals 45 percent. Land rental activities take place between households residing in the village. The share of rented irrigated land in the total cultivated irrigated land is 20 percent (table 10.7). Only few households participate in dry land and forestland renting. Thus, in the discussion below land refers to irrigated land only. Table 10.7 compares the land rented from other households and the contracted land for different household groups. Contracted land is the land contracted from the village collective or villagers' group; rented land is the area rented from other households within the village. The percentage of land rented for one-season rice production is much higher (17 percent) than that for two-season rice production (3 percent).

Table 10.7. Percentage of Rented and Contracted Irrigated Land Area by Crop Type

| Household Groups | One-Season Rice | | Two-Season Rice | | |
	Rented	Contracted	Rented	Contracted	Total
Group 1	23	58	2	16	100
Group 2	0	92	0	8	100
Group 3	24	60	4	12	100
Group 4	19	62	5	15	100
All groups	17	68	3	12	100

Land lease contracts normally last only one year or half a year. In 2000, institutional barriers were reduced in the village, and farmers became free to rent-in or -out their contracted land. However, in previous years it was risky to rent out contracted land, as it remained possible for the village collective or villagers' group to reallocate the rented land to other households because of quota obligations or other reasons.

Group 2 (no oxen) does not rent land to expand agricultural production. The other three groups, however, rent between 24 and 28 percent of their cultivated land, which is used mainly for growing one-season rice (table 10.7). Each group has few households renting out land. Only Group 2 is a net "seller (renting out)" in the land market; the other groups are net "buyer (renting in)" in the land market. The biggest "landlord" (renting out land) in the village is the group of households whose entire family or labor force participates in off-farm activities, especially in migration ("absent landlords"); they rent out all or a large share of their land. As mentioned above, these households could not be interviewed and therefore were not included in our sample. Group 2 only supplies 3 percent of contracted land to the land market; the rest of the rented land in the village comes from "absent landlords."

Oxen Rental Market

As an important production factor, oxen are used in the land preparation for rice and (to a lesser extent) vegetable production. Keeping oxen is a time-consuming activity. Children or elderly people normally take care of them. The oxen rental market is functioning to a certain extent, with 12 percent of the households (two cases in each group) hiring oxen in one-season rice production. Group 2 (no oxen ownership) is the main group hiring oxen. Groups 3 and 4 are the main suppliers of oxen in the village.

Credit and Savings

More than half of the households in *Shangzhu* village stated that they obtained credit in 2000, and that most credit was received in cash. Only 34 percent of the borrowed amount is obtained from households in the same village, especially from friends. The remaining 66 percent mainly came from relatives and friends outside the village. Banks, credit cooperative agencies, some shops, and individuals also lent money or lent in kind, but there were few such cases. The picture of the village credit market derived from the survey is unbalanced in terms of money borrowed and lent. The reason is that most of the households are not willing to be considered as moneylenders. Moreover, some of the households who are most likely to lend money, the "absentee landlords," were not interviewed.

Resource Use

Agricultural production in the village affects soil quality and environmental quality in a number of ways. Firstly, green manure crops planted during the previous year are important for the yield of the current year and can reduce chemical fertilizer (nitrogen) application substantially. Using green manure needs more labor but less capital than chemical fertilizer. The green manure area has decreased gradually each year. In 2000, two-season rice had a larger share of its area planted with green manure in the previous year than one-season rice (see table 10.8). Reduction of green manure planting application is an important reason for soil problems, such as natural compaction or soil blocking (Kuiper et al. 2001; Wei 1999). Rice production with green manure will help to improve the soil. Secondly, one-season rice needs less chemical fertilizer than two-season rice because fertilizer is applied only once per crop; fertilizer use in one-season rice is therefore close to half that in two-season rice (table 10.9). Shifting from two-season rice to one-season rice may therefore be beneficial for soil quality and environmental quality. However, application of pesticides and herbicides in oneseason rice is much higher than that in two-season rice, because the planting period of one-season makes the crop more susceptible to diseases. Thirdly, animal manure is another important soil-friendly input that substitutes for chemical fertilizer. There is no manure market, which

Table 10.8.   Percentage of Area with Green Manure Planting in Previous Year by Type of Land and Crop

Household Groups	One-Season Rice Rented	One-Season Rice Contracted	Two-Season Rice Rented	Two-Season Rice Contracted	Total Rice Area (mu)
Group 1	11	16	100	72	75.2
Group 2	n.a.	25	n.a.	0	68.6
Group 3	5	16	33	46	218.4
Group 4	44	31	48	45	305.8
All groups	20	22	60	41	668.0

Table 10.9.   Use of Chemical Fertilizer, Pesticides, Herbicides, and Manure per Household Group (*yuan/mu*)

Household Groups	Manure One-Season	Manure Two-Season	Fertilizer One-Season	Fertilizer Two-Season	Herbicides and Pesticides One-Season	Herbicides and Pesticides Two-Season
Group 1	18.7	69.7	33.7	89.5	5.7	0.2
Group 2	10.3	29.3	37.9	56.7	11.5	0
Group 3	8.5	21.5	29.5	70.2	9.7	2.7
Group 4	9.3	17.7	26.9	67.9	8.7	2.9
All groups	10.6	27.3	30.1	71.4	8.9	2.3

makes manure application closely linked to household livestock production. A decrease in livestock production therefore means a reduction in manure availability and possibly also in manure application.

THE SOCIAL ACCOUNTING MATRIX
OF *SHANGZHU* VILLAGE IN 2000

A village SAM represents the transactions among production activities, institutions, and the outside village. It shows the flows of inputs, outputs, and income between sectors, flows of income between production activities and households, expenditures of households on consumption and investment, and goods and services transfers between institutions. The rows of a village SAM show incomes of each account and the columns present the expenditures made by each account. The choice of the accounts and their subdivision are dependent on the research purposes and the types of policy experiment the researchers want to perform.

The structure of the SAM for *Shangzhu* is given in Table 10.10. It is to a certain extent similar to the SAM used by Taylor and Adelman (1996). It comprises seven main entries (activities, commodities, factors, institutions, government, saving and investment, and outside village). There are a few deviations from the SAM used by Taylor and Adelman (1996). Firstly, it treats migrants as part of household labor endowments, and migration

Table 10.10. The Structure of the Village SAM for Shangzhu

Expenditures Receipts	1. Activities	2. Commodities	3. Factors	4. Institutions	5. Government	6. S-I	7. ROW	8. Total
1. Activities								*Total Production and Transactions*
a. Farming and Livestock								
b. Manure Activity								
c. Fuel Wood Livestock		A. Village Production and Factor Transactions						
d. Factor Renting								
e. Agricultural Labor Work								
f. Non-farm Activities								
g. Migration								
h. Transaction								
i. Leisure								
2. Commodities								
a. Agricultural Products								*Total Demands*
b. Manure								
c. Fuel Wood	B. Village I/O Table			D. Household Consumption	E. Taxes in kind to Government	F. Seeds for Next Year	I. Goods and Services Exports	
d. Rented Factors								
e. Agricultural Labor Work								
f. Non-farm Activities						G. Capital Investment		
g. Migration								
h. Transaction								
i. Leisure	C. Transaction							

Account	Activities	Goods	Factors	Household	Government	Savings & Investment	Rest of World	Total
j. Agricultural Inputs								
k. Livestock Feed								
l. Other goods								
3. Factors								
a. Labor	J. Value Added in Village Production							*Factor Income*
b. Land								
c. Capita								
4. Institutions								
a. Household Groups			K. Payments to Households				L. Family Remittances	*Household Income*
5. Government				M. Taxes in Cash				*Government Income*
6. Savings and Investment		N. Seeds from Last Year		O. Saving				*Total Savings*
7. Rest of World		Q. Imports		Q. Imports	R. Transfers			*Imports*
8. Total	*Total costs*	*Total Supply*	*Total Factors*	*Household Expenditures*	*Government Expenditures*	*Total Investments*	*Exports*	

as an activity and commodity (service) in the SAM. Taylor and Adelman (1996) only include the remittances from migration as factor incomes from outside the village, so migrants are not available for activities in the village. However, migration of rural households has to be treated as part of a household's livelihood because institutional constraints such as the urban registration system (*hukou*) often discourage migrants from settling permanently in urban areas (de Brauw et al. 2001). Secondly, we disaggregated all the activity, commodity and factor accounts at household group level. The resulting SAM shows differences between household groups in factor market participation, productive activities, and consumption. Savings and investment are used as the balance account in order to balance the rows and columns in the SAM. Household expenditure is most likely to be overstated and less accurate, hence the savings and investment account is preferred to balance income and expenditure of household groups.

In table 10.A.4 (in appendix), each subaccount within every main entry is presented in detail. Activity accounts represent two major parts, one is the production activities and the other is the factor transaction activities (for example, land, labor, and oxen rental activities). Production activities are divided into rice production, vegetable production, perennial crop, livestock, manure production, and fuel wood collection. Rice production has been subdivided into four types of production (one-season rice with and without green manure planting in the preceding year, and two-season rice with and without green manure planting in preceding year). For the aim of this study, fuel wood collection and manure activity are included as separate activities. Commodity accounts are divided into products (agricultural and manufactured), services and rented factors. Factor accounts are divided into irrigated land, dry land, forestland, low-educated labor, high-educated labor, and capital, as they are the factors households have in the village. The institution account distinguishes five household groups, with household Group 5 representing the group that is absent in the village and provides much of the rented land in the village, while receiving income from land rents. The last two accounts in table 10.10 have not been subdivided. The government account includes the village committee and township government. In order to simplify the analysis, we did not separate them. The rest of the world account refers to the world outside the village; it is linked to activities inside the village through trade.

Total village GDP is 3,133,590 *yuan*. The relative importance of different production activities (derived from the SAM) is given in table 10.11. One-season rice (15 percent), perennial crops (19 percent), and livestock production (12 percent) are the most important sources of farm income; two-season rice and vegetable production provide much smaller contributions. Agricultural off-farm work contributes only 3 percent to the total village GDP. A very important sector is the off-farm sector, which accounts

Table 10.11. Village GDP Distribution among Activities

Sector	Percentage	Sector	Percentage
One-season rice	10.1	Agricultural work by low-educated labor	1.8
One-season rice and green manure	4.4	Agricultural work by high-educated labor	1.6
Two-season rice	1.8	Nonagricultural work by low-educated labor	0.5
Two-season rice and green manure	2.2	Nonagricultural work by high-educated labor	8.4
Vegetables	4.8	Self-employment by low-educated labor	0.8
Perennial crops	18.9	Self-employment by high-educated labor	4.1
Livestock	11.8	Low-educated labor migration	3.5
Manure production	1.2	High-educated labor migration	24.2
Sum (on-farm)	55.2	Sum (off-farm)	44.9

for 45 percent of GDP. Income from migration activities, particularly from educated labor, is the most important component of off-farm GDP.

MODEL AND SIMULATIONS

To examine the impact of off-farm employment on factor and variable input use and on agricultural production within the village, we apply a SAM multiplier model. Household models can capture household responses to outside shocks, but do not cover the interactions among households. Especially when household linkages within a village are strong, such indirect effects can be very important. Villagewide models that capture the linkages among households are needed to conduct policy analysis in such cases (Taylor and Adelman 1996). SAMs and village SAM multiplier models have been applied, for example to villages in Mexico, Zambia, and India, to examine the village-level implications of relevant policy options and recommend appropriate development strategies (Taylor and Adelman 1996; Holden et al. 1998; Parikh and Thorbecke 1996; Adelman et al. 1988). In this study, we use a village SAM multiplier model to examine the impact of off-farm employment on factor and variable input use, farm production, and farmland production capacity change.

A village SAM multiplier model can be used to analyze the impact of remittances from non-farm employment, self-employment, or migration on agricultural production and input use of household groups. Such changes reflect the income effect of off-farm employment, which is an important element of off-farm employment. However, this approach is not suitable for analyzing the effect of reduced labor availability or reduced consumption of absent household members involved in off-farm employment and migration. All production and consumption relationships in a village SAM multiplier model are linear, and substitution effects (e.g., between labor and other inputs in farm production) are not taken into account. Village SAM multiplier models give insights into linkages between different production sectors and into income/expenditure effects within the village that may arise (e.g., from changes in the renting in and out of production factors). They capture the direct and indirect income and demand effects, but not the local price variations that result from income and demand changes. Multi-market or computable general equilibrium (CGE) models can be used to analyze such local price changes. In this chapter, we focus on the income and expenditure effects of off-farm employment on different household groups within a village.

Three accounts in the village SAM are considered to be exogenous; these are the rest of world (i.e., outside the village), government, savings, and investment. Other accounts are treated as endogenous. The first step is to convert the SAM into a coefficient matrix by dividing each endogenous element in the matrix by its column sum. The resulting coefficient matrix A_n represents the average expenditure propensities of the endogenous accounts.

Fixed price multipliers can be obtained as follows (Parikh and Thorbecke 1996):

$$dy_n = (I-A_n)^{-1}dx = M_A dx$$

In this equation dy_n is the change in production or incomes of village activities (all endogenous accounts), dx represents an exogenous change in the demand for village goods (export) and services (labor exports), A_n is the coefficient matrix of average expenditure propensities, and I and M_A are the identity and the multiplier matrix, respectively. Multiplier analysis shows how the production or incomes of endogenous accounts will be affected by a change in exogenous demand or government investment.

We took 156,680 *yuan*, which equals 5 percent of the village GDP, as the total injection (used as policy simulation) in all simulations. Five scenarios are presented:

1. With rapid economic growth and urban expansion, farmers will get more opportunities to work in the urban sector; laborers with a high education level are most likely to find a job and get high payment. Hence, the first scenario is a 23 percent increase in income from high-educated labor migration to the urban sector.
2. Assuming migration activities will keep the same pattern as before, both low-educated and high-educated labor migration will increase proportionally. Hence, we injected additional income flowing proportionally to low-educated and high-educated labor migration income in the second scenario.

3. Development of the local economy is one of the important ways to promote rural development, because local non-farm activities have strong linkages with farm production. Hence, instead of migration, we injected the income into local nonagricultural employment and self-employment. We assume that only high-educated labor has access to this type of employment in the third scenario.

4. In the fourth scenario, we assume that both high- and low-educated laborers have access to local nonagricultural employment and self-employment. The injection is again distributed proportionally.

5. Poor rural infrastructure is a main cause of poverty in the village economy. Hence, an alternative scenario is for the government to invest 99,100 *yuan* (around 60 percent of the injection) into local nonagricultural employment involved in infrastructure construction; this injection is assumed to be proportionally distributed to high-educated labor and to low-educated labor in scenario 5. In addition, the village road construction resulting from this public investment is assumed to raise the demand for rice, vegetables, perennial crops, and livestock products from outside the village by 7,842; 225; 28,160; 21,339 *yuan*. The total injection is 5 percent of GDP.

Premultiplying these injections with the multiplier matrix, we obtain the total (i.e., direct and indirect) effect of the injections on farm production, agricultural factor and variable input use, factor market participation, and total income.

The simulation results for production of major crops and livestock are shown in table 10.12. The figures in the table show the percentage changes as compared to the base situation represented in the SAM, as is the case with the remaining tables. For most crops, production is expanded most by the household groups that experience the largest direct income gains. Additional income from off-farm activities is mainly spent on food (annual crops). As can be seen from comparing the results for scenarios 1 and 2, income from low-educated migration has a strong positive impact on annual crop production activities and livestock production of group 1. A comparison of scenarios 3 and 4 shows that the effects of increasing low-educated local off-farm employment on annual crop and livestock production are much smaller than those of increasing low-educated migration. Infrastructure investment (scenario 5) raises production of perennial crops and livestock by increasing the external demand for these products. Local off-farm employment (scenarios 3 and 4) benefits especially Group 2 (without oxen), which is less involved in agricultural production and obtains a relatively large share of its income from local nonagricultural employment (table 10.4).

Comparing the general effects of the five scenarios on stimulating agricultural production (group average), infrastructure investment gives the

Table 10.12.   Simulation Results for Farm Production

Farm Production		Scenario 1	Scenario 2	Scenario 3	Scenario 4	Scenario 5
One-season Rice	Group 1	2.50	6.11	2.29	3.41	5.41
	Group 2	4.13	4.59	8.25	7.70	8.93
	Group 3	7.30	6.64	3.74	3.54	4.77
	Group 4	4.97	4.67	6.36	6.50	7.75
	All groups	5.24	5.50	5.16	5.25	6.60
Two-season Rice	Group 1	2.61	6.38	2.39	3.56	5.33
	Group 2	4.77	5.30	9.52	8.89	9.20
	Group 3	7.66	6.97	3.93	3.71	4.65
	Group 4	5.18	4.87	6.63	6.78	7.77
	All groups	5.54	5.82	5.51	5.59	6.59
Other Annual Crop Production	Group 1	2.48	7.42	2.20	3.77	5.30
	Group 2	4.86	5.53	10.69	9.96	10.23
	Group 3	8.91	8.18	4.65	4.39	4.56
	Group 4	6.04	5.86	7.51	7.62	8.04
	All groups	6.29	6.79	6.22	6.32	6.80
Perennial Crop Production	Group 1	0.16	0.59	0.14	0.28	7.03
	Group 2	0.44	0.51	1.01	0.94	7.49
	Group 3	1.58	1.43	0.74	0.70	6.68
	Group 4	0.42	0.39	0.55	0.56	7.27
	All groups	0.76	0.77	0.61	0.61	7.07
Livestock Production	Group 1	1.84	5.70	1.62	2.85	5.62
	Group 2	1.82	2.08	4.15	3.86	8.41
	Group 3	5.56	5.06	2.84	2.67	5.47
	Group 4	4.01	3.77	5.26	5.38	7.90
	All groups	3.91	4.25	3.81	3.94	6.85

Note: Data in table are percentage changes with respect to the base scenario.

best results for stimulating all types of agricultural production (6.6–7.1 percent). Turning to the other four scenarios, the migration of low- and high-educated labor scenario shows a stronger impact on all production activities than the other three scenarios, whereas the local off-farm employment by high-educated labor scenario shows the smallest effects. The latter scenario benefits Group 2 mostly, which is the group with the smallest links with the other groups.

Table 10.13 presents the changes in the use of inputs that affect soil and environmental quality. In contrast to manure production, the use of fertilizer, pesticides, and herbicides is not subdivided by household groups in the village SAM. We therefore present aggregate results for the latter. Because manure is an output from livestock production, the increase of manure production is directly related to the changes in livestock production resulting from increased off-farm incomes.

Table 10.13.   Simulation Results for Input Use

Input Use		Scenario 1	Scenario 2	Scenario 3	Scenario 4	Scenario 5
Manure	Group 1	2.53	6.43	2.30	3.52	5.37
Production	Group 2	4.47	5.01	9.27	8.65	9.42
	Group 3	8.16	7.44	4.23	4.00	4.65
	Group 4	5.49	5.18	6.93	7.06	7.88
	All groups	5.78	6.07	5.68	5.76	6.67
Chemicals	Fertilizer	5.43	5.61	5.29	5.38	6.60
	Pesticides and herbicides	5.50	5.62	5.50	5.56	6.72

Note: Data in table are percentage changes with respect to the base scenario.

At the aggregate level, scenario 5 causes the largest increase of manure production, and also of fertilizer, pesticide, and herbicide use. The other four scenarios show similar tendencies as for agricultural production. The increase in manure production is slightly higher than that of fertilizer, pesticide, and herbicide use in all four scenarios, which seems to indicate that land productivity and environmental quality slightly improve.

It should be noted, however, that these results show the effects of additional incomes earned by off-farm employment. The impact of reduced labor availability is not analyzed in the SAM multiplier model. Because manure application is a relatively labor-intensive activity, the results in table 10.13 are likely to change when changes in opportunity costs of labor are taken into account. This requires, however, a change in the modeling approach from a fix-price to a flex-price model, which is outside of the scope of this chapter.

The simulation results for village factor market participation are shown in table 10.14. One single account in the SAM is used for the agricultural labor hiring activity, so we cannot distinguish between household groups, only between low- and high-educated workers. The infrastructure investment scenario (scenario 5) shows the largest impact on local agricultural labor markets both for low-educated and high-educated labor. All five scenarios show a slightly higher impact on high-educated agricultural employment than on low-educated agricultural employment, except for the low- and high-educated migration scenario (scenario 2).

The infrastructure scenario also shows the largest impact on land renting and oxen renting. We may therefore conclude that factor market development is stimulated mostly by local infrastructure investment. Migration (scenarios 1 and 2) stimulates in particular land renting-in by household Group 3, the group that is most involved in migration (see table 10.4). Nonagricultural wage employment and self-employment (scenarios 3 and 4), on the other hand, stimulates particularly land renting-in by Group 4, one of the two groups that earn most from this type of employment (Group

Table 10.14. Simulation Results for Factor Market Participation

Factor Market	Development Scenario	1	2	3	4	5
Agricultural labor hiring	Agricultural employment by low-educated labor	3.18	3.53	2.56	2.64	5.61
	Agricultural employment by high-educated labor	3.47	3.28	3.77	3.75	6.48
Land renting-in	Group 1	2.51	6.13	2.30	3.43	5.40
	Group 3	7.35	6.69	3.77	3.56	4.76
	Group 4	5.01	4.71	6.42	6.56	7.75
	All groups	5.45	5.67	4.74	4.92	6.25
Land renting-out	Group 2	5.44	5.53	5.24	5.31	6.59
	Absentee landlords	5.44	5.53	5.24	5.31	6.59
Oxen renting-in	Group 1	2.52	6.37	2.29	3.50	5.38
	Group 2	4.33	4.82	8.74	8.16	9.11
	All groups	3.36	5.64	5.30	5.67	7.12
Oxen renting-out	Group 3	4.23	4.92	8.36	7.89	8.89
	Group 4	4.23	4.92	8.36	7.89	8.89

Note: Data in table are percentage changes with respect to the base scenario.

2 does not rent-in land; see table 10.7). Oxen renting-out is stimulated much more by nonagricultural wage employment and self-employment than by migration (see last two rows of table 10.14). The same scenarios (3 and 4) also have a much larger impact on oxen renting in by Group 2, the group that is involved most in nonagricultural wage employment and self-employment (table 10.4).

In scenarios 1 and 3, we simulated the effects of income increases for high-educated labor, which Group 1 does not possess. The changes in participation in land, oxen rental markets, and agricultural labor hiring for Group 1 in these two scenarios therefore present the indirect effects on this group of income increases in the other groups. Both land renting-in and oxen renting-in increase significantly for Group 1, the group that depends most on agriculture for its income (table 10.2).

Table 10.15 shows the simulation results for household income levels of the four groups. As expected (see table 10.4), Group 3 benefits most from migration (scenarios 1 and 2), while Groups 2 and 4 benefit most from non-farm wage employment and self-employment (scenarios 3 and 4). Migration of low- and high-educated labor (scenario 2) has the highest impact on average income, while infrastructure investment (scenario 5) has the smallest impact. The results of scenarios 1 and 3 show that, as a result of village market exchanges, Group 1 benefits around 2 percent in income

Table 10.15. Simulation Result for Household Incomes

Income Scenario	Scenario 1	Scenario 2	Scenario 3	Scenario 4	Scenario 5
Group 1	2.18	8.04	1.84	3.74	5.01
Group 2	4.90	5.62	11.20	10.43	10.49
Group 3	10.13	9.16	4.75	4.47	3.91
Group 4	7.01	6.55	9.15	9.41	8.52
All Groups	7.03	7.49	6.93	7.12	6.78

Note: Data in table are percentage changes with respect to the base scenario.

from the 5 percent income injection that goes to the other three groups. Total income gains in the village are strongest for scenario 3 (increase in local off-farm employment for high-educated labor). The indirect effect on village income under this scenario equals 7.49 – 5.00 = 2.49 percent.

CONCLUSION

In this chapter, we have examined the impact of off-farm employment on farm production, factor market development, and factor use and variable input use in farm production for *Shangzhu* village, a remote village in Northeast *Jiangxi* Province. Four household groups are distinguished within the village, based on the number of educated household members (as a resource for earning off-farm income) and oxen ownership (as a resource for earning agricultural income) as criteria.

More than 30 percent of the households use exchange labor in one-season rice production, while 14 percent use hired labor. The large shares of exchange and hired labor used by households show that seasonal agricultural labor markets exist in the village. Contrary to the other three groups, the group that does not own oxen (Group 2) uses little exchange labor and relatively much hired labor. All household groups except Group 2 are net agricultural labor sellers.

The land rental market is more developed than the agricultural labor market; the percentage of households in the village participating in land rental activities equals 45 percent. Land rental activities take place between households residing in the village. The share of rented irrigated land in the total cultivated irrigated land is 20 percent. Only a few households participate in dry land and forestland renting. Group 2 (no oxen) does not rent land to expand agricultural production. The other three groups, however, rent between 24 and 28 percent of their cultivated land, which is used mainly for growing one-season rice. Only Group 2 is a net "seller (renting-out)" in the land market, the other groups are net "buyer (renting-in)" in the land market.

The oxen rental market is functioning to a certain extent, with 12 percent of the households hiring oxen in one-season rice production. Group 2 (no oxen ownership) is the main group hiring oxen. Out of the three groups owning oxen, the ones with at least one educated member (Groups 3 and 4) are the main suppliers of oxen services in the village.

Contrary to the markets for agricultural labor, land, and oxen services, the market for credit is not limited to the village. More than half of the households in the village receive credit, but only 34 percent of the borrowed amount is obtained from households within the same village. The remaining 66 percent mainly comes from relatives and friends outside the village; banks, credit cooperative agencies, and shops play a negligible role.

We examined the impact of increased incomes from off-farm employment on farm production, factor use, and variable input use, and income change for the four household groups by means of a village SAM multiplier analysis. Five different scenarios are distinguished, with a total injection equal to 5 percent of the village GDP for each scenario. We find that infrastructure investment (scenario 5) gives the best results for stimulating all types of agricultural production. It has, however, the smallest impact on household incomes of all five scenarios. Out of the other four scenarios, the migration of low- and high-educated labor scenario (scenario 2) shows a stronger impact on all production activities than the other three scenarios, while it has the largest impact on household incomes.

Government investment in infrastructure also has the largest impact on local agricultural labor markets (for both low-educated and high-educated labor) as well as on land renting and oxen renting. Factor market development within the village is therefore stimulated most by local infrastructure investment. Migration stimulates in particular land renting-in the household group that is most involved in migration, while local off-farm employment stimulates particularly land renting by one of the two groups that earn most from this type of employment (the other group does not rent-in land). Oxen renting is stimulated much more by local off-farm employment than by migration.

The simulation results further reveal that off-farm employment tends to have a small positive effect on land productivity and environmental quality, because manure application increases more in all scenarios except the infrastructure investment scenario than chemical input use. It should be remembered, however, that manure application is a relatively labor-intensive activity. The impact of reduced labor availability caused by off-farm employment cannot be analyzed with a SAM multiplier model.

The analysis shows that indirect income/expenditure linkages within the village are considerable. Total income gains within the village range from 6.8–7.5 percent, implying that indirect income/expenditure gains amount to 1.8–2.5 percent. Moreover, the household group with no educated household members (Group 1) sees its income increase by around 2 percent when the off-farm incomes of households with educated members increase (scenarios 1 and 3).

Although SAM multiplier models can provide useful insights into linkages between households and into the strength of income/expenditures linkages, they do not take into account potential changes in village prices for production factors and other commodities that are only traded within the village. A village CGE model is needed to incorporate such general equilibrium effects into the analysis.

NOTE

1. The authors would like to thank Arie Kuyvenhoven, Hans Opschoor, Max Spoor, and the participants of the *Nanjing* seminar for their comments on a draft version of this chapter. Special thanks are due to Marijke Kuiper for her helpful comments in the village SAM building. All remaining shortcomings are of the authors. Financial support by the Netherlands Ministry of Development Cooperation (DGIS-SAIL program), the European Union (INCO-DC program), and Nature Science Fund of China (70403007) is gratefully acknowledged.

BIBLIOGRAPHY

Adelman, I., E. Taylor, and S. Vogel. "Life in a Mexican Village: A SAM Perspective." *Journal of Development Studies* 1 (1998): 5–24.

Benjamin, D. and L. Brandt. "Property Rights, Labor Markets, and Efficiency in a Transition Economy: The Case of Rural China." *Working paper*, Department of Economics. Toronto: University of Toronto, 2002.

De Brauw, A. "Migration and Investment in Rural China." Illinois-selected paper at the American Agricultural Economics Association Meeting, August 5–8, Chicago, 2001.

De Brauw, A., J. Huang, S. Rozelle, L. Zhang, and Y. Zhang. "The Evolution of China's Rural Labor Markets during the Reforms." *Journal of Comparative Economics* 30, no. 2 (2002): 329–53.

De Brauw, A. and S. Rozelle. "Household Investment through Migration in Rural China." *Working paper*, University of California at Davis, 2003.

De Brauw, A., J. E. Taylor, and S. Rozelle. "Migration and Source Communities: A New Economics of Migration Perspective from China." *Working paper*, University of California at Davis, 2001.

De Janvry, A., M. Fafchamps, and E. Sadoulet. "Peasant Household Behavior with Missing Markets: Some Paradoxes Explained." *Economic Journal* 101 (1991) 1400–1417.

Deininger, K. and S. Jin. "Land Rental Markets as an Alternative to Government Reallocation? Equity and Efficiency Considerations in the Chinese Land Tenure System." *World Bank Policy Research Working Paper*, no. 2930. Washington, D.C.: World Bank, 2002.

Hoff, K., A. Braverman, and J. E. Stiglitz, eds. *The Economics of Rural Organization. Theory, Practice, and Policy.* Oxford: Oxford University Press, 1993.

Holden, S., J. E. Taylor, and S. Hampton. "Structural Adjustment and Market Imperfections: A Stylized Village Economy-wide Model with Nonseparable Farm Households." *Environment and Development Economics* 4 (1998): 69–87.

Huang, G. "Farmland Resource and Its Sustainable Use in Jiangxi Province." *ACTA Agriculture Jiangxi* 11 (*in Chinese*), 1999.

Huang, J. "Land Degradation in China: Erosion and Salinity." A Report Submitted to World Bank. March, 2000.

Huang, J. and S. Rozelle. "Environmental Stress and Grain Yields in China." *American Journal of Agricultural Economics* 77 (November 1995): 853–64.

Huang, X., N. Heerink, N., R. Ruben, and F. Qu. "Rural Land Markets and Economic Reform in Mainland China." Pp. 95–114 in *Agricultural Markets beyond Liberalization*, edited by A. Tilburg, H. Moll, and A. Kuyvenhoven. Dordrecht and Boston: Kluwer Academic Publishers, 2000.

Kuiper, M., N. Heerink, S. Tan, Q. Ren, and X. Shi. "Report of Village Selection for the Three Village Survey." Report Department of Development Economics. Wageningen: Wageningen University, 2001.

Kuiper, M., X. Shi, and N. Heerink. "Report of the Household Classification." Department of Development Economics. Wageningen: Wageningen University, 2002.

Kung, J. K. S. "Off-Farm Labor Markets and the Emergence of land Rental Market in Rural China." *Journal of Comparative Economics* 30 (2002): 395–414.

Kung, J. K. S. and Y. Lee. "So What If There Is Income Inequality: The Distributive Consequences of Nonfarm Employment in Rural China." *Economic Development and Cultural Change* 50, no. 1 (2001): 19–46.

Lohmar, B., Z. Zhang, and A. Somwaru. "Land Rental Market Development and Agricultural Production in China." Illinois-selected paper at the American Agricultural Economics Association Meeting, August 5–8, Chicago, 2001.

Li, Z. and X. Lin. *Changes of Soil Organic Matter Content and the Way to Improve It in Cultivated Land of Red Soil Region: Research on Red Soil Ecosystem.* Press of Chinese Agricultural Science and Technology, Beijing (*in Chinese*), 1998.

Niu, W. and M. W. Harris. "China: The Forecast of Its Environmental Situation in the 21st Century." *Journal of Environmental Management* 47 (1996): 101–14.

Parikh, A. and E. Thorbecke. "Impact of Rural Industrialization on Village Life and Economy: A Social Accounting Matrix Approach." *Economic Development and Cultural Change* 2 (1996): 351–77.

Parish, W., X. Zhe, and F. Li. "Non-farm Work and Marketization of the Chinese Countryside." *China Quarterly* 143 (1995): 697–30.

Rozelle, S. "Stagnation without Equity: Patterns of Growth and Inequality in China's Rural Economy." *China Journal* 35 (1996): 63–96.

Rozelle, S., J. E. Taylor, and A. de Brauw. "Migration, Remittances, and Productivity in China." *American Economic Review* 89, no. 2 (1999): 287–91.

Sadoulet, E. and A. de Janvry. *Quantitative Development Policy Analysis.* Baltimore: John Hopkins University Press, 1995.

Shi, X., N. Heerink, and F. Qu. "Factors Driving Off-farm Employment Participation Choices—An Empirical Analysis for Jiangxi Province, China." Paper prepared for

International Symposium on "China's Rural Economy after WTO: Problems and Strategies," *Hangzhou*, P. R. China, June 25–27, 2004.

State Environment Protection Administration (SEPA). "China National Planning for Eco-Environment Construction." *China Water Conservation* 2 (1999): 14–18.

Taylor, J. E. and I. Adelman. *Village Economies.* Cambridge: Cambridge University Press, 1996.

Turner, M., L. Brandt, and S. Rozelle. "Local Government Behavior and Property Rights Formation in Rural China." *Working Paper*, Department of Economics, Toronto: University of Toronto, 2001.

Wei, J. "On Implementation of 'Fertile Soil Program' in Red Soil Exploitation." *Journal of Jiangxi Agricultural University* 21, no. 2 (1999) (*in Chinese*).

World Bank. "China Environmental Strategy Paper." Washington, D.C.: World Bank, 1992.

Wu, H. X. and X. Meng. "Do Chinese Farmers Reinvestment in Grain Production?" *China Economic Review* 72 (1997): 123–34.

Wu, H. X. and X. Meng. "The Direct Impact of Relocation of Farm Labor on Chinese Grain Production." *China Economic Review* 72 (1996): 97–104.

Yao, S. "China's Rural Economy in the First Decade of the 21st Century: Problems and Growth Constraints." *China Economic Review* 13 (2002): 354–60.

Yao, Y. "The Development of the Land Lease Market in Rural China." *Land Economics* 76, no. 2 (2000): 252–66.

Zhang, L., S. Rozelle, and J. Huang. "Off-farm Jobs and On-farm Work in Periods of Boom and Bust in Rural China." *Journal of Comparative Economics* 29 (2001): 505–26.

Zhang, L., J. Huang, and S. Rozelle. "Employment, Emerging Labor Markets, and the Role of Education in Rural China." *China Economic Review* 13 (2002): 313–28.

Zhao, Q. "Land Degradation and Control." *China Land Science* 5, no. 2 (1991): 22–25 (*in Chinese*).

Zhao, Y. "The Role of Migrant Networks in Labor Migration: The Case of China." *Working Paper*, no. E2001012, China Center for Economic Research. Beijing: Beijing University, 2001.

11

Land Fragmentation and Smallholder Rice Farm's Production Costs in *Jiangxi* Province, China

Shuhao Tan, Gideon Kruseman, and Nico Heerink[1]

Land fragmentation is the practice of farming a number of spatially separated plots of owned or rented land by the same farmer (McPherson 1982). It is a common phenomenon in many agricultural societies. According to Simmons (1988), 80 percent of agricultural holdings in the world are fragmented. China is one of the countries confronting the most severe land fragmentation in rural areas since the land reform was initiated about two decades ago (Qu et al. 1995; Nguyen et al. 1996; Wan and Cheng 2001; Tan et al. 2006). Under this reform, farmland was allocated equally to individual households on a land per capita and/or per laborer basis, while taking into account differences in soil quality and distance of plots to homestead, so as to provide an adequate base for a family's livelihood. In subsequent years, periodic land readjustments in nearly 80 percent of Chinese villages further reduced the size of each per capita plot share.

The new Land Management Law adopted in 1998, officially limited further land readjustments. However, in a survey among 1,621 households in 17 of China's major agricultural provinces, 64 percent of farmers still expected further readjustments (Ye et al. 2000). Nevertheless, even if recurrent land readjustments can be ended, land fragmentation may be persistent if no public policies are implemented that will reduce it under the existing land tenure arrangements.

Land fragmentation can also have beneficial side effects. For instance, it can facilitate risk management through diversification, such as to spread out fresh food supplies over time by producing a more diversified portfolio of crops (Fenoaltea 1976; Heston and Kumar 1983) and it may enable households to allocate their own labor over the seasons (Bentley 1987). In addition, land fragmentation induced by land reform in some countries has to a great extent realized food security and equity among farm households by distributing land plots in terms of soil quality and family size (Hu 1997; Kompas 2002; Blarel et al. 1992). Land fragmentation, however, is more often believed to be one of the major problems existing in rural land management, especially in developing countries (Wan and Cheng 2001; Rusu 2002; Tan et al. 2004). Many researchers (Jabarin and Epplin 1994; Najafi and Bakhshoodeh 1992; Birgegard 1993; Wan and Cheng 2001; Nguyen et al. 1996) have argued that the adverse effects of land fragmentation overshadow its possible benefits, primarily because it enlarges economic costs and reduces agricultural efficiency. It is further regarded as an obstacle to adopt modern agricultural technologies, to construct or maintain the rural infrastructure and thus to advance agricultural modernization. In sum, according to these studies, the existence of land fragmentation means a loss of efficiency to society.

Despite the fact that it is a widespread and important problem in agriculture, quantitative research on the impact of land fragmentation on production costs is scarce. An exception is the study by Jabarin and Epplin (1994) for northern Jordan.[2] The main finding of this study is that land fragmentation measured by average plot size has a significant, but small, negative impact on production costs. An increase of 1 *dunum*[3] in average plot size will induce a 0.51 JD[4] decrease in production costs measured by JD per ton of wheat. However, this research had some limitations, as it only used average plot size to indicate land fragmentation at farm level. This means that much information about land fragmentation which may also be relevant for production costs, for instance the distance of plot to homestead and the inequality in plot sizes, is lost. Furthermore, it did not see production costs as an outcome of household decision-making. Instead of using a household model, the study simply analyzed the relation between plot size and production costs.

Production cost is not only one of the most important factors affecting competitive advantage of a product, but equally so influencing producers' decision-making (Norton and Alwang 1993). To our knowledge, no research has been conducted to empirically estimate the impact of fragmentation on production costs in China. This chapter intends to fill this gap, by examining the impact of land fragmentation on the variable cost of rice production in *Jiangxi* Province, where rice farming is the most important source of income and of poverty reduction. Results of this research may provide an important input for policy discussions on land fragmentation.

The remainder of this chapter is structured as follows. The next section presents the theoretical farm household model used for our empirical analysis. It is followed by two sections describing the data collection and model specification and the results of the empirical analysis, respectively. The last section presents the conclusions.

THE FARM HOUSEHOLD MODEL

Production cost is an outcome of household's decision-making, which is restricted by the constraints confronting the household. Within the framework of a farm household model the impact of land fragmentation on small rice farms' production costs will be empirically examined.

Farm household theory states that households try to attain their goals and aspirations using their limited resources (including fragmented land resources) to choose between alternative (off- and on-farm) productive activities. The basic agricultural household model[5] consists of a utility function that households try to maximize:

$$\max U = u\,(c,\,l;\,\xi) \tag{1}$$

where c is a vector of consumption goods, l is leisure and ξ is representing a set of household characteristics, like age and education of household head, etc. The utility function is subject to the following budget constraint:

$$p_c c = p_q q\,(L,\,A,\,K,\,B;\,\zeta) - p_b B - w_L(L^{in} - L^{off}) - w_A(A^{in} - A^{out}) - w_k(K^{in} - K^{out}) \tag{2}$$

where p_c and p_q are vectors of prices of commodities that are consumed and produced, and p_b is a vector of variable input prices. L is on-farm labor, including own labor and hired-in labor. A is land, K is capital, and B is a vector of material inputs used in the production. w_L and w_A are prices related to factor payments (labor and land here), whereas w_k is the price of capital. $q(.)$ is a production function, ζ represents the farm's characteristics, including the degree of land fragmentation, L^{in} and L^{off} is hired-in and off-farm labor, A^{in} and A^{out} are rented-in and rented-out land, and K^{in} and K^{out} are rented-in and rented-out capital, respectively. This model is subject to the following resource constraints:

$$
\begin{aligned}
L - L^{in} + L^{off} + l &= L^{TOTAL}\\
A - A^{in} + A^{out} &= A^{con}\\
K - K^{in} + K^{out} &= K^{assets}
\end{aligned} \tag{3}
$$

where L^{TOTAL} is the time endowment of the household. L^{TOTAL} can be allocated to off- and on-farm work as well as leisure. Land area input in the production balances the household contracted land A^{con} and rented-in land, minus rented-out land. Total capital available for production equals own capital endowment K^{assets} plus rented-in capital, minus rented-out capital.

Institutional constraints govern the ability of a household to participate in markets and its access to services and infrastructure:

$$x \le x^{max}\,(\psi,\,\xi,\,\zeta),\ x \in \{L^{in},\,L^{off},\,A^{in},\,A^{out},\,K^{in},\,K^{out},\,B\} \tag{4}$$

where x is a vector of choice variables and x^{max} is a function of household characteristics (ξ), farm characteristics (ζ) and institutional characteristics (ψ), like access to credit, road infrastructure, etc. Reorganizing and substituting equation (3) into equation (2) gives:

$$p_c c = p_q q(L, A, K, B; \zeta) - p_b B - w_L(L + l - L^{TOTAL}) - w_A(A - A^{con}) - w_k(K - K^{assets}) \quad (5)$$

The Lagrangian of this problem can be formulated as:

$$\Im = u(c, l; \xi) - \lambda[\, p_c c - p_q q(L, A, K, B; \zeta) + p_b B + w_L(L + l - L^{TOTAL}) +$$
$$w_A(A - A^{con}) + w_k(K - K^{asset})] - \mu_x(x^{max} - x) \quad (6)$$

where λ is the Lagrange multiplier of the budget constraint and μ_x is the multiplier related to the labor, land, capital, and material inputs. The first order conditions with respect to the choice variables c, l, L, A, K, and B are

$$\frac{\partial \Im}{\partial c} = \frac{\partial u(c, l; \xi)}{\partial c} - \lambda p_c = 0 \quad (7)$$

$$\frac{\partial \Im}{\partial l} = \frac{\partial u(c, l; \xi)}{\partial l} - \lambda w_L = 0 \quad (8)$$

$$\frac{\partial \Im}{\partial L} = -\lambda \left[-p_q \frac{\partial q(A, L, K, B; \zeta)}{\partial L} + w_L \right] = \mu_x \quad (9)$$

$$\frac{\partial \Im}{\partial A} = -\lambda \left[-p_q \frac{\partial q(A, L, K, B; \zeta)}{\partial A} + w_A \right] = \mu_x \quad (10)$$

$$\frac{\partial \Im}{\partial K} = -\lambda \left[-p_q \frac{\partial q(A, L, K, B; \zeta)}{\partial K} + w_x \right] = \mu_k \quad (11)$$

$$\frac{\partial \Im}{\partial B} = -\lambda \left[p_q \frac{\partial q(A, L, K, B; \zeta)}{\partial B} - p_b \right] = \mu_b \quad (12)$$

Solving this model gives a set of reduced-form equations:

$$M = f(w_A, w_k, w_L, p_b, p_c, p_q, \xi, \zeta) \quad (13)$$

where M represents the choice variables c, l, L, A, K, and B.

This assumes that prices are exogenous. However, farm households confront institutional constraints and market imperfections, which may result in the so-called nonseparability of household production and consumption. They therefore face shadow prices and effective prices that are not identical to market prices. We can assume that the prices depend on village location and household-specific variables captured by household, farm, and village characteristics. This can be displayed as:

$$W = w(v, \xi, \zeta) \quad (14)$$
$$P = p(v, \xi, \zeta)$$

In equation (14) W is a vector of prices of land, capital, and labor, P is a vector of input, consumption goods, and output prices, and v is a vector of village-specific variables.

Production costs C are a function of household choice variables, exogenous prices, and farm characteristics.[6]

$$C = (w_L L + w_A A + w_K K + p_b B)/q \; (L, A, K, B; \zeta) \tag{15}$$

Combining equations (13)–(15), the agricultural production cost function can therefore be denoted as:

$$C = g(v, \xi, \zeta) \tag{16}$$

where C represents production costs per ton grain, and $g(.)$ the reduced-form equation.

EMPIRICAL ANALYSIS

This study is conducted under the framework of the Sino-Dutch SERENA project.[7] *Jiangxi* was selected as research area for two reasons that are relevant to this study: (1) rice production plays an important role in its economy; and (2) land fragmentation in this province is more prominent than at the national level.

Jiangxi is an inland province in Southeastern China (see figure 11.1). It is located on the Southern bank of the *Yangtze* River, between 24′29″–30′04″ north and 113′34″–118′28″ east. It covers a total area of about 0.17 million square kilometers. The topography of *Jiangxi* is dominated by mountainous (36 percent) and hilly land (42 percent). The annual average temperature of *Jiangxi* is around 18°C, and annual rainfalls ranges from 1,341–1,940 mm. The total population of *Jiangxi* was 42.22 million in 2002. GDP was 245 billion *yuan* (US$25.59 billion), and per capita net income of rural household was 2,306 *yuan*, which was 7.3 percent lower than the national level. Many rural residents are still living under the poverty line.[8]

As part of the field research undertaken in the framework of the SERENA project, three villages, *Banqiao, Shangzhu*, and *Gangyan* were selected as case studies[9] (see figure 11.1). A summary of the major characteristics of the three selected villages is given in table 11.1.

Banqiao is the smallest one with around 900 persons living in 220 households. It is located in a hilly region, with upland area accounting for 60–70 percent of its total land. Market access is good in this village. It is within 10 km distance of a major city. However, the conditions of roads from its hamlets to the main road are poor. Irrigation conditions are adequate, thus paddy fields can be easily irrigated with water from the reservoir against payment of irrigation fees. In the upland areas, the farming systems are rain-fed.

Figure 11.1.   Location of *Jiangxi* Province and Case Study Villages

Shangzhu is a remote village, as it takes about two hours from the location of its village committee to the capital city of the county. Its sixteen hamlets are scattered in a mountainous area, with some of them even difficult to reach. There are 472 households with 2,028 persons in this village. Its main crops are rice and bamboo, with rice planted on terraces in the valley areas, while bamboo and fir (a kind of cash tree) are grown in the hilly areas. The terraces are well-constructed with stone and are several hundred years old.

Gangyan is the largest village, with 730 households and 3,200 persons. It is located in a plain area, and has a medium distance (about 30 km) from the village to the county capital city. Its main crops are rice and vegetables. Tractors are used in this village.

SAMPLING AND DATA COLLECTION

In the survey around 23 percent of the total households were randomly interviewed in the selected three villages. These samples were proportionately distributed in the hamlets of each village in terms of the total number of included households. The distribution of the sample in each village is shown in the table 11.2.

Table 11.1. Summary Description of the Three Villages

		Banqiao	*Shangzhu*	*Gangyan*
Location	Prefecture	*Yingtan*	*Yingtan*	*Shangrao*
	County	*Yujiang*	*Guixi*	*Yanshan*
	Township	*Honghu*	*Tangwan*	*Wanger*
	Distance	Close to city (10 km)	Remote	20 km from city
Population	Persons	900	2,028	3,200
	Households	220	472	730
	Hamlets	4	16	7
	Village groups	4	8 22	—
Land (*mu*)	Farmland	1,700	2,759	3,880
	Paddyland	1,234	2,359	3,780
	Highland	500	400	100
	Farmland/capita	1.89	1.36	1.21
	Upland/total land	60–70%	97%	"plain"
Agriculture	Main crops	Rice, peanuts, fruit trees	Rice, bamboo, fir	Rice, vegetables
	Green manure planting	Yes, but low production (cattle & soil quality problem)	By more than 80% of households	Depends on hamlets
	Manure	Limited use	Limited use	Limited use
	Technology: seed variety	High-yielding varieties	High-yielding varieties	High-yielding varieties
	Technology: plowing	Animal plow	Animal and human force plow	Animal plows and machines
	Farm infrastructure condition	Good	Rain-fed or irrigated with conserved water	Good
Land tenure	Quality/distance classes	4	3	3–4
	Allocation basis	Family size & labor force	Family size	Family size
	Frequency of adjustment	For some hamlets: never adjusted	Small adjustments	Small: 3–5 years; Large: 5–10 years (depends on hamlet)
	Collective management	—	—	Hamlet management of some forest-land

Source: Based on the fieldwork.

Table 11.2.   Sample Size Per Village

	Number of Households	Share of Households (%)	Total Sample Size	Share of Sample Households (%)
Banqiao	220	15.5	52	23.6
Shangzhu	472	33.2	112	23.7
Gangyan	730	51.3	174	23.8
Total	1,422	100.0	339	—

The resulting 339 households were surveyed for the agricultural season 2000. Information from 2,490 plots was collected. Among the 339 households, 265 planted early rice, 211 planted one-season rice, and 262 planted late rice. In total 323 households cultivated rice. These households provide the sample that is used for our analysis. Data were collected at different levels, with farm characteristics and household characteristics at the household level, land fragmentation indicators at plot level, and planting area, yield, and production costs at crop level (that is, all plots planted with rice).

Production costs cover the cost of labor, seed, chemical fertilizer, herbicides and pesticides, and oxen and tractor. Labor costs include all labor used in activities such as nursery, land preparation, seeding, weeding, fertilizing, transplanting, harvesting (including the transportation from plot to the homestead), and field monitoring. The agricultural labor market in the research area faced major imperfections (Kuiper 2004); the shadow wage rate is therefore assumed to be 25 percent lower than the average market price in rice planting activities (20 *yuan*/day), namely 15 *yuan* per day.

Seed cost includes those for seeds proper and for plastic film (used only in early rice nursery). Two kinds of seeds are commonly used, namely local variety and high-yield-variety (HYV). Local seeds are normally reserved by households after harvesting. Its price can be represented by the output price after the harvest. HYVs are bought in the market. Likewise, fertilizers, herbicides, and pesticides are bought in the market, thus market prices are used in calculating their cost.

Oxen were mainly owned by farms in *Banqiao* and *Shangzhu*. In *Gangyan*, 17 percent of households rented-in oxen from the hamlet or the village. The tractor was not popularly used in rice production; only one household in *Banqiao* rented in a tractor for plowing and 20 households rented-in tractors for plowing in *Gangyan*. Like the shadow wage rate, the estimated shadow price of own oxen is about 25 percent lower than that of rented-in oxen. As tractors were rented-in by 20 of 21 households, we used the average rented-in tractor price as its shadow price for the household that used its own tractor.

The resulting variable cost in rice production and the percentage of each category are calculated in table 11.3. It indicates that labor cost accounts for

Table 11.3.   Structure of Variable Cost of Rice Production (*yuan/ton*)

	Labor	Fertilizer	Seeds	Herbicide and pesticide	Oxen and tractor	Total
Variable costs	727	160	33.6	43.6	104	1,068
Percentage of each category	68.07	14.98	3.15	4.08	9.74	100

Source: Calculated from the survey data.

a large share (68 percent) of the total cost, whereas seeds, and herbicides and pesticides account for very small shares, 3 and 4 percent, respectively. The costs for oxen or tractor account for about 10 percent of total costs and fertilizers about 15 percent. This suggests that in rice production in Southeastern China, labor cost is the largest item in production costs. Other categories are relatively smaller, even though they are important and can to some extent be substituted by labor.[10]

Choice of Variables

Production costs depend on farm size and technology adopted in the production process. Factors that may affect technology used in rice production (such as land fragmentation, farm household characteristics, and village characteristics) are included in this analysis. An appropriate land fragmentation indicator system that can reflect a relatively complete picture of land fragmentation and can be used to derive reasonable policy outcomes is desired. However, many indicators can be used to measure land fragmentation. The most popularly used are one or more of the basic three: the number of plots, plot size, and plot distance. The Simpson index (SI) is also used in some studies (see for example, Blarel et al. 1992).

In this chapter, we use farm size, the SI and plot distance to represent land fragmentation. The *SI* is a general index of land fragmentation. It is defined as:

$$SI = 1 - \sum_{i=1}^{n} a_i^2 / (\sum_{i=1}^{n} a_i)^2 \tag{17}$$

where *n* is the number of plots and a_i is the area of each plot. *SI* is located between zero and one, with a higher value of *SI* indicating a larger degree of land fragmentation. The value of the SI is determined by the number of plots, the average plot size and the plot size distribution. It does not capture farm size and distance to the plots. Farm size is used to capture the scale effect of economies, and the average distance of plots to the homestead captures the spatial distribution of plots within a farm.

Descriptive statistics for the variables used in the empirical models are presented in table 11.4. Of special interest are the dependent variables and the land fragmentation indicators. There is a large spread in total produc-

Table 11.4.   Descriptive Statistics of Variables Used in the Analysis

	Minimum	Maximum	Mean	Std. Deviation
Production costs (PC)				
Total cost (*yuan*/ton) (TC)	326	5,463	1,068	516
Labor cost (*yuan*/ton) (LC)	166	4,716	727	450
Fertilizer cost (*yuan*/ton) (FC)	19.6	507	160	75.6
Seed cost (*yuan*/ton) (SC)	3.78	164	33.6	20.7
Herbicide and pesticide cost				
(*yuan*/ton) (HC)	0.00	180	43.6	23.6
Oxen and tractor cost				
(*yuan*/ton) (OC)	18.2	655	104	66.4
Explanatory variables				
Simpson index (SI)	0.00	0.91	0.73	0.17
Farm size (*mu*) (FS)	1.00	34.2	10.4	5.68
Average distance of plots				
to homestead by walking				
(minutes) (DT)	3.00	61.3	16.1	7.70
Age of household head				
(years) (AH)	23.0	75.0	47.1	10.4
Education years of household				
head (years) (EH)	0.00	13.0	4.79	2.80
Household size (persons) (HS)	1.00	14.0	4.46	1.52
Share of good irrigated				
land (%) (GI)	0.00	1,000	28.7	30.3
Contracted forestland (*mu*) (CF)	0.00	31.0	2.26	3.66
Household owns cattle,				
1 = yes, 0 = otherwise (CO)	0.00	1.00	0.66	0.48
Available savings (*yuan*) (AS)	0.00	40,000	2,587	5,406
Total credit (*yuan*) (TC)	0.00	30,000	1,722	3,810

Source: Calculated from the survey data.
Note: All 322 households planting rice are included in the regression, except one household in *Shangzhu*
   which reported fertilizer expenditures were as high as 6,300 *yuan* per ton grain.

tion costs. It varies from 326 to 5,463 *yuan*/ton across households, with
a mean value of 1,068 *yuan*/ton. Likewise, there are large spreads in the
individual cost categories.

Land fragmentation degree also differs greatly from household to household. The SI varies from 0 to 0.91 (0.73 on average).[11] The average distance of
plots to the homestead varies from three minutes to more than one hour, while
it takes sixteen minutes from a plot to the homestead on average. The farm size
is 10.4 *mu* on average, varying from 1 to 34 *mu*. The other indicators also show
substantial variations, as required for use in the regression analyses.

MODEL SPECIFICATION AND ESTIMATION METHOD

The reduced-form equation (16) gives information on the variables that
should be included in the model, but not on the functional form that

should be used. We therefore tested different functional forms for normality (Jarque-Bera test), misspecification (Ramsey RESTE test) and goodness of fit (R-squared, F-test) in order to select the most appropriate one. The semi-logarithmic functional form passed all the tests, while both the linear and double-logarithmic functional forms failed to do so. See appendix table 11.A.1 for details. We therefore select the following model for the empirical analysis:

$$\ln (PC_i) = \alpha_{0i} + \alpha_{1i}SI + \alpha_{2i}FS + \alpha_{3i}DT + \alpha_{4i}AH + \alpha_{5i}EH + \alpha_{6i}HS + \alpha_{7i}GI$$
$$+ \alpha_{8i}CF + \alpha_{9i}CO + \alpha_{10i}AS + \alpha_{11i}TC + \upsilon_{1i} \qquad (17\text{–}21)$$

with PC_i = TC, LC, FC, SC, HC, and OC, respectively; α_{0i}, ..., α_{11i} are unknown coefficients, and υ_{1i} is a disturbance term with standard properties.

The definitions of the variables are shown in table 11.4. The reduced-form of the farm household production cost function (16) indicates that production costs are a function of village-specific variables, household characteristics (age and education of household head, and household size) and farm characteristics that include land fragmentation, oxen ownership, available savings, and available credit. In our case, we use two village dummies to measure the impact of village-specific factors.

The expected impact of each variable on the total production costs is given in table 11.5. A larger value for the SI is expected to increase production costs, because technologies are more difficult to use on the fragmented plots. An increase in farm size is expected to decrease production costs due

Table 11.5. Anticipated Signs of Variables Included in the Model

Explanatory Variables	Production Cost (yuan/ton)
Land fragmentation indicators	
Simpson index	+
Farm size (mu)	−
Average distance of plots to homestead by walking (minutes)	+
Household characteristics variables	
Age of household head (years)	−/+
Education years of household head (years)	−
Household size (person)	−
Farm characteristics variables	
Share of good irrigated land	−
Contracted forestland (mu)	+
Oxen ownership	−
Available savings (yuan)	−
Available total credit (yuan)	−
Village characteristics variables	
Dummy variable, =1 if household lives in *Shangzhu* village	+
Dummy variable, =1 if household lives in *Gangyan* village	+

to the potential economies of scale effect. Larger average distances to the plots may increase production costs, since greater distance means travel time losses, keeping other factors fixed.

The farm household characteristics may affect decision-making on technology adoption and therefore influence production costs. Older farmers are expected to face higher production costs because they may spend more time to manage the plots compared with their young counterparts. If they are more experienced in managing their fields, however, age is expected to reduce production costs. A higher education of the household head and a larger household size are expected to decrease production costs, because more educated farmers or larger households may adopt better management methods and be more able to manage the crop in a timely manner. Keeping other factors constant, farms with a higher share of well-irrigated land face lower production costs, since such land is more productive. Farms with larger areas of forestland, on the other hand, will generally face higher production costs. This is because, given the farm size, the larger forestland means smaller land for rice production. Oxen ownership can reduce production costs, because their shadow price is lower than that of hired-in oxen. Both available savings and received credit are expected to reduce production costs as they can reduce cash constraints and enable farmers to more efficiently manage their fields.

Village characteristics are represented by two dummy variables. Farmers in *Shangzhu* and *Gangyan* villages are expected to have higher rice production costs because of poorer market access and fewer contacts with agricultural extension agents.

The above-mentioned effects apply to total production costs, but may not apply to individual cost categories. Farmers may switch from one input to another (for example from fertilizer or herbicides to labor) in response to changes in land fragmentation or other explanatory variables. We therefore analyze both the impact on total production costs and on individual cost categories.

EMPIRICAL RESULTS

Equations (17)–(21) are estimated with ordinary least squares. The results for these equations are presented in table 11.6 (for total production costs) and in table 11.7 and Appendix table 11.A.2 (for different cost categories). We will first discuss the results for total production costs.

Total Production Costs

The F-statistic in the total production costs equation is high enough to reject the null hypothesis that the listed variables cannot explain the differences in production costs between farms (see table 11.6). Farm size is found to have a significant negative impact on total production costs. Larger farm

Table 11.6. Regression Results for Total Production Costs

	Coefficient	t-statistic
Land fragmentation		
Simpson index	0.16	1.23
Farm size (*mu*)	–0.015***	–3.60
Average distance (minutes)	0.008***	3.19
Household characteristics		
Age (years)	0.003*	1.65
Education (years)	–0.020**	–2.54
Household size (persons)	–0.007	–0.54
Farm characteristics		
Share of good irrigated land	–0.001	–0.72
Forestland (*mu*)	0.003	0.59
Own oxen	–0.076*	–1.84
Available savings (*yuan*)	–1.86E-6	–0.52
Available credit (*yuan*)	–8.27E-6	–1.52
Village dummies		
Shangzhu village	0.41***	6.98
Gangyan village	0.10*	1.82
Constant	6.65***	37.84
Number of observations	322	

Notes: (a) Dependent variable is in logarithm.
　　　(b) *Significant at 10% level; **Significant at 5% level; and ***Significant at 1% level.

size reduces production costs. A 1 *mu* increase in farm size causes a 1.5 percent decrease in production costs. As expected, the distance of plots to the homestead has a significant positive impact on production costs. One minute's additional travel time to the plots causes a 0.8 percent increase in production costs. The SI, however, is not found to have a statistically significant impact on production costs per ton. A possible explanation is that farms with a higher SI can facilitate the spreading of (natural) risk, labor and crops, which counterbalances the negative impact on management and technology adoption.

In addition, household characteristics are found to have a significant impact on production costs. Age has a weakly significant positive impact on the total production costs, while education of the household head has a negative impact on production costs per ton. The latter finding confirms that farmers with a higher education level are more skillful in producing rice. The size of the household, however, does not affect total production costs.

As expected, if farmers own oxen, the cost in rice production decreases, either because the shadow price of own oxen is lower than that of rented oxen, or because owning oxen permits greater flexibility and efficiency in preparing the land compared to hiring them from others. The other farm characteristics, however, do not have a significant impact on production costs. Production costs in *Shangzhu* are 440 *yuan* (41 percent) higher than in *Banqiao*, keeping other factors fixed. The difference is much smaller (108 *yuan* or 10 percent) between *Gangyan* and *Banqiao*. The main reason for this is probably that farmers in Shangzhu village face higher input prices due to more difficult access to markets than farmers in the other two villages. Moreover, they may have less access to extension services.

Production Costs Categories

The empirical results for total production costs show that farm size and distance, not the SI, significantly affect the production costs of growing rice. The available data allow us to examine which cost categories are most significantly influenced by land fragmentation, and how. The main results are presented in table 11.7; the full regression results are presented in Appendix table 11.A.2. All the F-statistics are high enough to reject the null hypothesis that the variation in each production costs category cannot be explained by the listed variables at a 1 percent testing level.

The results for the SI show a very interesting pattern. It has a significant positive impact on labor cost per ton, but it decreases fertilizer and oxen costs. A 1 percent increase in the SI causes a 0.31 percent increase in labor use, but a 0.25 percent decline in fertilizer and a 0.28 percent decrease in oxen and tractor costs. An increase in the SI therefore induces a shift from fertilizer and oxen and tractor use toward higher labor use. Management of more fragmented plots therefore requires more labor because of inconvenience in the management of fragmented plots or because farmers avoid household labor bottlenecks by spreading peak labor requirements. On the other hand, adoption of modern technologies is less on more fragmented plots. As a result, total production costs are not significantly affected.

Farm size is observed to have a significant negative impact on labor, seed, and herbicide and pesticide costs. A 1 percent increase in farm size will cause a 0.21 percent, 3.59 percent, and 0.15 percent decrease in labor, seed, and herbicide and pesticide costs, respectively. Costs of fertilizer and oxen, however, are not affected. The average distance to the plots is found to have significant positive impacts on all categories except for seed costs.

Table 11.7. Regression Results for Land Fragmentation Indicators for Cost Categories in Rice Production

	Labor	Fertilizer	Seed	Herbicides and Pesticides	Oxen and Tractor
	Estimated coefficients				
Simpson index	0.429***	−0.35**	−12.07**	−0.123	−0.386**
Farm size	−0.02***	0.001	−0.345**	−0.014**	−0.004
Average distance	0.009***	0.008***	0.090	0.007*	0.007**
Observations	322	322	315	319	322

Notes: (a) Except for seed, all the dependent variables are in logarithms.
　　　(b) *Significant at 10% level; **Significant at 5% level; and ***Significant at 1% level.

A 1 percent increase in the average distance gives rise to a 0.14 percent increase in labor, 0.13 percent increase in fertilizer and 0.11 percent increase in herbicides and pesticides and oxen and tractor costs, respectively. The result for labor confirms that plots at larger distances from the homestead require more time for traveling. Fertilizer cost is observed to be significantly higher on farms with larger average distances, because farmers prefer to use more chemical fertilizer to substitute manure, which is much more difficult to transport. More herbicides and pesticides are used on farms with larger average plot distances. A possible explanation is that farmers can spend less time on detecting weed and pest outbreaks on far-away plots and therefore apply more chemicals for prevention. Oxen and tractor is less used on farms with higher land fragmentation degree, but more used on farms with larger average distance of plots.

CONCLUSION

In this chapter a farm household model framework is developed to address the impact of land fragmentation on smallholder rice farm production costs. This is motivated by the fact that under the household responsible system, the small size of farms and their fragmented plots are likely to influence farm production decision-making on technology use and to increase economic costs, thus weakening the competitive advantage of agricultural commodities in the international markets (Cai 2003). The model specification is based on a reduced-form equation derived from the farm household model. It explicitly accounts for the impact of household characteristics, farm characteristics including land fragmentation, and village-specific variables on the total production costs, as well as on individual cost categories.

Estimation results indicate that land fragmentation—as measured by farm size, SI and average distance to plots—has a significant impact on production costs (see table 11.6). Farm size and average distance to plots have negative and positive effects on production costs, respectively. The SI, however, does not show an impact on the total production costs. This is explained through a more detailed examination of the impact of land fragmentation on individual cost categories. A higher SI increases labor costs but reduces fertilizer, seed, and oxen costs. Apparently, farmers with highly fragmented plots switch to more labor-intensive methods and use fewer modern technologies, making the net impact on total production costs not significant. The increase in labor input on fragmented plots may be due to the inconvenience in the management of fragmented plots or to avoidance of household labor bottlenecks through spreading peak labor requirements over the plots. The detailed examination further shows that a larger farm size entails reduced labor, seed, and herbicide and pesticide costs per ton, but does not affect the costs of fertilizers, oxen, and tractors. A larger average plot distance increases all cost categories except seed.

To sum up, fragmented landholdings have higher production costs for rice farms. Keeping other factors constant, an increase in farm size and a reduction of the average distance to plots decrease the total production costs per ton. Changes in the average plot size and its distribution, as measured by the SI, cause a shift in input use but do not affect total production costs.

Stimulating grain production and reducing the rural-urban income gap have become major goals in Chinese policy making in recent years. Adoption of new agricultural technologies plays an important role in this respect. Our results show that the high degree of land fragmentation serves as an important bottleneck in realizing these goals, as it puts an upward pressure on agricultural production costs and hampers the adoption of modern technologies. Elsewhere we have argued that major reforms are needed in the system of land allocation to deal with the problem of land fragmentation (Tan et al. 2006). The current system of land allocation divides land within a village into different land classes (based on soil quality and irrigation and drainage conditions), and gives households equal rights to land in each class.[12] If land would be assigned on the basis of land values instead of physical units, this could substantially reduce the degree of land fragmentation. In addition, provision of tradable land use rights to all farmers, so that they can freely transfer their agricultural land in the market, may help to further reduce land fragmentation and thereby reduce production costs and stimulate the adoption of modern agricultural technologies.

NOTES

1. The authors are grateful to Xiaobo Zhang and the book editors Max Spoor and Nico Heerink for their comments on an earlier version of this chapter, and for the financial support provided by the Netherlands Ministry of Development Cooperation (SAIL program) and the European Union (INCO-DC program).

2. Although there are many cases of land fragmentation in some transitional countries, such as Bulgaria, Vietnam, and Armenia and so on, no research on production costs has been conducted there.

3. Area unit in Middle East; 1 *dunum* = 0.1 ha.

4. Jordanian *dinar*, official currency in Jordan. Official exchange rate in the 1990s is $1 = 0.68–0.71 JD.

5. Our discussion of the household model builds upon Sadoulet and de Janvry (1995).

6. Note that total production cost C is closely related to total factor productivity (TFP), a well-known efficiency measure in (agricultural) economics, which is

defined as:

$$TFP = p_q q/(w_L L + w_A A + w_K K + p_b B); \text{ hence } TFP = p_q/C.$$

7. The project "Strengthening Education and Research on Environmental and Resource Economics at Nanjing Agricultural University (SERENA)," financed by the Netherlands Ministry of Development Cooperation (SAIL program), was a collaboration between the Development Economics Group, Wageningen University, The Netherlands, the College of Land Management, Nanjing Agricultural University (NAU), and the Institute of Social Studies (ISS), The Hague, The Netherlands.

8. By the end of 2000, the population under the poverty line was estimated at about 0.9 million. Their average per capita annual net income was less than 1,300 *yuan* (US$157) in the three years from 1997–1999, for rural residents from 563 townships of *Jiangxi*.

9. The three villages are located in a soil degradation prone area. A stratified sample was used to select the villages, with the strata reflecting differences in market access, economic development level, and geography. Landholdings in all three villages show substantial land fragmentation, while households in all three villages earn a considerable share of their incomes outside their farms.

10. Labor cannot substitute for seeds and pesticides, but can to some extent substitute for fertilizer (to use more manure, which is a more time-consuming activity), herbicides (to weed with the use of labor), oxen, or tractor.

11. Land fragmentation indicators behind the Simpson index show that on average the number of plots per household is 7.4 (varying from 1 to 17), and the average plot size varies from 0.36 to 6.84 *mu* with an average of 1.5 *mu*.

12. The assignment of land to each household is based on the number of household members and/or the number of labor force members in each household.

BIBLIOGRAPHY

Bentley, J. "Economic and Ecological Approaches to Land Fragmentation: In Defense of a Much-Aligned Phenomenon." *Annual Review of Anthropology* 16 (1987): 31–67.

Birgegard, L.-E. "Natural Resource Tenure: A Review of Issues and Experiences With Emphasis on Sub-Saharan Africa." Swedish University of Agricultural Sciences, International Rural Development Centre, 1993.

Blarel, B., P. Hazell, F. Place, and J. Quiggin. "The Economics of Farm Fragmentation: Evidence from Ghana and Rwanda." *World Bank Economic Review* 6 (1992): 233–54.

Cai, X. "Small Scale Management and Transaction Costs." *Current Economic Research* 1 (2003): 54–57.

Fenoaltea, S. "Risk, Transaction Costs, and the Origin of Medieval Agriculture." *Exploration in Economic History* 13 (1976): 129–51.

Heston, A. and D. Kumar. "The Persistence of Land Fragmentation in Peasant Agriculture: An Analysis of South Asian Cases." *Exploration in Economic History* 20 (1983): 199–220.

Hu, W. "Household Land Tenure Reform in China: Its Impact on Farming Land Use and Agro-Environment." *Land Use Policy* 14 (1997): 175–86.

Jabarin, A. S. and F. M. Epplin. "Impact of Land Fragmentation on the Cost of Production Wheat in the Rain-Fed Region of Northern Jordan." *Agricultural Economics* 11 (1994): 191–96.

Kuiper, M. "Moving Together by Moving Away? Village Level Impact of Asymmetric Access to Rural-Urban Migration in China." Paper presented at the EAAE seminar Agricultural Development and Rural Poverty under Globalization, Florence, September 8–11, 2004.

Kompas, T. "Market Reform, Productivity and Efficiency in Vietnamese Rice Production." apseg.anu.edu.au/pdf/apseg_seminar/ap01_tkompas.pdf, 2002.

McPherson, M. "Land Fragmentation: A Selected Literature Review." *Development Discussion Paper*, no. 141. Harvard Institute for International Development, Harvard University, 1982.

Najafi, B. and M. Bakhshoodeh. "The Effects of Land Fragmentation on the Efficiency of Iranian Farmers: A Case Study." *Journal of Agricultural Science and Technology* 1, no. 1 (1992): 15–22.

Nguyen, T., E. Cheng, and C. Findlay. "Land Fragmentation and Farm Productivity in China in the 1990s." *China Economic Review* 7 (1996): 169–80.

Norton, G. W. and J. Alwang. *Introduction to Economics of Agricultural Development.* Toronto: McGraw-Hill, 1993.

Qu, F., N. Heerink, and W. Wang. "Land Administration Reform in China: Its Impact on Land Allocation and Economic Development." *Land Use Policy* 12 (1995): 193–203.

Rusu, M. "Land Fragmentation and Land Consolidation in Romania." www.landentwicklung-muenchen.de/cd_ceec_conference/case_studies, 2002.

Sadoulet, E. and A. de Janvry. *Quantitative Development Policy Analysis.* Baltimore: Johns Hopkins University Press, 1995.

Simmons, S. "Land Fragmentation in Developing Countries: The Optimal Choice and Policy Implications." *Explorations in Economic History* 25 (1988): 254–62.

Tan, S., N. Heerink, A. Kuyvenhoven, and F. Qu. "Impact of Land Fragmentation on Rice Producers' Technical Efficiency in Southeast China." Paper presented on International Symposium on China's Rural Economy: Problems and Strategies. *Hangzhou*, China, 25–27 June, 2004.

Tan, S., N. Heerink, and F., Qu. "Land Fragmentation and Its Driving Forces in China." *Land Use Policy* 23 (2006): 272–85.

Wan, G. and E. Cheng. "Effects of Land Fragmentation and Returns to Scale in the Chinese Farming Sector." *Applied Economics* 33 (2001): 183–94.

Ye, J., L. Prosterman, X. Xu, and X. Yang. "Research on 30 Years Chinese Rural Land User-Rights Investigation—Results and Policy Suggestions from 17 Provinces." *Management World* 2 (2000): 163–72.

12

Intensification of Rice Production and Negative Health Effects for Farmers in the Mekong Delta during Vietnam's Transition

Nguyen Huu Dung and Max Spoor

Rice production has played a significant role in the livelihood of people and the Vietnamese economy for several thousands of years. Currently, rice accounts for 64 percent of the sown crop area, and is grown by more than two-thirds of the Vietnamese households. Comparable to intensive farming with high output levels in other Asian countries, rice production in Vietnam is also characterized by a dependence on agro-chemicals (chemical fertilizers and pesticides). Benefiting from the new institutional environment and market reforms, the growth of rice output has been—on average—more than double the growth of the Vietnamese population since 1989. Yet in spite of this enormous success, there are serious concerns over future rice productivity, and the environmental sustainability of the rice system, due to negative environmental effects of the very intensive use of agro-chemicals (Greenland 1997; Mallon and Irvin 1997; Dung and Dung 1999).

The main objective of this chapter is to examine the response of rice farm households to changes in the economic and institutional environment, with a focus on the usage of agro-chemicals and its impact on farm profitability and farmers' health. The data for analysis has been collected during two farm household surveys in the Mekong Delta (MD) during the 1996–1997 and 2000–2001 dry seasons. The target region is the MD, as it is the biggest rice-growing region in Vietnam, the "rice basket" of the country and the main source of rice exports. In each survey a total of 180 farmers in six villages were interviewed (30 farmers in each site). The two surveys were carried out with the help of a grant and the academic support from the Economy and Environment Program for Southeast Asia (EEPSEA), managed by the Environmental Economics Unit (EEU), Ho Chi Minh City University of Economics. The first farm household survey was held between

January and April 1997, in six subdistricts (or villages) in four provinces of the MD, including *Tien Giang* (*Nhi my*, *Cai Lay* district), *Dong Thap* (*Tan Phu Trung*, *Chau Thanh* district), *An Giang* (*Vinh My*, *Chau Doc* district, and *Long Dien B*, *Cho Moi* district), and *Can Tho* (*Thanh Xuan*, and *Dong Phuoc*, *Chau Thanh* district).

A supplementary field survey was conducted from October 2000 to January 2001, in order to collect additional production data, and look into environmental aspects of agro-chemicals from another (but similar) sample of 180 individual farmers in irrigated rice production systems. The study sites comprised of six villages from four provinces, namely, *Tien Giang* (*Nhi my*, *Cai Lay* district), *Dong Thap* (*Tan Phu Trung*, *Chau Thanh* district, and *Tan Binh*, *Thanh Binh* district), *An Giang* (*Vinh My*, *Chau Doc* district, and *Long Dien B*, *Cho Moi* district), and *Can Tho* (*Thanh Hoa*, *Phung Hiep* district). Of the six villages, four villages were also surveyed in 1997, which enabled us to reduce a large number of questions in the questionnaire, to save time and cost, while being able to make some comparisons with the previous survey.

The 2000–2001 survey was again conducted in cooperation with officials from the local Departments of Extension Services, and of Plant Protection, the People's Committees, the Farmers' Association, and the Rice Research Department of *Can Tho* University. In this survey, 98 households were the same as those interviewed in 1997. However, only 76 households with complete data are available. In the analysis we denote this group as the "same household group" (SHH group) within the total sample.

Multistage sampling techniques were employed in both surveys for choosing study sites and interviewing farmers. The production data in the two surveys concentrated on the Winter-Spring (WS) season, while information of other seasons was also recorded. There are two reasons explaining why the WS season received more attention. First, the rice cultivation calendar is almost similar for all households, and the climate also similar. Hence, the influence of agro-chemicals on yields can be examined through comparing the data from the two surveys. Second, the WS season provides the highest rice yield among growing seasons, higher profits to farm households, and a large part of the production surplus for export.

This chapter will develop its analysis in the following manner. In the next two sections the changes in the economic environment for farm households during the transition will be examined, followed by an analysis of the responses of rice households to this changing environment. Thereafter, the impact of the national integrated pest management programs (IPM) is presented, followed by an analysis of the farmers' health impairments from pesticides' exposure, which has been a direct consequence of the intensification of rice cultivation in the MD. The final section presents some brief conclusions.

CHANGES IN ECONOMIC ENVIRONMENT
FOR FARM HOUSEHOLDS

In Vietnam, economic policy reforms were implemented in 1981, accelerating further since 1986, with the introduction of *Doi Moi*. Reforms in agriculture have caused gradual but persistent changes in markets for rice and inputs, and fundamentally transformed the economic incentives to farm households.

Property Rights' Reform and Market Liberalization

Before the introduction of various stages of land reform, agricultural land, and other means of production were owned, managed, and allocated by agricultural cooperatives, since their foundation in the late 1950s and early 1960s in North Vietnam. Farm households had no economic incentives and no individual rights to access land, except for a small share of land provided as subsidiary plots and home gardens. Since land could be reallocated to other households by the management of the collectives, there was no long-term security of tenure on "assigned land" to individual households, and farm households lacked incentives to invest in order to sustain land productivity. In addition to this insecurity of land tenure, the elimination of private ownership of other production assets (such as tractors or threshing machines) and a limited supply of fertilizers and pesticides led to stagnation in production. This was particularly noticeable in the late 1970s with a substantial food crisis appearing in the north during the 1979–1980 season. This crisis was further inspired by the forced and failed collectivization policies implemented in the Mekong Delta in the south (see also chapter 5).

Following the introduction of the household contract system in 1981 (Directive 100CT, April 1981), farm households were allowed to sign production contracts with the cooperative to produce a certain level of output in return for access to land and agricultural inputs. Though land was still under the management of cooperatives or collective farms, the revenue gained from "above-contract" output, which could be freely sold, provided at least an initial incentive for farmers to increase rice yields. The first concrete manifestation of a new land policy was resolution No. 10 of April 1988, which decentralized responsibility for agricultural management from collectives to farm households. The policy allowed the cooperatives to distribute the land use rights to farm households for a 10–20 years' term, on the basis of a renewable lease.

Farm households in general and rice farm households in particular were made responsible for making the main decisions regarding the allocation and use of resources, production, and the marketing of their products, given the prices of rice and inputs. A more recent land law of 1993 gave greater stability of tenure, making land use rights marketable and decreasing state control over land usage. Long-term security of land tenure and associated rights promoted farm households' efforts to invest in agricultural land.

While other (such as investment) policies have been implemented during the past decade, price and market liberalization were the main elements of the economic reform in the agricultural sector. In the transition from central planning to a market-oriented economy, Vietnam is moving to a system of market allocation of resources, having liberalized prices and markets since 1989. In agriculture, the process of market liberalization has been more extensive and most evident in the rice sector. The prices of rice, inputs, and other resources such as water, fuel, and machinery, became determined by the market, and farm households were allowed to freely sell their products. Private traders could buy rice and other agricultural products directly from the farmers, and the compulsory government purchase of agricultural products was eliminated as a result of resolution No. 10 of 1988. Increased security of land tenure, along with price and market liberation of input and output has offered important incentives for farm households to increase their output.

The initial output and yield improvements were impressive, but this should be seen as a "catching up" effect, starting at relatively low levels. In the first half of the 1980s, yields increased on average by 7.3 percent per annum, while output rose with 6.8 percent per annum (see table 12.1). In the following fifteen years output increases remained at a level between 4–6 percent per annum, while yields rose with an average of 2.7–3.1 percent per annum. In the most recent period the output increases are flattening out somewhat (2.3 percent per annum for the period 2000–2004), while yields still rose by 3.1 percent per annum (see FAOSTAT 2005 database).

Agro-chemical Markets

The market for fertilizers in Vietnam largely depends on imports since domestic production of chemical fertilizers is far less than domestic demand. In the year 2000, domestic demand of fertilizers was 2.28 million

Table 12.1. Annual Rice Yield and Production Growth in Vietnam, and Other Selected Countries

	Yield Growth (%)				Production Growth (%)			
Countries	1980–1985	1985–1990	1990–1995	1995–2000	1980–1985	1985–1990	1990–1995	1995–2000
Indonesia	3.9	1.8	0.2	0.4	6.3	3.1	2.0	0.5
Myanmar	2.1	–0.9	0.3	2.4	1.5	–0.5	5.7	2.3
Philippines	3.4	3.0	–1.2	2.9	3.0	2.5	1.3	3.7
Thailand	1.8	–1.0	4.7	–0.7	3.3	–3.0	5.6	1.3
Vietnam	6.8	2.9	3.1	2.7	7.3	4.2	6.0	5.6

Source: Calculated based on data from FAOSTAT database (2001).

tons, of which 2.06 million tons, or 90 percent of total consumption, was imported (table 12.2). Before the market reforms of 1989, the State Trading Cooperation distributed the imported fertilizers into the trade system, from provincial to district level, and from district level to the cooperatives. Fertilizers were distributed according to planned quantities, which were verified by the People Committees at local levels. Then, based on the fertilizers available, the cooperatives decided how much fertilizer should be applied on the fields and redistributed to farm households. While fertilizer quantities were variable, prices of chemical fertilizers were determined by the State Pricing Committee, and not likely to be changed under specific circumstances such as high demand in a specific location. Prices were subsidized by the State in support of agricultural (and food) production, especially to promote food self-sufficiency. Thus, before 1989 fertilizers were not distributed through the market, but by an administrative allocation system. In the late 1980s, the system of imports and distribution increasingly showed to be inappropriate to meet domestic demand. Until then, fertilizers had to be imported both from nonsocialist countries and under trade agreements with the CMEA (the Soviet dominated Council for Mutual Economic Assistance). Since 1991, the latter source disappeared. In order to meet domestic demand, fertilizers were imported from all possible suppliers. Both central and local agricultural trading companies were allowed to import and market fertilizers. However, with new private trading companies having less experience in the world fertilizer market and lacking sufficient foreign currency, and with the resulting unbalanced imports between agricultural regions, the Vietnamese government still had an important role in fertilizer markets. For example, the import quantity was controlled by quota although fertilizers were freely sold in the domestic market and prices were not controlled. Domestic prices of fertilizers were influenced and varied seasonally and geographically according to the performance of trading and distribution agencies both at provincial and national levels.

From 1996, the Ministry of Agriculture and Rural Development (MARD) and the National Chamber of Commerce were assigned to be responsible for estimating domestic demand and controlling import quota of chemical fertilizers. Some private companies were allowed to import fertilizers, subject to fulfillment of certain criteria. The fertilizer import restrictions were finally abolished in May 2001, following the Decree No. 46/2001/QD-TTg on "Vietnam's Export-Import Management Mechanism for 2001–2005." This Decree eliminated both the rice export quota and the fertilizer import quota. Both State and private companies/enterprises holding a license to trade food or agricultural commodities could participate in rice export and fertilizer imports. In addition, the direct nomination of importers and exporters of these products was also brought to an end. That is to say, the markets for rice and fertilizers were fully liberalized from May 2001 onward.

Table 12.2. Fertilizer Consumption and Imports (1992–2002)

Types of Fertilizers	Year										
	1992	1993	1994	1995	1996	1997	1998	1999	2000	2001	2002
Fertilizer consumption (× 1,000 tons of nutrients)											
Nitrogen	541.3	565	874.9	813.7	995.3	922.9	1,186.1	1,224.2	1,332	1,136	1,305.4
Urea	448.8	508.8	754.2	674.8	818.3	725.8	925.8	892.4	1,098.5	784	887
Phosphate	183.5	165.3	241.6	322	380.2	386.8	399.8	456.4	501	492	532
Potash	41.6	23.8	68.4	88	109	162	271	377	450	400	444
Total	766.4	754.1	1,184.9	1,223.7	1,484.5	1,471.7	1,856.9	2,057.6	2,283	2,027.8	2,281.4
Fertilizer import (× 1,000 tons of nutrients)											
Nitrogen	484.3	504.7	806.8	734.1	900.8	805.5	1,076.3	993.2	1,293	1,020.2	1,132.6
Phosphate	84.9	50.8	97.2	159	201.4	192.6	160.9	165.9	314	790	—
Potash	41.6	23.8	68.4	88	109	162	271	377	452.9	294.6	342.4
Urea	410.8	462.8	709.7	623.8	762.6	666	896.4	869.4	1,036	337.3	406
Total	610.8	579.3	972.4	981.1	1,211.2	1,160.1	1,508.2	1,436.1	2,059.9	1,652.1	1,881.0

Source: www.fertilizer.org/IFA/statistics (2005), and FAOSTAT Statistical Database (2005).

Like with fertilizers, the public sector had a monopoly in the pesticides market until 1988. There had not been any policy on pesticides regulation, except for some plant protection techniques. Before 1988, domestic prices and costs of pesticides were administratively fixed by the government, through the official buying and selling cooperatives. The market of pesticides in Vietnam only started to develop in 1989 following the market reforms for inputs and outputs that were announced in November 1988. From 1991, markets of pesticides were widely spread at the national scale. Pesticides became well available and there was a strong competition in the market, with prices not fluctuating much (in comparison with the case of fertilizers). However, the difference with the market of fertilizers is that pesticides are imported by both private and State companies, while import quota are absent. What is clear from table 12.3 is that the consumption of pesticides grew rapidly in the 1990s, from a level of 20,300 Mt in 1991 to 32,700 Mt in 1996. It has been declining since then, possibly under the influence of IPM programs (see below).

PRODUCTION RESPONSES BY RICE FARM HOUSEHOLDS

Benefiting from the new institutional environment established since the mid-1980s, the rice sector has made considerable strides in the last decade. From 1986 to 2000, rice production increased from about 16 million metric tons to over 32 million metric tons, and yields increased from 2.81 to 4.18 tons per hectare. The growth in rice production has resulted from a substantial improvement of agricultural techniques and an increase in cropping intensity (i.e., rice-sown area). The intensification in rice production has resulted in food self-sufficiency and having substantial surplus for export since 1989, making Vietnam into one of the worlds' most important rice exporters. Intensification in rice production has taken several forms, which are discussed in the following section, using panel data from the two household surveys in the 1996–1997 and 2000–2001 seasons.

Agricultural Land and Rice Production

Agricultural land in Vietnamese statistics includes land for annual crops, perennial crops, and for aquaculture. Distribution of agricultural and rice lands in the study sites are presented in table 12.4. In general, the land for agriculture per household is quite small for all farm households and across villages, with an average of around 1 hectare. Average landholdings have increased between the two surveys from 0.96 to 1.22 ha, which indicates a gradual land consolidation process in the survey areas.

Table 12.3. Pesticides Use (1991–2001)

							Year				
	1991	1992	1993	1994	1995	1996	1997	1998	1999	2000	2001
Pesticides consumption (1,000 Mt)											
Total	20.3	23.1	24.8	20.4	25.6	32.7	30.4	20.5	33.7	16.3	19.0
Insecticides	16.9	18.0	18.0	15.2	16.5	17.4	17.8	9.7	8.6	8.4	9.7
Fungicides and Bactericides	2.6	2.5	3.6	3.3	3.4	8.2	7.7	5.5	4.5	4.6	5.4
Herbicides	0.8	3.7	3.7	2.8	4.9	7.2	6.7	5.3	3.4	3.3	3.9
Pesticides used per hectare (kg a.i./ha)	0.67	0.77	0.82	0.68	0.85	1.08	1.01	—	—	—	—

Source: Data for 1991–1997 from Plant Protection Department, Government of Vietnam; data for 1998–2001 from FAOSTAT (2005).
Note: a.i. = Active ingredients.

Given the small average land acreage, it is even more important to know how households allocate their land to production. Almost 94 percent of agricultural land in all villages is primarily devoted to rice production, and this figure has remained the same between 1997 and 2001. Households usually dedicate the remaining land for home gardens or small ponds for fishery.[1] But some households with rather large agricultural landholdings use some of this land for perennial crops. Nevertheless, most agricultural land is for rice production, with increasing crop intensity. In general, rice land in the study sites has favorable conditions for crop cultivation according to their land classification.[2] Around 78 percent belongs to land classes 1 and 2, and only 9.4 percent belongs to classes 4 and 5 together.

Increasing Cropping Intensity and Rice Monoculture

Cropping intensity has increased in all regions of the country, especially in the MD, and contributed 38.4 percent to the growth of rice production during the period 1985–1995. There is a substantial increase in cropping intensity in the study area from the 1996–1997 to the 2000–2001 season. The growth in intensity was driven by increased cultivation during the dry season, made possible by the improved availability of water through irrigation. This implies that increased cropping intensity has led to intensive farming of rice in areas most suitable to it and thereby leaving some land formerly devoted to rice to other crops. Farm households cultivate rice from two to three seasons or even three-and-a-half seasons a year. As can be seen in table 12.4, triple rice crops were practiced more and more from 1996–1997 to 2000–2001. Nearly 60 percent of farm households growing three rice crops, shows that they are implementing an intensive rice monoculture system.

Rice is the main crop cultivated in the study sites during the survey periods, though some other crops were grown in rotation with rice. As reported by most farmers, there are reasons explaining why the rotation of other

Table 12.4. General Characteristics of the Study Sites

Characteristics	1996–1997		2000–2001	
	Mean	Range	Mean	Range
Agricultural land per household (hectare)	0.96	0.2–4.5	1.22	0.2–3.9
Rice land per household (hectare)	0.88	0.2–4.3	1.04	0.2–3.6
Rice land/Agric. land ratio (%)	93.76	50.0–100.0	93.55	41.7–100.0
Double rice crops (% of households)	70.60	—	42.10	—
Triple rice crops (% of households)	29.40	—	57.90	—
Household size (persons)	5.84	1.0–14.0	5.25	2.0–13.0
Household labor (persons)	3.26	1.0–9.0	2.81	1.0–8.0
Years of rice farming (years)	23.20	3.0–55.0	22.50	2.0–56.0

Source: surveys of 1996–1997 and 2000–2001.

crops in rice fields is seldom practiced. Among these, the most significant factors are unfavorable field conditions, fewer markets for alternative crops, the availability of rice markets, and the family tradition. However, as we will see, in intensive rice monoculture systems, there are serious concerns about the limits for yield increases and the environmental consequences of intensive use of agro-chemicals.

High-Yielding Varieties (HYVs)

The spread of HYVs has contributed significantly to the increase in rice production during the 1980s and 1990s. With the increasing availability of new and further improved rice varieties in recent years, farmers soon recognized the value attached to seed in terms of yield, pest resistance, and short growth duration. In both surveys, the HYVs were omnipresent in all study sites (actually all farm households grew HYVs). There are three factors that seem to account for the wide adoption of HYV in the study sites: (1) the availablity of HYVs suitable for a particular area; (2) the strategy of farmers to maximize annual crop production or net economic gains per unit area of land; and (3) the relatively low yields of traditional rice varieties. Taken together, these factors reflect both farm households' behavior and the influence of rice research and extension.

By cultivating these HYVs, farmers in well-irrigated villages could grow up to three crops a year. These rice crops with short gestation periods provide larger total yields than for example double crops in which in one season a HYV is used and the other a local rice variety. The traditional rice varieties are not suitable for two or more rice crops a year in the study sites. However, the HVYs grown in the study sites are not hybrid seeds, they are inbred rice varieties, which can be kept and planted in following seasons.

Chemical Fertilizers and Rice

Intensification in rice production also took place in the form of greater use of chemical fertilizers and pesticides per hectare. The introduction of HYVs was accompanied by more dependence on agro-chemicals. This is understandable since the Green Revolution technology is generally considered to include chemical control for pests, and fertilizers for plant nutrients (Greenland 1997: 17).

At the farm level, primary nutrients, such as nitrogen, phosphorus, and potassium (N, P_2O_5, and K_2O) play a vital role in rice growth. They are used in large amounts and supplied under numerous forms of chemical fertilizers, while imposing significant costs to farmers and the environment. Technically, the amount of fertilizers to be applied depends on many factors, such as rice varieties, growing seasons, soil, and water conditions. Economically speaking, the market price of fertilizers is an important factor.[3] Farm households have to consider all these factors and eventually decide the final amount of fertilizers applied in specific fields. All farm households used chemical fertilizers as the main source of nutrient supply to the rice crop. No organic matter and manure, except rice straw, is applied to the fields. Nitrogenous fertilizers account for the largest proportion in fertilizer composition. All farm households follow a split application of fertilizers, which enhances their technical efficiency.

There is actually a reduction of fertilizers used per crop per hectare in the four-year period between the two surveys. The mean value of the fertilization rate was 6.86 kg of N, P, K nutrients lower in the dry season 2000–2001, compared to that of 1996–1997 (table 12.5). Farmers have tended to increase the amount of potassium fertilizers, but reduced the use of nitrogenous and phosphorous components. All these changes, except the change in total fertilizer use, are significantly different between the two surveys. The improved balance of N, P, K nutrients could provide better yield and cost saving.

In order to understand production decisions by farmers in terms of fertilizer use, the farm-gate prices of all fertilizers, and relative prices of fertilizers and rice are important. The expected economic returns to fertilizer use normally increase when there is a decrease in this relative price ratio. In such situations we would expect farm households used more fertilizers than if the relative prices had remained the same. Table 12.6 reports the farm-gate prices of N, P, K fertilizers and their relative prices with regard to rice, experienced by households surveyed in 1996–1997 and 2000–2001.

For all households in the SHH, fertilizer relative to rice prices were slightly lower in the 2000–2001 WS season. In 2000–2001 a kilogram of N, P, K nutrients "cost" about 4.27 kg of rice, while that in 1996–1997 was equivalent to 4.40 kg of rice. Though the relative price of fertilizers to rice shows that farm households bought fertilizers cheaper in WS season 2000–2001, this was not homogenous across villages. Moreover, there are many factors causing the variations in prices between households, such as the composition of nutrients, types of fertilizers (e.g., straight or compound fertilizers), terms of payment, and purchasing places.

Table 12.5.   Changes in Fertilizers Used per Crop per ha in Rice Production

| Input quantity | Unit | All Households | | | |
		1996–1997	2000–2001	Mean Difference	t-ratio
Fertilizer	kg of N, P, K	178.5	171.6	−6.86	−1.48[NS]
Nitrogen	kg of N	111.73	99.48	−12.25	−4.55***
Phosphorus	kg of P_2O_5	48.45	43.01	−5.43	−2.31**
Potassium	kg of K_2O	18.29	29.10	10.80	5.97***

Source: Surveys of 1996–1997 and 2000–2001.

Note: *Significant at 10%; **Significant at 5%; and ***Significant at 1% respectively; NS: not significant.

Table 12.6. Farm-Gate Current Prices of Rice and Fertilizers of the Surveyed Households

Season N = 76 (SHH)	Price of Fertilizers (VND/kg of N, P, K)	Rice Price (VND/kg)	Price Ratio of Fertilizers and Rice
1996–1997	5,601.7	1,279.3	4.40
2000–2001	5,562.1	1,313.6	4.27

Source: Surveys of 1996–1997 and 2000–2001.

In the survey areas, an IPM program was introduced already before 1997 in order to diminish the overuse of agro-chemicals, and convince the farmers to use more environmental (and health) friendly methods of plague management. IPM farmers were trained not only in pest control, but also how to manage rice crops in a more sustainable manner through better nutrient management. IPM can be best described as a key component of integrated farming practices that are based on farmers' knowledge and use of pest control measures and skills, taking into account dynamics of pest population, the rice yield, profits from production, safety, and risks to the environment.

Under this assumption, farmers are categorized into two groups: IPM and non-IPM farmers in the subsequent analysis, in order to examine the consequences of agro-chemicals used in rice production. It was expected that IPM farmers would make different decisions on the amount used and timing of fertilizer application. However, the amount of N, P, K nutrients and the total amount of fertilizers used by IPM farmers and by non-IPM farmers did not show a statistically significant difference.

Total Fertilizers Applied per Hectare per Year

Despite the fact that the fertilization rate per crop per hectare in 2000–2001 was slightly lower than that in 1996–1997, the total fertilizers applied per hectare per year increased during the four-year period (table 12.7). The total volume of fertilizers used in rice production per hectare per year is estimated to be 321 kg of N, P, K nutrients for double rice crops, and 556 kg for triple rice crops in 2000–2001. Hence, the overall increase in fertilizer use is due to the increased cropping intensity. Cropping intensity in the survey sites increased from 2.23 to 2.57 for all households.

Table 12.7. Fertilizers Used in Rice Production per Hectare (kg of Nutrients) and Cropping Intensity

Categories	Cropping Intensity	Double Rice	Triple Rice	Average/Year
1996–1997	2.23	344.47	554.70	399.10
2000–2001	2.57	321.34	556.00	452.91

Source: Surveys of 1996–1997 and 2000–2001.

THE IMPACT OF NATIONAL IPM PROGRAMS

Pest management in rice fields is inseparable from sound farm management. In crop fields, pest management practices comprise of farmers' activities to prevent the rice crops being infected and to control for all harmful organisms from seedlings until harvesting. In traditional rice systems farmers had a number of practices that promoted healthy plants and kept pest damage low without using pesticides. Switching from traditional rice varieties to HYVs, farmers in the study sites have made available a range of pest management practices, from natural controls based on cultural, physical, or mechanical techniques to the use of biological control agents and chemical pesticides. During the surveys, however, chemical pesticides were the main technique to control pests applied by all farmers. IPM was introduced to farmers in some of the study villages very soon after the national IPM program started in Vietnam (Dung 1994). The proportion of farmers adopting IPM to control rice pests has increased through time. By the year 2001, 70 percent of farmers practiced IPM on their fields, which is more than double compared to that practice IPM in 1996–1997 (table 12.8).

The farm-gate prices of pesticides again were different among farm households, depending on the composition (herbicides, fungicides, and pesticides), but also depending on the distance to markets and possibly personal connections and knowledge (major elements of transaction costs). For all households in the SHH, the relative price of pesticides to rice was higher in 2000–2001 than in 1996–1997. In 2000–2001 a pesticide active ingredient costs about 0.36 kg of rice, while in 1996–1997 it was equivalent to 0.30 kg of rice (table 12.9). However, the increase is not homogenous among farm households.

Table 12.8.   Adoption of Integrated Pest Management (IPM)

IPM-adoption	1996–1997		2000–2001	
	Number	Percentage	Number	Percentage
Non-IPM farmers	119	67.2	47	29.6
IPM farmers	58	32.8	112	70.4
Total	177	100.0	159	100.0

Source: Surveys of 1996–1997 and 2000–2001.

Table 12.9.   Farm-Gate Current Prices of Rice and Pesticides of the Surveyed Households

Season N = 76 (SHH)	Price of Pesticides (VND/gram a.i.)	Rice Price (VND/kg)	Price Ratio of Pesticides and Rice
1996–1997	380.05	1,279.3	0.30
2000–2001	464.26	1,313.6	0.36

Source: Surveys of 1996–1997 and 2000–2001.

Pesticides: The Main Instrument to Control Rice Pests

Pesticides of various kinds have been used on a large scale as early as the introduction of HYVs to the MD in general, and the study sites in particular. Farmers typically apply insecticides, fungicides, and herbicides during the course of the growing season. All farmers use pesticides to control rice pests, including several products that have either been banned or restricted because of their toxicity. In terms of pesticide toxicity, farmers have used relatively more pesticides in categories I and II[4] than they did in the dry season 1996–1997, accounting for more than 50 percent of total pesticides used in the study sites. Of the 891 types of pesticides allowed to be traded in the market in 2000, 75 and 96 brand name pesticides were used by rice farmers in the dry seasons of 1996–1997 and 2000–2001, respectively.

Herbicides are vital inputs in rice production in the MD. Given the current direct seeding techniques in rice farming, using herbicides is almost a must for farmers to eradicate weeds at the very early stage of crop growth. Less than 5 percent of farmers did not use herbicides to control weeds at the beginning of the growing seasons, because their fields were flooded some months before and weeds were destroyed. Most insecticides used were in categories I and II, which are classified as extremely and moderately hazardous, respectively. However, there was a significant decrease in the use of restricted pesticides[5], which are belonging to these categories (table 12.10). This change can be explained by the regulatory policies in pesticide market and the expansion of the IPM program in rice production.

Another major group of pesticides that farmers applied to control rice diseases was fungicides. Similar to insecticides, there is an increase in the number of fungicide types used by rice farmers, from twenty-eight types in 1996–1997 to forty-one types in 2000–2001.

Despite the widespread use of pesticides, the per hectare consumption of pesticides in rice production per crop per hectare has declined during the four-year period in all study villages. Pesticide use by active ingredient is

Table 12.10. Trend in Restricted Insecticide Use for Surveyed Households

Pesticides	WHO Category	Dry Season 1996–1997		Dry Season 2000–2001	
		% Farmers	Gram a.i/ha	% Farmers	Gram a.i./ha
Methyl Parathion	Ia	4.5	180	2.0	165
Metaphos	Ia	—	—	—	—
Azodrin	Ib	5.6	317.5	—	—
Monitor	Ib	17.4	424	—	—
Thiodan	II	2.8	29.8	—	—
Furadan	Ib	2.8	350	3.0	300

Source: Surveys of 1996–1997 and 2000–2001.
Note: a.i. = Active ingredients.

Table 12.11. Pesticide Used per Crop per Hectare in Rice Production

	Non-IPM Farmers	IPM Farmers	t-ratio	All Farmers
	1996–1997 Survey (N = 177)			
Pesticide use (grams a.i.)				
Total	1,035.69	898.86	–1.95	990.85
Herbicide	336.42	266.91	–1.53	313.64
Fungicide	275.99	347.25	1.60	299.34
Insecticide	423.29	284.71	–2.50	377.88
Category I & II	453.17	387.14	–1.27	431.53
Category III& IV	582.57	511.62	–1.29	559.32
	2000–2001 Survey (N = 159)			
Pesticide use (grams a.i.)				
Total	756.64	748.53	–0.16	750.93
Herbicide	192.78	187.78	–0.18	189.26
Fungicide	219.17	296.66	2.05	273.76
Insecticide	344.69	264.10	–1.69	287.92
Category I & II	287.64	296.24	0.22	293.70
Category III & IV	469.01	452.29	–0.34	457.23

Source: Surveys of 1996–1997 and 2000–2001.
Note: a.i. = Active ingredients.

shown in table 12.11. Taking all three types of pesticides together, the average of pesticides used per hectare per crop fell from 990.85 grams a.i. in 1996–1997 to 750.93 grams a.i. in the 2000–2001 dry season. In other words, per crop per hectare consumption of pesticides has reduced by approximately 25 percent for all survey farm households. Pesticide doses used by IPM farmers in the two surveys went down relatively less than those reported by the non-IPM farmers. The comparison is, however, complicated by the fact that the groups are rapidly changing (with a strong increase in IPM farmers).

Though the pesticide rate per crop per hectare in 2000–2001 is substantially lower than that in 1996–1997, the total amount used per hectare per year is only 7 percent lower during the four-year period (table 12.12). Thus, the increase in cropping intensity lessens the positive effects of diminishing use of pesticides per crop per hectare. Furthermore, it should be noted that the amounts of pesticides used in triple rice crops were higher in the 2000–2001 survey than in the 1996–1997 survey.

Table 12.12. Pesticides Used in Rice Production per Hectare (grams a.i.) and Cropping Intensity

Categories	Cropping Intensity	Double Rice	Triple Rice	Average/Year
1996–1997	2.23	2,066.73	2,288.63	2,124.40
2000–2001	2.57	1,420.23	2,380.06	1,975.60

Source: Surveys of 1996–1997 and 2000–2001.
Note: a.i. = Active ingredients.

Cost and Benefit of Rice Production

In economic terms, the production performance of IPM farmers was much better than that of non-IPM farmers in both surveys, as shown in table 12.13. Gross return of the former was significantly lower than that of the latter. The most significant issue is that the IPM program successfully helped farmers to increase their family income. Thus, the relative family incomes to rice prices show that farm households attained higher income from rice production in the WS season 2000–2001 than in the season 1996–1997 (t-ratio = 2.05). Rice yields and household incomes from rice increased during the four-year period, reflecting better economic performance of rice production of these farm households.

Farmers' Health Impairments from Pesticide Exposures

Despite the success of rice farm households in terms of productivity and income from rice, intensification of rice production that relies heavily on HYVs

Table 12.13. Cost and Benefit of Rice Production in the Mekong Delta (VND/ha), Dry Season, All Farmers

Item	Non-IPM Farmers	IPM Farmers	t-ratio	All Farmers
	1996–1997 Survey (N = 177)			
Gross return	7,483,273	8,191,798	4.07***	7,715,445
Total cost	4,427,851	4,526,957	0.91^{NS}	4,460,327
Benefit	3,055,422	3,664,841	3.38***	3,255,119
Household income	3,887,123	4,618,096	4.16***	4,126,651
Relative household income	3,028.45	3,596.06	4.66***	3,214.45
Benefit/cost ratio	0.72	0.84	2.35**	0.76
	2000–2001 Survey (N = 159)			
Gross return	7,874,794	8,253,739	1.81*	8,141,724
Total cost	4,573,206	4,556,555	–0.14^{NS}	4,561,477
Benefit	3,301,587	3,697,184	1.95*	3,580,247
Household income	4,176,433	4,591,856	2.08**	4,469,058
Relative household income	3,185.45	3,473.11	2.25**	3,388.08
Benefit/cost ratio	0.74	0.84	1.78*	0.81

Source: Calculated from 1996–1997 and 2000–2001 surveys.
Notes: Costs and benefits for the 2000–2001 survey are expressed in constant prices of 1996 (using the consumer price index).
Gross return = Yield in kg × rice price per kg .
Total cost = Costs of pesticides, fertilizers, seeds, costs of laborers (hired laborers + the imputed value of family labor) + other costs (land preparation, irrigation fee or fuel for running irrigation pumps, and land tax).
Benefit = Gross return – total cost.
Household income = Gross return – total cost + imputed value of family labor.
Relative household income = Household income ÷ rice price per kg.
*Statistically significant at 10%; **Statistically significant at 5%; and ***Statistically.- significant at 1% respectively; NS: not significant.

and agro-chemicals has caused spillover effects to water quality and human health impairments. These effects pose a significant cost to society and to rice farmers, severely affecting the long-term sustainability of rice production. In the following part, we examine these health effects and farmers' health cost due to pesticide exposure. Results of the 1996–1997 survey revealed that 69.7 percent of the farmers in the study sites were "rather sure" that they showed acute poisoning symptoms from pesticide exposure while 16.8 percent were "completely sure" (Dung and Dung 1999: table 13). The survey data also showed that each person can simultaneously get more than one acute poisoning symptom. Among poisoning symptoms caused by exposure, the impact of chemical pesticides on eye, neurological system (headache, dizziness), and skin effects are most discernible to farmers, as seen in table 12.14.

Eye Effects

Table 12.14 presents determinants of farmers' health impairments. In the five senses of human beings, the eye provides the most help to people in

Table 12.14.   Logit Regression Results for Health Impairments of Rice Farmers

Variables	Eye Effect	Headache	Skin Effect
Constant	–1.74*	0.33	–0.37
	(0.98)	(1.93)	(0.68)
Age (years)	0.0033	0.025*	–0.012***
	(0.0079)	(0.014)	(0.0058)
Smoking (dummy)	—	0.13	—
	—	(0.44)	—
Drinking alcohol (dummy)	0.73***	1.25***	0.30**
	(0.23)	(0.43)	(0.17)
Weight/height	–0.056**	–0.095*	–0.036***
	(0.026)	(0.05)	(0.018)
TOCA1 (gram a.i. per ha)	0.000033	0.00033	–0.000092
	(0.00018)	(0.00045)	(0.00015)
TOCA3 (gram a.i. per ha)	0.001***	0.00073*	0.0011***
	(0.00018)	(0.0004)	(0.00015)
NA1	0.195***	0.12	0.15***
	(0.061)	(0.12)	(0.047)
NA3	–0.058	–0.185	0.086**
	(0.057)	(0.11)	(0.042)
Log-likelihood	–443.2	–101.53	–681.34
Chi-square	63.15***	23.1***	138.53***

Source: Dung and Dung (1999: table 15).

Notes:

*Statistically significant at 10%; **Statistically significant at 5%; and ***Statistically.- significant at 1%, respectively.

Figures in parentheses are standard errors.

TOCA1 = Total dose of categories I and II pesticides.

TOCA3 = Total dose of categories III and IV pesticides.

NA1 = Number of times in contact with TOCA1 per season.

NA3 = Number of times in contact with TOCA3 per season.

terms of perception. Farmers generally pay little attention to bad effects of pesticides on the eye and other organs. Incidence of eye irritation increases significantly with the drinking of alcohol and the severity of exposure to herbicides and fungicides, that is categories III and IV pesticides (TOCA3). The weight-to-height ratio has a negative impact, as expected, on eye abnormalities. In addition, the number of contacts with insecticides, i.e., pesticides of categories I and II (NA1) contributes significantly to an increase in eye irritation, whereas the number of contacts with categories III and IV pesticides (NA3) does not have a significant effect.

Neurological Effects

The incidence of headache symptoms is significantly related to drinking habits, age, and nutritional status (weight for height). Exposure to herbicides and fungicides (TOCA3) has a significantly positive effect on the occurrence of this symptom, but the impact of the other three pesticides contact variables does not differ significantly from zero.

Skin Effects

Skin problems are popularly discerned in rice farmers who are often exposed to pesticides. The logit regression estimates indicate that the incidence of skin problems is positively and significantly related to the dose of herbicides and fungicides (TOCA3) as well as to the number of contacts with insecticides (NA1) and herbicides and fungicides (NA3). In addition, the age of the farmer and the nutritional status have significant negative effects on skin problems, while drinking alcohol has a significant positive effect.

FARMER'S HEALTH COSTS FROM EXPOSURE TO PESTICIDES

Based on the above information, a health damage function was constructed to calculate the health costs of pesticide-related visible acute health impairments. Farmers' health costs are estimated as a function of pesticides exposure via total dose of active ingredients used by farmers, and other characteristics of farmers such as: nutrition status (weight-to-height ratio), age, and dummy variables indicating whether the individual smokes cigarettes and drinks alcohol or not. These costs included opportunity costs of lost work days and restricted activity days, expenses for medication, physical examination fees, and costs of protecting equipment.

Regression estimates based on data from the 1996–1997 WS rice crop survey presented in table 12.15 show that the total dose of pesticides significantly affect health costs (consisting of opportunity costs of days lost and days of

Table 12.15. Regression Results for Health Costs

Dependent Variable	Log of Health Costs	
	Coefficient	Standard Error
Constant	0.65***	0.20
Log of age (years)	1.41***	0.41
Weight/height	−0.026	0.027
Smoking (dummy)	0.02	0.27
Drinking alcohol (dummy)	0.72**	0.25
Log of total dose of all pesticides used (gram a.i. per ha)	0.385**	0.138
R^2	0.15	

Source: Dung and Dung (1999: table 16).
Note: *Statistically significant at 10%; **Statistically significant at 5%; and ***Statistically significant at 1%, respectively.

restricted activity, recovery costs, and costs of protective equipment). Health costs increase by 0.385 percent for every 1 percent increase in total doses. Drinking habits and age also have a significant positive effect on farmers' health cost, as expected. Somewhat surprisingly, however, smoking habits and nutritional status do not have a statistically significant effect on health costs.

CONCLUSION

Policy changes toward land security and market liberalization of agro-chemicals and rice have offered major incentives for rice farm households to increase their land productivity and family income via intensification. Rice monoculture resulting from intensification suggests that so far agricultural research and extension and market development have generally focused on rice. Though more and more farm households practice better nutrient management and IPM, increasing cropping intensity outweighs the positive effects of diminishing use of fertilizers and pesticides per crop per hectare. While rice yields and family incomes from rice have increased, evidence of farmers' health impairment and high health costs due to pesticides exposure shows that the quest of sustainability of rice production in the MD is uncertain.

NOTES

1. The homestead area is separated from the agricultural land and not included in the surveys.
2. Following the agricultural land tax legislation in 1993, land for annual crops was classified into six classes with different tax rates.
3. The prices of fertilizers are totally different, depending on the nutrient concentration in commercial fertilizers. As a result, using the total amount of com-

mercial fertilizers to make comparisons is inappropriate. All types of commercial fertilizers used by farm households are therefore converted into N, P, K nutrients in our analysis.

4. According to the World Health Organization (WHO), pesticides are classified into four categories of toxicity of the chemical compound and on their formulation, namely, extremely hazardous (Ia) and highly hazardous (Ib), moderately hazardous (II), slightly hazardous (III), and unlikely to present hazard in normal use (IV).

5. Pesticides, especially insecticides, posing extreme hazards to human health and the environment are banned or restricted to use in Vietnamese agriculture since 1992.

BIBLIOGRAPHY

Antle, J. M. and P. L. Pingali. "Pesticides, Productivity, and Farmer Health: A Philippines Case Study." *American Journal of Agricultural Economics* 76 (1994): 605-7.

Dung, N. H. *Profitability and Price Response of Rice Production System in the Mekong Delta, Vietnam*, MSc Thesis Chiangmai University. Chiangmai, Thailand, 1994.

Dung, N. H. and T. T. T. Dung. *Economic and Health Consequences of Pesticide Use in Paddy Production in the Mekong Delta, Vietnam*, Research Report, Economy and Environment Program for Southeast Asia (EEPSEA), 1999.

Dung. N. H., T. C. Thien, N. V. Hong, N. T. Loc, D. V. Minh, T. D. Thau, H. T. L. Nguyen, N. T. Phong, and T. T. Son. *Agro-chemicals, Productivity and Health in Vietnam*, Research Report, Economy and Environment Program for Southeast Asia (EEPSEA), 1999.

FAOSTAT. *Statistical Database*. Rome: Food and Agricultural Organization, 2001. www.fao.org.

Greenland, D. J. *The Sustainability of Rice Farming*. Wallingford, UK: CAP International and IRRI, 1997.

General Statistical Office (GSO). *Statistical Yearbooks*. Hanoi: Statistical Publishing House, 1996-2002.

Heong, K. L., M. M. Escalada, N. H. Huan, and V. Mai. "Use of Communication Media in Changing Rice Farmers' Pest Management in the Mekong Delta, Vietnam." *Crop Protection* 17, no. 5 (1998): 413-25.

Mallon, R. and G. Irvin, "Is the Vietnamese Miracle in Trouble?" *Institute of Social Studies working paper*, no. 253. The Hague: Institute of Social Studies, 1997.

Minot, N. and F. Goletti. "Rice Market Liberation and Poverty in Vietnam." *Research Report*, no. 114. Washington, DC: International Food Policy Research Institute, 2000.

Oskam, A. J., H. V. Zeijts, G. J. Thijssen, G. A. A. Wossink, and R. A. N. Vijftigchild. *Pesticide Use and Pesticide Policy in the Netherlands: An Economic Analysis of Regulatory Levies in Agriculture*. Wageningen: Wageningen University, 1992.

Pingali, P. L., C. B. Marquez, and F. G. Palis. "Pesticides and Philippine Rice Farmer Health: A Medical and Economic Analysis." *American Journal of Agricultural Economics* 76, no. 3 (1994): 587-92.

Rola, A. C. and P. L. Pingali. *Pesticide, Rice Productive, and Farmers' Health: An Economic Assessment.* Washington, DC: World Resource Institute, and Los Baños, Laguna: International Rice Research Institute, 1993.

Sadoulet, E. and A. de Janvry. *Quantitative Development Policy Analysis.* Baltimore: Johns Hopkins University Press, 1995.

Wossink, G. A. A., G. H. Peters, and G. C. van Kooten. "Introduction to Agro-chemicals Use." Pp. 1–40 in *Economics of Agro-chemicals: An International Overview of Use Patterns, Technical and Institutional Determinants, Policies and Perspectives: Selected Papers of the Symposium of the International Association of Agricultural Economists Held at Wageningen, The Netherlands, 24–8 April 1996,* edited by G. A. A. Wossink, G. H. Peters, and G. C. van Kooten. London: Ashgate Publishing, 1998.

13

Public Investment, Agricultural Research, and Agro-chemicals Use in Grain Production in China

Jing Zhu and Yousheng Li

Agricultural production has been growing rapidly in China during the past several decades. Grain production, as an example, has increased from approximately 300 million in the late 1970s to more than 500 million metric tons in the late 1990s. However, while providing the country with sufficient food and a solid basis for overall economic development, this expansion also brought with it the problem of excessive usage of chemical fertilizer and pesticides. In the late 1970s, chemical fertilizer consumption in China was at a level of less than 14 million metric tons. This figure then doubled to nearly 30 million metric tons in the early 1990s, and reached at 35.4 million metric tons in the year 2001—an increase of 226 percent in twenty-four years (FAOSTAT database). In addition, pesticides of various kinds have been used on a large scale in China since the 1950s. According to Huang et al. (2000) annual pesticides production reached more than 500,000 metric tons after the mid-1990s. China is now the second largest consumer, producer and exporter of pesticides in the world, following the United States.

Intensive application of chemical fertilizer and pesticides comprises hazards to both human health and environment. When applying pesticides, only about 20–30 percent is absorbed by the plant; the rest is left in the environment. Furthermore, although intensive pesticide application has increased grain yield, it may also contaminate the products of field crops, and pose a serious danger to the agro-ecosystem (Rola and Pingali 1993). Pesticides use may also harm the health of farmers applying them (see also chapter 12). In recent years serious environmental problems have been identified in Chinese agriculture, including leaching of nitrate and other agro-chemicals into groundwater, surface water pollution from soil erosion

251

and nutrients, and pesticides in runoff. There is currently a great concern about the impact of these chemicals on the environment and the sustainability of agricultural production.

Efforts have been made in the past to reduce the extent of usage and the harmfulness of agro-chemicals to the environment. Less persistent and highly efficient compounds were introduced in pesticide production. Several policies and regulations on pesticide production and utilization were formulated in the early 1980s, including the ban on some extremely hazardous products. However, the annual fertilizer and pesticides consumption in China was still increasing rapidly, reaching 43.4 million metric tons and 1.3 million metric tons respectively by the year 2002. China has become the first largest consumer of chemical inputs for agriculture in the world (National Bureau of Statistics 2003).

The intensive usage of agro-chemicals can be attributed to various factors. As far as pesticides are concerned, farmers' lack of knowledge of crop diseases and application of inappropriate technology often result in the overuse of pesticides in developing countries (Mumford 1982; Norton and Mumford 1983; Pingali and Carlson 1985). The technological deficiencies of the chemicals themselves are also a factor of importance, that is, pesticides that are high in toxicity and residue, or compound fertilizers that are not balanced. Nevertheless, from the farmers' point of view, increasing output and thus income, is the final objective of an intensive application of inputs. Given the limited available arable land in China, any output increase can only be achieved by an increase in crop yield. In consequence, the intensity of use of these chemicals has been rapidly increased.

Institutional reform, in particular the introduction of the HRS (household responsibility system) in the late 1970s and early 1980s, was fundamental for Chinese farmers to increase yield substantially. Furthermore, public investment in agricultural research, irrigation, education, and rural infrastructure, such as irrigation, has been a major driving force to the growth of agricultural production and yield improvement in China (Fan 2000; Fan and Pardey 1997). However, with the impact of the HRS being fully realized during the 1980s, and the insufficiency of investments by the public sector in following years, Chinese farmers relied more and more on inputs to realize yield increase. As a result, the application of chemical fertilizer and pesticides intensified.[1]

The objective of this chapter is to analyze how public sector investment in agricultural research and irrigation can be utilized as a policy tool to reduce the application of chemical fertilizer and pesticides application in crop production in China. As grain is predominant in China, the research mainly focuses on the grain production sector. By using provincial-level data for the period 1979–1997, this study analyzes the crop-specific contribution of public investment in agricultural research and development (R&D) and irrigation to yield increases for the three major grain crops—wheat, maize, and rice—and how the government can use investment in agricultural research investment as a policy tool to reduce chemical inputs by grain producers.

The chapter is organized as follows. The second section presents the regression results of a model explaining crop yields from public investments and private input use. These results are used in the third section to simulate the impact of increased public investments in research on chemical input use per unit output, and in the fourth section to estimate the potential reduction in chemical inputs of these crops through input substitution at given output levels. The conclusion and policy implications are presented in the last section.

PUBLIC INVESTMENT, PRIVATE INPUTS, AND GRAIN YIELDS

Grain production is the dominant crop production in China. According to the Agricultural Production Cost-Benefit Survey data,[2] the average yield of the three main grain crops has increased substantially in the past decades. Wheat showed a 40 percent yield increase from 198 kg per *mu* in 1979 to 277 kg per *mu* in 1997; maize an 18 percent yield increase from 297 kg per *mu* to 350 kg per *mu*, and rice a 13.4 percent increase from 373 kg per *mu* to 423 kg per *mu* for the same period. Material input cost rose from around 25 *yuan* per *mu* to 45 *yuan* per *mu* in the same period (measured in 1980 constant prices). In the late 1990s, about 40 percent of total material cost of grain production was composed of the cost of chemical fertilizer and pesticides, 25 percent was represented by machinery, animal and irrigation costs, and 10 percent by seed.

In our study the yield of each grain crop is expressed as a function of two categories of inputs:

$$Y = f\left(\sum X_i, \sum Z_k \right) \tag{1}$$

where Y stands for the crop yield;
X_i for all types of private material inputs; and
Z_k for different forms of public investment.

To grain producers, it is private material inputs, X_i, that are within their realm of control to be increased in order to promote a higher crop yield. The increase in public investment, Z_k, on the other hand, will improve the production infrastructure for the producers, and lead to a higher level of output with the same private material inputs, or to reduced private investment while keeping the same output level. In both cases, the producer's input needed for per unit of output will be reduced. This study will discuss both cases.

The double-log, or Cobb-Douglas, functional form is used to estimate the production functions of the three major crops of grain, namely wheat, corn, and rice. The production function for each crop is then expressed as follows:

$$\ln Y_{it} = \alpha_i + \beta_i \ln X_{it} + \sum_{k=1}^{2}\sum_{g=1}^{6} \gamma_{ikg} D_g \ln Z_{tkg} + \sum_{j=1}^{28} \gamma_{ij} D_j + \varepsilon_{it} \tag{2}$$

where Y is the yield of a given crop, i.e., wheat, corn, or rice, denoted by i;

X is the total value of physical inputs for the production per *mu*;

Z is the public input in agricultural research and irrigation, respectively, denoted by k;

D_g is a regional dummy;

D_j is a provincial dummy;

g and j stand for region and province, respectively; and

t is the year.

By dividing China into six agro-ecological regions based on their characteristics of natural endowments and resulting cropping systems (Carter and Zhong 1988), the difference in efficiency of public sector inputs between these regions is captured by setting up interaction variables $Z_{kg} D_g$, with $g = 1, 2, 3, \ldots 6$, between regional dummies and public investment in agricultural research and irrigation. Also included are a set of provincial dummy variables, D_j, with $j = 1, 2, 3, \ldots 28$, to represent the time-persistent, different social, economic, and natural endowments of each province which have not been accounted for by variables in the equation. Regional and provincial dummies are only included when the crop in question is grown in that region or province.

The private material inputs are expressed as the total material cost per *mu* (= 1/15 ha) for each crop under investigation, measured at *yuan* per *mu*. This approach is different from most previous research in the sense that, due to data limitations, value is used instead of physical volume of the input. It should further be pointed out that labor is not included in the model, as labor is mostly in surplus supply in the Chinese agricultural sector and hence, in general, does not constitute an input constraint to output.

There are many forms of public investment, such as agricultural R&D, irrigation, rural education, and infrastructure (including road, electricity, and telecommunication) that contribute to agricultural production growth. Due to data limitations, in this research we choose agricultural research and irrigation as a proxy to estimate the effects of public investment on yield growth of the three crops in China.

The irrigation variable is expressed as the ratio of irrigated area to total arable land. It is not included in the production function for rice, as virtually all paddy rice fields in China are irrigated. The agricultural R&D variable is measured in the form of a "stock" defined as a function of past government expenditures on agricultural R&D. This stock variable is calculated using the lag structure estimated by Fan, Zhang, and Zhang for Chinese agriculture (Fan et al. 2002).[3]

Pooled time-series and cross-province data from 1979 to 1997 and for 28 provinces were used in the estimation of each function.[4] Yield and material cost data used are from the Agricultural Production Cost-Benefit survey conducted under the State Development and Planning Commission. The research stock variable is constructed from annual research expenditures taken from Fan (2000) and Fan et al. (2000). Irrigated area data are from various issues of the *China Statistical Yearbook*.

Estimates of each production function—wheat, corn, and rice—are reported in table 13.1. Results of provincial dummies are not presented here for reasons of simplicity.

The results of the estimated yield functions show that the material inputs have positive and significant coefficients as expected, ranging from 0.20–0.31 for the three crops. The agricultural R&D stock variable also exhibits significant and positive coefficients for most of the six regions for all three crops. In contrast, the coefficients for the ratio of the irrigated area are insignificant in all regions for both wheat and corn.[5]

PUBLIC RESEARCH INVESTMENT AND INPUT USE RATIOS

Based on the estimation of the yield functions, two scenarios are simulated to estimate the effect of increased public investment in agricultural research on the material input cost reduction. It is assumed that public investment in agricultural research increases at an annual rate of 5 percent in Scenario I and 10 percent in Scenario II during the following years, with all private material inputs remaining at the same level as in 1997 for wheat, corn, and rice. The total sown area in the future remains at the 1997 level. The increased yield resulting from increased investment in agricultural research would lead to a higher output in the same area, and because private material input remains constant, private cost per unit of output would be reduced. Table 13.2 presents the results calculated for Scenario I and Scenario II, showing the percentage of material input cost reduction with yield increasing over the years through a 5 and 10 percent annual growth in R&D investment.

It can be noted from table 13.2 that, by increasing output, a 5 percent annual growth of public research expenditure helps to bring down the costs per unit of output in terms of private material input, by 0.6 to 1.7 percent in 2005 (compared with 1998), to up to 2.9 to 8.1 percent by the year 2020; a 10 percent annual increase in research expenditure will roughly double these gains. The largest cost reduction would be in wheat, followed by corn and rice.

If all types of material inputs are assumed to reduce proportionately, the application of chemical fertilizer and pesticides could also be brought down by approximately the same ratio respectively. As can be seen from

Table 13.1.　Regression Results for Wheat, Corn, and Rice Yields (1979–1997)

Explanatory Variable	Dependent Variable		
	Wheat	Corn	Rice
Constant	4.2874*	4.6495*	4.9564*
	(6.38)	(7.41)	(9.89)
Total material cost	0.3115*	0.1968*	0.2627*
	(4.64)	(4.34)	(5.91)
R&D stock, Region 1	0.106*	0.138*	0.064
	(6.99)	(2.81)	(1.06)
R&D stock, Region 2	0.120*	0.097*	0.052*
	(6.57)	(5.09)	(3.24)
R&D stock, Region 3	0.119*	0.072	0.058
	(5.97)	(0.97)	(0.36)
R&D stock, Region 4	0.038*	0.066*	0.075*
	(1.66)	(3.37)	(4.48)
R&D stock, Region 5	0.012	—♣	0.011
	(0.80)		(1.55)
R&D stock, Region 6	0.076*	0.036*	0.066
	(2.61)	(0.82)	(1.35)
Irrigation, Region 1	0.0836	0.0980	
	(0.60)	(0.71)	
Irrigation, Region 2	–0.0677	0.0415	
	(–0.51)	(0.30)	
Irrigation, Region 3	0.0808	–0.3313	
	(0.22)	(–0.89)	
Irrigation, Region 4	–0.1275	–0.5526	
	(–0.41)	(–1.44)	
Irrigation, Region 5	0.2129	—♣	
	(0.49)		
Irrigation, Region 6	–0.0289	0.2094	
	(–0.10)	(0.81)	
Provincial dummy			
Degrees of freedom	360	318	353
R2-adjusted	0.84	0.65	0.74

Source: Calculated in this study.
Notes:
Numbers in parentheses are t-values. * Statistically significant at 10 percent level.
Coefficients for provincial dummies are not presented here.
Region I:　Pastoral: Inner Mongolia, Ningxia, Gansu, Qinghai, Xinjiang, Xizan;
Region II:　Spring Wheat: Heilongjiang, Jilin, Liaoning;
Region III:　Winter Wheat: Beijing, Tianjin, Hebei, Henan, shandong, shanxi, shaanxi;
Region IV:　Wheat-Rice: Jiangsu, Anhui, Hubei;
Region V:　Double Rice: Zhejiang, Shanghai, Fujian, Jiangxi, Hunan, Guangdong, Guangxi; and
Region VI:　Southwest Rice: Sichuan, Yunnan, and Guizhou (See Carter and Zhong 1988: 67–70).
♣: corn is not a major crop grown in this region.

table 13.3, this implies that for one ton of output of wheat, maize, and rice, taking the 1997 data as base, chemical fertilizer use will be reduced by 11–36 *yuan* for the three crops, and the use of pesticides by 1–3 *yuan*. Although the percentage of reduction of agro-chemicals usage is smaller for

Table 13.2. Reduction of Average Material Inputs Application through Output Increase (in percent compared with 1998)

	Scenario I			Scenario II		
	Wheat	*Corn*	*Rice*	*Wheat*	*Corn*	*Rice*
2000	0.22	0.27	0.01	0.34	0.32	0.07
2005	1.69	0.25	0.62	2.84	0.75	1.34
2010	3.72	0.99	1.04	6.70	2.49	2.80
2015	5.94	2.33	1.96	10.96	5.37	4.74
2020	8.13	4.22	2.88	15.00	8.93	6.49

Source: Calculated in this study.
Note: Scenario I and II assume that agricultural research expenditure grows at 5 and 10 percent annually respectively, while other factors remain unchanged after 1997.

Table 13.3. Reduction of Application of Chemical Fertilizer and Pesticides through 10 Percent Annual Increase in Agricultural Research in 2020 (yuan per ton, 1997 prices)

Crop	Chemical Fertilizer	Pesticides
Wheat	36.0	3.0
Maize	16.7	1.0
Rice	11.3	2.6

Source: Calculated in this study.

rice than that for wheat and maize, the absolute cost savings for pesticide use per ton of rice is more than twice that for maize, and only 13 percent lower than in the case of wheat. The reason for this is the much higher application of pesticides in rice production, which is two to three times higher than in wheat or maize production.

PUBLIC RESEARCH INVESTMENT AND INPUT SUBSTITUTION

Whereas the material inputs per unit of grain production could be reduced through higher output, it can also be reduced if output is kept constant and public investment is increased. In the late 1990s, China experienced a situation where domestic grain supply exceeded demand. In such circumstances, it would be more desirable to invest in public inputs that will reduce cost with output being kept at a certain level.[6]

Model Specification

In the production function of our study above, inputs have been aggregated into the categories of private material inputs and public sector inputs, expressed as $Y = F (C, R)$, with Y the yield per unit of land, C the private cost, and R the research stock (as the proxy for public investment). The two inputs in the production function are substitutable, and there exists a trade-off between the two inputs if the output is to be kept constant. If we increase the amount of public expenditure on agricultural research, the quantity of research stock will increase accordingly, which may enable the producers to reduce their private inputs while keeping production on the same output *isoquant*. In other words, for the same amount of output, private material inputs can be reduced due to the increasing investment in public inputs. As chemical fertilizer and pesticides comprise a substantial part of private material inputs, this increase in efficiency and return to private material inputs will result in a decreased application of chemical fertilizer and pesticides, promoting more sustainable agricultural production.

The Marginal Rate of Technical Substitution (*MRTS*), which measures the rate at which one input has to be substituted for another in order to keep output constant, can be expressed as:

$$MRTS_{cr} = MP_r / MP_c \tag{3}$$

where MP_r and MP_c are marginal products of the research stock and private material cost, respectively. The $MRTS_{cr}$ indicates how much private material input can be saved by increasing one unit of the research stock. Both the private material input and the research stock variable are in absolute volume terms. The MRTS can also be measured in percentage terms, derived from the above equation, and is expressed as

$$MRTS_{cr}^* = E_r / E_c \tag{4}$$

where $MRTS_{cr}^*$ stands for the modified RTS, E_r, and E_c for the output elasticity of the research stock and private material input, respectively. It measures the percentage of private material input that can be saved if the research stock increases by 1 percent, without reducing the output.

The $MRTS_{cr}^*$ for each crop in each region is calculated with the values of E_r and E_c estimated in table 13.1 and is listed in table 13.4. The national $MRTS_{cr}^*$ is also calculated as a weighted average (weighted by output in each region), which is used for simplicity in the following simulations for each crop.

As can be seen from table 13.4, for each percent increase in research stock, the private material inputs in wheat, corn, and rice in China as a whole can

Table 13.4. Modified Rate of Technical Substitution

Region	I	II	III	IV	V	VI	National
Wheat	0.34	0.39	0.38	0.12	0.04	0.24	0.31
Corn	0.70	0.49	0.37	0.34	—	0.18	0.40
Rice	0.24	0.20	0.22	0.28	0.04	0.25	0.14

Source: Calculated in this study.

Table 13.5.   Reduction of Application of Chemical Fertilizer and Pesticides through an Increase in Agricultural Research, 1997 (yuan/mu)

5 percent increase in agricultural research expenditure			1 yuan increase of agricultural research expenditure per mu		
Wheat	Corn	Rice	Wheat	Corn	Rice
1.12	1.34	0.53	2.17	1.49	0.89

Source: Calculated in this study.

be reduced by 0.31 percent, 0.40 percent, and 0.14 percent, respectively, if the output is to be kept constant. In other words, keeping the same output, a 1 percent increase in public investment in research can reduce farmer's material inputs and therefore the need of use of chemical fertilizer and pesticides by 0.31 percent for wheat, 0.40 percent for maize, and 0.14 percent for rice, assuming all types of material inputs are decreased at the same ratio.

These results are used to estimate the reduction in expenditures on chemical inputs that can be obtained through investments in research (see table 13.5). Taking the data of 1997 as base, a 5 percent increase in research expenditure will reduce the input of chemical fertilizer and pesticides for wheat, corn, and rice by 1.12, 1.34, and 0.53 *yuan* per *mu*, respectively. If one additional *yuan* per *mu* is to be invested in research for these crops as a whole, the private material input cost for wheat, corn, and rice will be reduced by 2.17, 1.49, and 0.89 *yuan* per *mu* respectively.

CONCLUSION

Agriculture is widely believed to be a sector of significant environmental effects, especially with regard to chemical input use. Agricultural production has been growing rapidly in China in the past few decades. However, the even more rapid increase of material inputs used in crop production, especially the application of chemical fertilizer and pesticides, poses threats to the environment and human health as well as to the sustainability of future agricultural production. In recent years serious environmental problems have been identified in Chinese agriculture, including leaching of nitrate, pesticides, and other agro-chemicals into the groundwater, surface water pollution from soil erosion and nutrients and pesticides in runoff, pesticide drift, and plastic film residues in soil. All of these problems highlight not only the fragility of the agricultural resource base, but also the importance of understanding the impact of government policies on farmers' input use.

One of the important reasons for the investment of material inputs is the deficiency of public sector inputs, which include agricultural research, extension, technological development, education, knowledge dissemination, and investment in rural infrastructure (such as irrigation). Higher levels of investments in these public inputs would improve the investment efficiency of the private material inputs by farmers.

This chapter examined the effects of public investment in agricultural research (as a proxy for overall public investment in agriculture) on the reduction of material inputs in grain production in a crop-specific manner. The research shows that yield increases caused by an annual increase of 5–10 percent of investment in agricultural research will result in a reduction of material inputs per unit output by 8–15 percent for wheat, 4–9 percent for maize, and 3–6.5 percent for rice, respectively, over the period 1998–2020. On the other hand, a 5 percent annual increase of public research investment will lead to a reduction of private material inputs of these three crops by 0.7–2 percent through input substitution if the same yield is to be maintained. If all types of material inputs are assumed to be reduced proportionately, this could be translated into a reduction of application of chemical fertilizer and pesticides by approximately 0.5–1.3 *yuan* per *mu*, or about 0.13–0.33 kg per *mu* for fertilizer application. The nation as a whole could achieve a reduction of chemical fertilizer consumption by 140,877 kg, 116,403 kg, and 61,070 kg for wheat, maize, and rice respectively (based on 1997 data). Although this effect is still very marginal compared with total fertilizer and pesticides use, it contributes to the reduction of negative environmental effects of grain production.[7]

In the past several decades, agricultural research has been one of the major driving forces to improvement of grain yields in China. However, compared with most developed and even many developing countries, the investment intensity in agricultural research and other agricultural production infrastructure in China is at the lower end. Increased investment in rural public goods such as agricultural research is urgently needed to provide grain producers with better production conditions and knowledge of improved farm management practices. With increased productivity, farmers' private material inputs into grain production can be brought down and the application of chemical fertilizer and pesticides reduced as a consequence. Although the impact is marginal in terms of the total use, it does contribute to the sustainability of agricultural production in the future.

NOTES

1. The overall investment in agricultural research in China is low in comparison with developed countries, and even with other developing countries; the "Green Box" support measures to agriculture production are also limited in China compared with those in developed countries.

2. The Agricultural Production Cost-Benefit Survey is conducted under the State Development and Planning Commission and covers more than 1,400 coun-

ties and approximately 60,000 farm households. The dataset is considered of higher quality and accuracy than other officially published data and is becoming widely used.

3. They use a two-step approach. First, econometric techniques are used to determine the length of the lag by including past government expenditures in the production function. The authors found that China's agricultural research lag is between years 7 and 17, that is, research will begin to affect production from the year 7, and will last until year 17. Second, once the length of the lag is estimated, they use the second polynomial distribution to calculate the coefficient of each year's expenditure in the production function.

4. More recent data was not available at the time of this study.

5. The insignificance of the irrigation variable can be explained by several factors. Firstly, the variable is taken as the share of irrigated area in the total cultivated area, which does not reflect the quality of irrigation. Secondly, most cost surveys are usually conducted in irrigated areas so the irrigation variable shows little impact on yields.

6. In fact, grain output has been on the decline after 1998 as a result of the government's advocacy of diversification of agricultural production. Since 2004 it has been on the rise again (see chapter 8).

7. In this study, constant return to scale of and constant elasticities of agrochemicals' use in production are assumed in the model specification. It would be helpful to test for return to scale and elasticity change in a follow-up study and compare to what extent this will change the impacts estimated in this study.

BIBLIOGRAPHY

Carter, C. and F. Zhong. *China's Grain Production and Trade.* Boulder, CO and London: Westview Press, 1988.

Fan, S. "Research Investment and the Economic Returns to Chinese Agricultural Research." *Journal of Productivity Analysis* 14, no. 92 (2000): 163–80.

Fan, S. and P. G. Pardey. "Research, Productivity, and Output Growth in Chinese Agriculture." *Journal of Development Economics* 53 (1997): 115–37.

Fan, S., L. Zhang, and X. Zhang. "Growth, Inequality and Poverty in Rural China: The Role of Public Investments." *Research Report*, no. 125. Washington, D.C.: International Food Policy Research Institute (IFPRI), 2002.

Huang, J., F. Qiao, L. Zhang, and S. Rozelle. "Farm Pesticide, Rice Production, and Human Health." *Project Report*, no. 11. Center for Chinese Agricultural Policy (CCAP), Chinese Academy of Agricultural Sciences. A Project Report Submitted to EEPSEA, Singapore, 2000.

Mumford, J. D. "Perceptions and Losses from Pests of Arable Crops by Some Farmers in England New Zealand." *Crop Protection* 1, no. 3 (1982): 283–88.

National Bureau of Statistics, *China Rural Statistical Yearbook*, China Statistics Press, 2003.

Norton, G. A. and J. D. Mumford. "Decision-making in Pest Control." Pp. 97–119 in *Advances in Applied Biology*, edited by T. H. Coaker. vol. VIII. London: Academic Press, 1983.

Pingali, P. L. and G. A. Carlson. "Human Capital, Adjustment in Subjective Probabilities, and Demand for Pest Control." *American Journal of Agricultural Economics* 67, no. 4 (1985): 853–61.

Rola, A. C. and P. L. Pingali. *Pesticides, Rice Productivity, and Farmers' Health: An Economic Assessment.* Los Banos and Washington, D.C.: International Rice Research Institute and World Resource Institute, 1993.

PART III

Agriculture and Sustainable Land Management

14

Sustainability Issues in East Asian Rice Cultivation

Herman van Keulen

This chapter reviews the major processes threatening the sustainability of high-intensity rice-based cropping systems in Asia, and discusses possible modifications in the system that will make them "more" sustainable, while maintaining the required production level to safeguard food security. The chapter is organized in the following manner. In the second section, traditional systems of paddy cultivation in Asia and their ecological functions are briefly discussed. The third section discusses the environmental impact of the Green Revolution in lowland rice production, with a focus on water use, nutrient supply (soil organic matter, nitrogen, and other nutrients) and the occurrence of weeds and pests and diseases. In the fourth section the impact of these environmental changes on the sustainability of rice cropping systems is discussed. The fifth section presents a number of options for sustaining soil quality, including green manure planting, crop diversification and site-specific nutrient management. Opportunities for short-term measures to promote environmental sustainability are discussed in the sixth section, while the chapter ends with some reflections on critical management issues in transforming lowland rice production.

TRADITIONAL SYSTEMS OF PADDY CULTIVATION

The origin of paddy rice cultivation is situated somewhere in the southeastern part of Asia, and is said to date back at least 7,000 years. Such systems originated in the river valleys and deltas and remained unchanged for thousands of years. Wetland rice is therefore the only major crop that has been grown for millennia in monoculture without major soil degradation (Bray 1986; Von Uexkuell and Beaton 1992).

Soil flooding and puddling maintained favorable soil properties for rice growth (Ponnamperuma 1972), and traditional rice-growing systems aimed at yield stability rather than high yields. Traditional long-duration varieties (130–210 days) characterized by low harvest indices and yields were grown and, in many areas, much of the straw remained in the field (Von Uexkuell and Beaton 1992). Bunds protected rice fields from soil erosion. Floodwater buffered soil temperature and allowed ample growth of N_2-fixing microorganisms (Roger 1996). Suspended particles and soluble nutrients from rainfall and irrigation water contributed to the indigenous nutrient supply, sufficient to cover the demand of these extensively grown crops. Current rainfall contributions to annual nutrient inputs to irrigated rice fields of Asia are estimated to be in the range of 1–10 kg N, 0.2–2 kg P, 3–10 kg K, and 5–20 kg S per hectare. In these traditional irrigated rice systems, where total net nutrient removal, as well as daily nutrient uptake rate was low, nutrient additions from natural sources were an important component of the overall nutrient balance, and even poor soils had the capacity to supply sufficient nutrients to sustain yield levels of 1–2 ton per hectare. Lowland rice ecosystems can consequently be seen as unique ecological regions—comparable for example with forests, except that they are the largest single land use focused on feeding the world. The traditional rice landscape with its patchwork of fields and surrounding areas is additionally a "hub" in biodiversity (Bambaradeniya et al. 2004; Cantrell 2000).

THE GREEN REVOLUTION AND BEYOND

The development and widespread adoption of high-yielding, early-maturing, semi-dwarf *Indica* varieties in the 1960s led to rapid intensification in the tropical lowlands of Asia. The new varieties such as IR8 had a short growth period, greater yield potential due to more efficient biomass partitioning, were short-statured and lodging-resistant, and responded well to N fertilizer. Use of external inputs, such as fertilizers, water, pesticides, and energy sharply increased and the diversity in rice varieties used in irrigated systems rapidly declined. The higher yield potential of modern varieties promoted private and public investments in irrigation infrastructure. Tillage and management intensity improved through extension programs, two to three rice crops per year became a reality and the soils thus remained submerged for longer periods. Average grain yield increased up to 4.9 ton per hectare in 1991 (Cassman and Pingali 1995a) and harvesting techniques changed. To facilitate land preparation for the next crop, farmers started to cut the entire crop and remove or burn the straw (Von Uexkuell and Beaton 1992).

During the last decade, trends of declining factor productivity have been noted in long-term rice monoculture experiments (Flinn and De Datta 1984; Cassman et al. 1995; Dawe et al. 2000). There is evidence that the declining productivity trends are caused by a gradual degradation of soil quality due to intensive cropping. Reduced indigenous soil N-supplying capacity was identified as a driving force, despite conservation or even increases in total soil organic matter contents (Cassman et al. 1995; Cassman and Pingali 1995a). Depletion of soil nutrient reserves, build-up of soil pests, physic-chemical changes in the soil due to prolonged submergence, or changes in composition of the soil micro flora have also been hypothesized as possible causes of the productivity decline, but universally accepted mechanisms have not yet been identified (Dobermann et al. 2000).

Some thirty years after the green revolution in Asian rice cultivation, sustainability of intensified rice production systems can be viewed from different perspectives, reflecting seemingly conflicting interests. The application of more recent concepts from agro-economy, however, seems to bridge the gap between economic and ecological goals.

Water

Rice produced on submerged soil is a major beneficiary of fresh water resources in Asia. Nearly 90 percent of the fresh water diverted for human use in Asia goes to agriculture, of which more than 50 percent is used to irrigate rice. Based on the most recently available data, Asian agriculture on the whole is not yet in the midst of a water crisis (Dawe 2004), but the total irrigated crop area in Asia continues to increase (Huke and Huke 1997), with associated consequences for agricultural water demand. As many of the world's most rapidly growing economies are in Asia, water demand from industries and households also continues to increase, so that serious consideration has to be given to the water issue (Rijsberman 2004). Heavy upstream diversions of water from some rivers have severely reduced the outflow to the sea during the dry season, leading to negative consequences for availability of water for irrigation and likelihood of saltwater intrusion (Geng et al. 2001). Groundwater has become the most important source for irrigation, particularly in South Asia, but groundwater tables are falling in many areas in Asia. Thus, there are legitimate concerns regarding the effect of future availability and price of water on farmers and farm management practices.

Lowland rice in Asia is traditionally established by transplanting, but as labor costs increase, there has been a move to direct seeding. Land preparation for both transplanted and direct-seeded rice typically consists of soaking the soil followed by ploughing and harrowing of saturated soil. This process, referred to as puddling, destroys soil structure, creates a soft

muddy layer of 10–15 cm depth, and reduces subsequent downward movement and loss of water during rice cropping. Rice fields are typically kept
continuously submerged with 5–10 cm of water throughout the growing
season, until just prior to harvest. Production of irrigated rice consequently
requires large quantities of water, because of the high water use for land
preparation and the large losses by seepage, percolation, and evaporation
when soil is submerged.

The use of irrigation water for production of lowland rice can potentially
be reduced by lowering the depth of standing water and by allowing the
soil to dry before the next application of irrigation water (i.e., alternate soil
wetting-drying). These techniques, however, entail a risk of yield reduction
due to drought stress on rice, and they are relatively difficult to implement
because of the strict requirements for water control. Even with the successful implementation of such water-saving irrigation techniques, water
requirements remain high for rice as compared to other cereal crops (Bouman and Tuong 2001). Savings in irrigation water and increases in water
productivity might be possible if lowland rice could be grown under *aerobic*
soil conditions similar to wheat and maize (Tuong et al. 2004). Varieties
and management technologies for production of lowland rice under aerobic soil conditions do not yet exist. Nonetheless, potential tradeoffs can be
examined for the situation in which localized or widespread water shortages necessitate the production of rice with less water.

Nutrient Supply

There is limited quantitative information on the impact of prolonged
submergence on soil physico-chemical properties and their effects on nutrient supply. Though most irrigated rice lands are probably not prone to
salinization, long-term use of poor-quality irrigation water may cause undesirable changes in soil chemistry. Due to precipitation of carbonates, soil
pH may increase (Marx et al. 1988). In some areas where groundwater is
the irrigation source, high net additions of Ca and Mg may result in reduced
K availability due to a wide $(Ca+Mg)/K$ ratio (Dobermann et al. 1995). We
do not have enough quantitative information at the moment about the
importance of such processes for sustaining soil quality.

Impact of Intensification on the Organic Phase

Crop intensification leads to increases in the total pool of organic matter
in the soil, as a result of higher production of roots and stubble, intensified
root exudation, and reduced mineralization under anoxic conditions (Olk
and Cassman 1995). Photosynthetic primary production in the floodwater
provides another soil organic matter source. An average fraction of 1–5

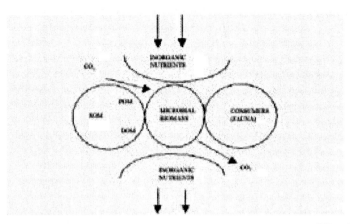

Figure 14.1. Central Role of Microbial Biomass in Soil Nutrient Immobilization and Release

percent of soil organic matter consists of living biomass (Anderson and Domsch 1980; Inubushi and Watanabe 1986). This consists mainly of heterotrophic microorganisms and represents an easily available pool of nutrients with a rapid turnover rate (figure 14.1; Lee 1994).

Traditional rice cultivation owed most of its sustainability to the continuous replenishment of the organic matter pool (Bray 1986). As a result of shorter turnaround time between intensified crops, farmers eliminate the entire crop from the field and often burn the straw (Von Uexkuell and Beaton 1992). The intensified cultivation of higher yielding, less photoperiod-sensitive varieties with shorter growth periods required roughly a doubling of nutrient supply. Mineral fertilizers can rapidly and efficiently satisfy this growing demand when bypassing removal in nutrient cycling or the retarding sequences of sequestration and remobilization (Broadbent 1984).

Tropical wetland soils are known for their rapid decomposition of organic debris (Kimura et al. 1990). Nevertheless, total soil organic carbon content seems to be conserved or even enhanced in long-term trials under intensive double or triple cropping for decades, though it is partially attenuated by decreasing bulk density (Cassman and Pingali 1995b). This may be an indirect evidence of fertilizer-induced, largely microbial soil organic matter production (Broadbent 1984). Although biomass pools in the photosynthesis-dominated floodwater subsystem are small, its autotrophic productivity can reach 600 kg/ha over a cropping period (Saito and Watanabe 1978).

Intensification of lowland rice crops implies extended periods of submergence. Thus, anoxic conditions prevailing in the bulk soil can both slow down the primary attack of extra cellular enzymes on particulate organic matter (Reichardt 1986) and modify the metabolic pathways of microbial mineralization (Schink 1988). Finally, humification processes involving the build-up of phenolic compounds depend strongly on the chemical composition of the organic input into the submerged system. In contrast to green manure, rice straw may release large quantities of phenolic compounds (Tsutsuki and Ponnamperuma 1987). Polymerization and mineralization of phenols are delayed under anaerobic conditions. This phenomenon favors accumulation of young, low-humified soil organic matter, rich in phenols (Ye and Wen 1991; Palm and Sanchez 1991; Becker et al. 1994b; Olk and Cassman 1995). The presence of N-containing aromatic components could give a mechanistic explanation for declining endogenous N supply in continuously flooded anaerobic fields (Cassman et al. 1995).

Nitrogen

Nitrogen occupies a special position as a plant nutrient, because of the large quantities that are needed by the crop for unrestricted growth, and because the soil-crop system is an open system for nitrogen. This implies that large quantities of the element move through the system, representing substantial risks for losses. In intensive systems, the major part of the nitrogen requirements of the crop evidently comes from fertilizer. But indigenous N-supply, that is, the nitrogen originating from the organic store in soil organic matter still forms a substantial proportion of total N-uptake. When looking at the nitrogen supply, therefore, both indigenous soil nitrogen and fertilizer have to be taken into account. As explained, soil organic matter contents in intensive rice systems are higher than in extensive systems. What happens, however, in the long run?

In the Long-Term Continuous Cropping Experiment (LTCCE) at the International Rice Research Institute (IRRI), where all aboveground biomass is removed at harvest, the concentrations of soil organic carbon and total N in the puddled topsoil have increased slightly since 1963 in control treatments without N-fertilizer inputs, while they are considerably higher in treatments that received high inputs of N-fertilizer (see table 14.1). Unfortunately, soil bulk density initially was not measured, so it is impossible to unequivocally establish whether quantities of organic N and C have been conserved. By 1978, when forty-eight consecutive crops had been cultivated, the soil physical properties presumably had stabilized, and the data then show that organic N and C contents remain constant, indicating equilibrium between inputs and outputs in the system. The C/N ratio also remains constant, and was not influenced by the fertilizer input. This homeostasis of C and N is attributed to the processes indicated above: (a) C-inputs from photosynthetic biomass in the floodwater; (b) biological N-fixation by blue-green algae in floodwater and heterotrophic bacteria in soil and soil rhizosphere; (c) slower rate of organic matter decomposition under anaerobic conditions, leading to modifications in organic matter composition.

Table 14.1. Soil Organic Carbon and Total N in the 0–20 cm Topsoil in the Long-Term Continuous Cropping Experiment at IRRI (Andaqueptic Haplaquoll with clay texture)

N-rate (kg/ha)		Organic Carbon (g/kg)			Total N (g/kg)			C/N ratio		
Dry Season	Wet Season	1963	1978	1983–91[a]	1963	1978	1983–91[a]	1963	1978	1983–91[a]
0	0	18.3	18.8	19.4	1.94	1.97	2.00	9.5	9.6	9.7
50	30	18.3	20.6	20.6	1.94	2.13	2.04	9.5	9.6	10.0
100	60	18.3	21.4	20.8	1.94	2.18	2.10	9.5	9.9	9.9
150	90	18.3	21.4	22.3	1.94	2.22	2.23	9.5	9.7	10.0
CV (%)		6	12	8	8	11	7	9	6	4

Source: Cassman and Pingali (1995a).
Note: [a] Average for 1983, 1985, and 1991; sampled in the same manner by the same agronomist, and kept for future reference; analyzed in 1992 by the same lab technician.

The observed yield decline in LTCCE at IRRI has received extensive attention, and especially the N-cycle in this context, as it was hypothesized that the slower rate of decomposition and the associated lower rate of N-mineralization could be at the base of the yield decline. To test this hypothesis nitrogen management was changed in the 1992 dry season, and detailed observations made in LTCCE (see table 14.2). The main conclusions from the analysis of this experiment and reanalysis of historical data were: crop yields in the zero-N treatment around 1990 were substantially lower than in the 1970–1972 period. This appeared to be associated with an appreciably lower N-supplying capacity of the soil ("supply from natural sources," including mineralization from soil organic matter). In 1992, maximum yield was much higher than in 1989–1991 at an appreciably higher N-fertilizer rate, of which moreover, the recovery ("uptake efficiency") was higher, presumably because special attention was paid to crop management, such as water management during and following N-application and pest and disease control (fungicide application). Agronomic efficiency therefore was also much higher. When these conclusions may be extrapolated, it would imply that under continuous high intensity rice cropping, indigenous N-supply gradually declines, so that to realize potential, N-fertilizer rates would have to be increased yields (Kropff et al. 1993).

Other Nutrients

There are numerous examples of depletion of soil nutrients other than soil N in intensive rice systems. In productive soils of the alluvial floodplains of South and Southeast Asia, P and K rarely limited rice productivity before intensification of these systems during the Green Revolution. In most early fertilizer trials with modern varieties, no significant responses to

Table 14.2. Yield Response and Partial Factor Productivity for N-Fertilizer Inputs in Three Periods of the Long-Term Continuous Cropping Experiment at IRRI

Period (cultivar)	Crop Duration (days)	N-Fertilizer Rate (kg/ha)	Grain Yield (kg/ha)	Yield Increase (Δ kg/ha)	Agronomic N-Fertilizer Use Efficiency (Δ kg grain/ kg N)	N-Fertilizer Factor Productivity (kg grain/ kg N)
1970–1972	127	0	6,300	—	—	—
		50	7,870	1,570	31.4	157
		100	8,870	2,570	25.7	89
		150	9,430	3,130	20.9	63
1989–1991	115	0	3,800	—	—	
		50	5,210	1,410	28.2	104
		100	6,100	2,300	23.0	61
		150	6,380	2,580	17.2	43
1992	115	0	4,250	—	—	—
		63	6,170	1,920	30.3	97
		127	7,760	3,510	27.7	61
		190	8,710	4,460	23.5	46

Source: Cassman and Pingali (1995a).

P or K additions were observed (Kawaguchi and Kyuma 1977; De Datta and Mikkelsen 1985; Bajwa 1994).

Depletion of extractable soil P to a level that significantly reduced N-use efficiency and grain yield was first shown in long-term experiments in the Philippines (De Datta et al. 1988). Similar effects were noted in long-term experiments in China. Across eleven sites in five countries, negative P balances averaged 7–8 kg /ha per crop in zero-P treatments, and it was estimated that P-fertilizer rates of 17–25 kg/ha would be required to maintain the P balance or to increase total soil P (see figure 14.2; Dobermann et al. 1996c).

Potassium deficiency has become a constraint in soils that were previously not considered K-limited (De Datta and Mikkelsen 1985; Von Uexkuell 1985; Mohanty and Mandal 1989; Chen et al. 1992; Dobermann et al. 1996a; Oberthuer et al. 1996). Modern rice varieties require similar quantities of K and N (20 kg each per ton grain yield). Most rice farmers in Asia do not apply K-fertilizer, while straw was increasingly removed from the field as a result of intensification, and the quantities of animal manure applied strongly declined. Although researchers did draw attention to the risk of negative K balances and soil K depletion many years ago (Ismunadji 1976; Kemmler 1980; Von Uexkuell 1985; De Datta and Mikkelsen 1985), until recently this did not lead to significant improvements in K management. In long-term experiments at eleven sites, the K balance was highly negative (−34 to −63 kg/ha per crop cycle; see figure 14.2) in all NPK combinations tested; even fertilizer K application at an average rate of 40 kg/ha in the +NK and +NPK treatments was insufficient to compensate for the K removal at most sites (Dobermann et al. 1996a). Examples of K depletion observed in farmers' fields include alluvial, illitic soils in India (Tiwari 1985), lowland rice soils of Java, Indonesia (Sri Adiningsih et al. 1991), and vermiculitic clay soils of Central Luzon, Philippines (Oberthuer et al. 1996).

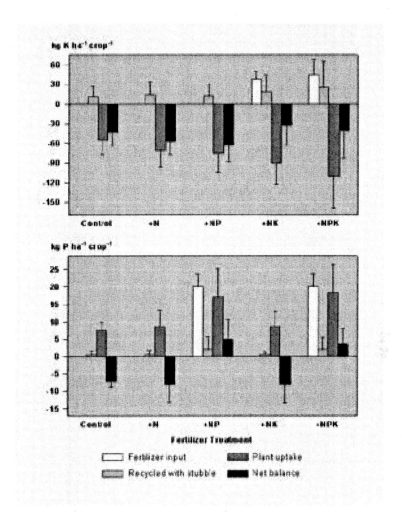

Values shown are averages and standard deviations (error bars) of long-term experiments at eleven sites in five countries sampled in 1993. Stubble was recycled at five sites and all straw was removed at six sites, reflecting standard farmer practices for each location (*Source:* Dobermann et al. 1996a, c)

Figure 14.2. Partial Net K and P Balances for One Rice Crop under Five Different Fertilizer Treatments.

Intensification also contributed to more widespread occurrence of S and Zn deficiencies in marginally productive lowland rice soils (Ismunadji et al. 1975; Blair et al. 1978; Von Uexkuell and Beaton 1992). The removal or burning of straw and the replacement of sulfur-containing fertilizers with non-S fertilizers (Yoshida 1981) has contributed to S depletion in several rice areas. More recently, however, increasing air pollution and S deposition associated with rapid industrial development seem to counteract this trend in some parts of South and Southeast Asia. Little is known about Zn balances in traditional and intensive irrigated rice culture. Zinc deficiency is usually associated with leached ultisols and oxisols with high pH and/or high organic matter contents, but Zn depletion may also occur on nonalkaline soils through ZnS formation (Oberthuer et al. 1996).

On marginally productive and highly weathered soils, the increased supply of N has intensified deficiencies in K, P, and Zn, resulting in the spread of a nutritional disorder known as iron toxicity (Ottow et al. 1981).

Weeds

Weeds are a constant problem in all rice-growing areas (Kropff and Moody 1992). Reduced availability of water and labor are driving forces to changing management practices in rice production. The shift from transplanted to direct seeding of rice aggravates the problem, because weeds and crop emerge together and it is more difficult to use early flooding for weed control. Herbicide use is increasing in Asia because herbicides are cheaper than hand labor and easy to apply. However, herbicides pose negative effects, such as changes in the weed flora resulting in an increase in hard-to-control weed species, environmental contamination, and selection for herbicide-tolerant weed biotypes. Thus, it is becoming increasingly important to develop integrated weed management systems in which various weed control measures are combined and herbicide use is minimized (Bastiaans et al. 1997; Labrada 2003; table 14.3).

Pests and Diseases

Pests (including insect-pests and diseases) of rice, evolved under the influence of host genes, are changing the rice environment. Thus, scientists are in a continuous war with ever changing races, pathotypes, and biotypes of rice pests. New and more potent genes, being added continuously using conventional or biotechnological tools, fight a losing battle. But these efforts are essential to add stability to production and avoid the recurrence of the great Bengal famine of the Indian subcontinent, the brown planthopper catastrophe of Indonesia and the Philippines, or the blast damage in the Republic of Korea and Japan during 1996.

Table 14.3. Integrated Weed Management Requires a Shift from Weed Control to Weed Management

Structure	Weed Control	Weed Management
Goal	Maximize crop yield and profits	Optimize long-term farm productivity
Objectives	Eradicate weeds from the crop	Maintain weeds below level of significant competition with the crop
Approach	Use one to two of the easiest, most effective methods suited to the crop	Balance the best available methods suited to the farming system
Action	Employ full tillage technology, apply full rates of herbicides	Employ minimum tillage, minimum effective rates of herbicide, integrated agronomic practices to increase competitive ability of the crop
Outputs	Near-perfect weed elimination, high crop yield	Substantial reduction of weed pressure, optimum farm profit
Application	Wide geographical regions	Adapted to specific locations/areas

Source: Adapted from Kon (1993).

Improvements in the control of crop diseases and pathogens could significantly increase yields. Rice provides 90 percent of the human food in Asia, and has become the most abundant species in Asia. A central conclusion of epidemiology is that both the number of diseases and disease incidence increase proportional to host abundance. This disconcerting finding illustrates the potential instability of a global strategy of food production in which rice accounts for so high a proportion of production and consumption. The relative scarcity of outbreaks of diseases on rice in Asia is a testament to plant breeding and cultivation practices. Breeders have been successful at improving resistances to abiotic stresses, pathogens and diseases, and at deploying these defenses in space and time so as to maintain yield stability despite low crop diversity in continuous cereal systems. However, it is unclear if such conventional breeding approaches can work indefinitely. Both integrated pest management and biotechnology that identifies durable resistance through multiple gene sources, should play increasingly important roles.

Nonetheless, the evolutionary interactions among crops and their pathogens mean that any improvement in crop resistance to a pathogen is likely to be transitory. Each defense sows the evolutionary seeds of its own demise. Maize hybrids in the United States now have a useful lifetime of about four years, half of what it was thirty years ago. Similarly, agrochemicals, such as insecticides, fungicides, and antibiotics, are also major selective agents. Insects often evolve resistance to insecticides within a decade. Resistant strains of bacterial pathogens appear within 1–3 years of the release of many antibiotics. The need to breed for new disease resistance and to discover new pesticides, however, can be reduced by crop rotation and the use of spatial or temporal crop diversity.

THE SUSTAINABILITY OF RICE CROPPING SYSTEMS
SINCE THE GREEN REVOLUTION

Irrigated rice systems in tropical Asia will remain the major source of food production in the region, but their sustainable management represents an enormous challenge. Due to increased cropping intensity and yields, the pressure on the soil resource base has increased tremendously. Above-mentioned observations on yield decline, change in organic matter quantity and quality or nutrient depletion seem to indicate that current intensive rice systems are less sustainable than the traditional rice culture practiced for thousands of years. Both increased nutrient demand and prolonged submergence seem to cause gradual changes in soil quality that need to be managed.

Some of the sustainability issues in irrigated rice such as negative nutrient balances clearly result from inadequate soil and crop management. Declining indigenous nutrient supply and negative nutrient balances are key factors that may reduce the capacity of the soil resource base to sustain high rice yields. The seed and fertilizer package approach used during the Green Revolution in Asia did not address such problems adequately. Nutrient management practices of most rice farmers in Asia focus on optimizing short-term gains rather than on sustaining soil quality over the long run (Von Uexkuell and Beaton 1992). Exploitation of native soil fertility prevails over maintenance or enhancement of soil fertility. The diverse nature of the soil resource base, particularly the large variation in indigenous nutrient supply, has not been taken into account adequately. The importance of returning at least part of the rice straw for soil organic matter conservation and the nutrient balance is well-known, but, for various reasons, is widely neglected in present-day crop management.

Therefore, some of the observed negative trends in productivity are probably reversible through site-specific nutrient management approaches that focus on optimizing nutrient use efficiency in combination with long-term soil fertility management (Dobermann et al. 1996d; 2004). It is likely that such a fine-tuning of system management will significantly improve productivity and sustainability of intensive rice systems.

Preservation of natural resources depends on a system's environmental sustainability. The latter is often viewed in terms of biodiversity (Schoenly et al. 1996). There are numerous examples of reductions in genetic richness and organismic diversity through agronomic land use, both among flora and fauna (Schoenly et al. 1996), as well as among microorganisms (Torsvik et al. 1990).

Microorganisms serve key functions, which are also crucial for agronomic sustainability. This refers to nutrient cycling as well as to incidence of pathogens and their antagonists. A preliminary comparison of a wetland soil left fallow with a continuously cropped soil of the same texture indicates that intensive irrigated rice cropping can substantially reduce microbial functional diversity (Reichardt et al. 1996).

Problems with agronomic sustainability such as declining yields have been observed in a few long-term experiments with good nutrient management in both predominantly anaerobic (rice-rice) or anaerobic-aerobic (rice-wheat) cropping systems. Identifying the causes for this remains a challenge. At this stage, it remains an open question at what productivity level intensive rice systems can become environmentally and economically sustainable.

SUSTAINING SOIL QUALITY AND ITS FUNCTIONS

Managing the Organic Phase in Rice Soil-Floodwater Systems

It has been estimated that over 100 million Mg of rice straw are produced annually in Southeast Asia, but only a very small fraction is presently being reincorporated into the soil. Also, enrichment of the organic phase with N_2-fixing green manure has ceased in many areas with ongoing intensification, since its main purpose of providing sufficient N was no longer needed (Becker et al. 1994a, b; George et al. 1998). When economic conditions would change, however, use of green manure might resume, possibly supplementing the use of inorganic N fertilizer as part of integrated nutrient management strategies (Kundu and Ladha 1995). Fallow periods of 40–60 days in intensive systems limit the use of leguminous green manure to fast growing, short-duration species such as the stem-nodulating *Sesbania rostrata* (Singh et al. 1991; Ventura and Watanabe 1993; George et al. 1993, 1998). As green manures are selected for their capacity to accumulate N from N_2-fixation, its performance is judged in terms of the agronomic efficiency of its N component (Morris et al. 1986; Becker et al. 1994b). This efficiency is comparable with that of inorganic fertilizer only at N levels below 100 kg/ha (Singh et al. 1991). Nitrogen input, however, should not be the only criterion to justify the use of green manure. In China, a number of K-rich green manure plants have proved successful as potash fertilizers, with the beneficial side-effect of enhanced protein content in the grain (Peng and Li 1992). Grown in situ, however, such green manure plants do not contribute to net addition of K to the soil. The complex effects of organic matter inputs on soil quality improvement are also reflected in soil reclamation practices. Green manuring has been shown to be effective in accelerating the reclamation of saline and sodic soils (Singh et al. 1991).

Management of organic matter has not kept pace with the recent intensification of rice systems. Success hinges on improved understanding of the regulation of the network of biogeochemical pathways. Part of the mineral fertilizers is also assimilated by a dynamic, metabolizing matrix of active biomass and organic matter that forms the system's food web (e.g., Nannipieri et al. 1994; Clarholm 1994). Improved management of soil organic matter should lead to better synchronization of nutrient release and crop demand. This might be achieved by making use of the dynamics of the biota in the rice soil/floodwater system, the most labile fraction of organic matter (Nannipieri et al. 1994; Dobermann et al. 2004).

Monocropping versus Diversification

Cassman and Pingali (1995a) discussed reduction in the intensity of flooded rice monocropping by diversifying into higher value non-rice crops in rotation with rice. In flooded rice systems with two to three crops per year, diversification means providing an aerated upland crop phase. Provision of an aerated phase between two flooded rice crops would reverse the buildup of phenol-rich humic compounds (Olk et al. 1996) that may cause a reduction in N availability (Cassman et al. 1995). An aerated phase, long enough to grow an upland crop, is justified if total productivity is maintained or even enhanced, provided the quality of the resource base is not adversely affected.

Recent research in favorable rain-fed lowland rice systems (George et al. 1992, 1993, 1994, 1995) indicates that total productivity can indeed be increased by proper management of dry season and dry-to-wet season transition crops, including grain and green manure legume crops. Without an enhancement in total productivity, a short aerated phase between successive flooded rice crops could provide the same benefits in terms of soil aeration, organic matter decomposition, and formation and microbial activity, as a full aerated crop cycle. On the other hand, an aerated soil phase may not always be possible, because of the heavy clay texture of rice lowlands in the humid tropics. In regions with coarse-textured soils where rice-vegetable rotation cropping is practiced, excessive use of agrochemicals for the dry-season crops is already threatening the quality of the entire resource base. Rice-wheat systems are facing soil quality problems, too (Nambiar and Abrol 1989; Ladha et al. 2003). Thus, diversification alone does not necessarily solve the problems associated with intensification in the irrigated rice lowlands, although it can be part of an overall solution.

However, crop diversification might well become an issue in the presently intensively cultivated irrigated rice systems with the prospect of water becoming a premium commodity. Development of high value rice-non-rice cropping systems, such as rice-soybean or rice-vegetables may become a real possibility with the ever-increasing urban demand for water.

Balancing Nutrient Inputs and Outputs

Substantial quantities of N, P, K, and S are removed from the soil with each crop. Maintaining nutrient balances is therefore a prerequisite for keeping the resource base sustainable. Our understanding of nutrient cycling in the intensive irrigated rice system has improved. This has not, however, led to large-scale improvements in nutrient management practices in farmers' fields. Farmers' decisions about fertilizer application are often more affected by socioeconomic factors (market availability, prices, availability of cash and labor) than by biophysical needs. Adoption of existing technology and recommended practices by the farmer is confounded by considerable field-to-field variability in indigenous soil nutrient supply. To sustain average yields exceeding 7–8 ton/ha the nutrient use efficiency from both indigenous and external sources will have to be increased. This implies that nutrient management recommendation domains will have to shift from large regions to farms, single fields, or even single parcels within a larger field.

Knowledge-based objectives and tactics for management differ for each essential nutrient (Dobermann et al. 1996d). Adjusting N fertilizer dose to variations in indigenous N supply is as important as timing, placement, and source of applied N (Peng et al. 1996; Cassman et al. 1996; Dobermann et al. 2004). As nutrients such as P and K are not easily lost or added to the root zone by the biological and chemical processes affecting N, their management requires a long-term strategy that emphasizes maintenance of soil nutrient supply to insure that crop growth and N use efficiency are not limited. Diagnosis of potential deficiencies is the key management tool for nutrients such as Mg, Zn, and S. Once identified as a problem, deficiencies can be alleviated by regular or irregular (single) measures as part of a general fertilizer/soil use recommendation (Dobermann et al. 1996d). Straw management is a key leverage point for maintaining a positive balance of most nutrients, particularly for N and K (Becker et al. 1994b; Dobermann et al. 1996b, d). Increasing combine or stripper harvesting may provide new opportunities for better crop residue recycling.

Implementing site-specific management will only be successful if the additional labor required is restricted to a minimum, if the economic gain is sufficient, and if suitable, easy-to-use decision aid tools become available. In many Asian countries, facilities for more sophisticated farmer support need to be built-up, such as soil testing laboratories and soil testing programs—perhaps with the involvement of the private sector—fertilizer recommendation services providing objective information about new fertilizer products, and use of mass media (radio, TV, newspapers) for extension of new technologies. As the transition to farm- or field-specific management will take time, other cost-effective ways to increase nutrient use efficiencies must be further explored in the shorter-term (Pingali et al. 1995).

OPPORTUNITIES FOR SHORT-TERM MEASURES TO PROMOTE ENVIRONMENTAL SUSTAINABILITY

Agronomic sustainability, which implies stable productivity, is reflected in characteristics such as annual yields, partial factor productivity, or nutrient balances that can be monitored for each crop. Such records, however, only provide a partial account of the total factor capacity of the resource base to sustain high yields. As they are production-targeted, they do not include aspects of the environmental quality of the flooded resource base. Possible gains from differences in crop management have been estimated on the basis of long-term crop and water management research at IRRI in the Philippines. These estimates refer to actual attainable yields and use of irrigation water for three annual cropping patterns (rice–rice, rice–rice–rice, and rice–maize), employing conventional water management with soil puddling and soil submergence for each rice crop (table 14.4). Rainfall is relatively abundant at IRRI and supplies most of the water requirement in the wet season, while only limited supplementary irrigation is needed for land preparation. The dry season crop, on the other hand, relies heavily on irrigation.

We consider two options for production of "aerobic rice," that is, the growth of lowland high-yielding rice on aerated soil without standing water (Bouman et al. 2002), but with sufficient irrigation to avoid water stress. In one, aerobic rice is grown in the dry season and rotated with conventional production of rice on puddled, submerged soil in the wet season. In the other, aerobic rice is grown in both the wet and dry seasons (table 14.4). Because aerobic rice is a relatively new concept, agronomic and water use data for aerobic rice are very limited. Irrigation water inputs for aerobic

Table 14.4. Estimates of Grain Yield and Irrigation Water Inputs for Rice Ecosystems with No Water Constraints to Plant Growth, Based on Conditions at IRRI in the Philippines

Annual Cropping Pattern		Mean Grain Yield (ton/ha)		Irrigation Water (mm)	
Wet Season	Dry Season	Wet Season	Dry Season	Wet Season	Dry Season
Conventional water management					
rice	rice	5[a]	7[a]	400–600	1,400–1,600
rice	rice	5[a]	7[a]	400–600	1,400–1,600
rice-rice	rice	10[a,b]	7[a]	700–900	1,400–1,600
rice	maize	5[a]	6[a]	400–600	600–700
Aerobic rice					
rice (submerged)	rice (aerobic)	5[a]	4–6[c]	400–600	600-700
rice (aerobic)	rice (aerobic)	4–5[c]	4–6[c]	200–300	600–700

Notes:
[a] Estimates based on actual average yields from long-term trials.
[b] Based on sum for two crops.
[c] Tentative estimates based on probable attainable yields.

rice systems in table 14.4 are estimates assuming that the aerobic rice is established and irrigated like an upland crop, such as wheat or maize. Grain yields for aerobic rice are based on "best estimates," which assume that yields of aerobic rice will never exceed and often be lower than those of rice with conventional water management.

Table 14.5 presents estimates of grain yield, irrigation water inputs, and N-fertilizer inputs with no water and nitrogen constraints to plant growth based on the same long-term crop and water management research at IRRI. From the estimates in tables 14.4 and 14.5, the partial factor productivity (PFP, defined as the output of paddy grain per quantity of input of a given factor; see Dawe and Dobermann 1999) was estimated for irrigation water and nitrogen fertilizer (see table 14.5). The PFP for irrigation water for double rice cropping with conventional water management reflects the high water requirement. Increased reliance on rainfall is one way to increase PFP for irrigation water, as illustrated by the higher PFP for triple rice cropping in which water for the second wet-season rice crop largely originates from rainfall. Rotation of wet-season rice with a dry-season upland crop is another way to increase annual PFP for irrigation water, as illustrated by the higher PFP for the rice–maize rotation in which maize uses less water than rice in the dry season.

The PFP for irrigation water in aerobic rice agro-ecosystems strongly depends on the attainable grain yield. The PFP for irrigation water can be increased if yields of aerobic rice closely approach the yields of rice with

Table 14.5. Estimates of Grain Yield, Irrigation Water Inputs, N-Fertilizer Inputs and Partial Factor Productivity for Rice Ecosystems with No Water and Nitrogen Constraints to Plant Growth, Based on Conditions at IRRI in the Philippines

Annual Cropping Pattern		Grain Yield (ton/ha)	Irrigation Water (mm)	N-Fertilizer[b] (kg/ha)	Partial Factor Productivity	
Wet Season	Dry Season				N (kg/kg[-1])	Water (kg/mm[-1])
Conventional water management						
rice	rice	12[a]	1,800–2,200	200–290	60–41	0.67–0.55
rice–rice	rice	17[a]	2,100–2,500	310–465	55–37	0.81–0.68
rice	maize	11[a,c]	1,000–1,300	250–290	44–38[c]	1.10–0.85[c]
Aerobic rice						
rice (sub-merged)	rice (aerobic)	9–11[d]	1,000–1,300	—	—	1.10–0.69
rice (aerobic)	rice (aerobic)	8–11[d]	800–1,000	—	—	1.10–0.80

Notes:
[a] Estimates based on actual average yields from long-term trials.
[b] Approximate range used in long-term experiments.
[c] Based on sum for two crops.
[d] Tentative estimates based on probable attainable yields.

conventional water management. Limited available information indicates that significant yield reductions can occur when lowland rice cultivars are grown on aerobic soils. In such cases, the PFP would be only slightly higher for aerobic rice than for rice grown with conventional water management. The rice plant might require modification before it can achieve near comparable yields in unsaturated soils and in submerged soils, and thus the potential for increased productivity of irrigation water can be realized.

The estimates for use and PFP of irrigation water in table 14.5 are based on researcher-managed conditions at IRRI. Actual values in farmers' fields strongly depend on soil type and management. The values presented in table 14.5 simply illustrate a comparative analysis among rice production options, and they should not be extrapolated to rice ecosystems throughout Asia. Tables 14.4 and 14.5, nonetheless, illustrate the risk of tradeoffs, whereby increased PFP for irrigation water is achieved at the expense of reduced rice yield. Efforts to reduce irrigation water requirements must not compromise food security.

Saving of water at the field scale largely results from the reduction in outflows, which include seepage, percolation, and runoff. These outflows are sometimes recaptured for other uses at some point downstream in the irrigation system or in the basin, and hence these outflows are not really "lost" at higher scales. Water saved at the field scale, consequently does not necessarily result in more available water to irrigate new lands or to be diverted to other uses. A multiscale approach is needed to understand the effects of on-farm water saving on water availability at higher scales (Tuong 1999).

Growing aerobic rice might reduce labor requirements, but would result in a substantially more competitive and diverse weed flora than in transplanted rice with conventional water management. Without the use of post-emergence graminicides, which are not always effective, grassy weeds will dominate this flora. Aerobic rice will therefore require full integration of all available weed management practices. More tools are required to complement good agronomic practices to manage weeds in rice and minimize herbicide use. This will include choice of weed suppressive germplasm and water, crop, and land management practices for preventive weed control in addition to precision in-crop weed control with herbicides (optimal doses and precise application). Allelopathy and biological control have shown potential to be added to the toolbox for weed control. Rice allelopathy research has suggested that some cultivars are able to suppress weed growth by more than 50 percent under field conditions. Research in biological control shows promising results for control of some of the major weeds in rice.

Nitrogen fertilizer rates reported in table 14.5 represent the approximate ranges used in long-term experiments at IRRI to eliminate N constraints to crop growth. The reported ranges in PFP for N fertilizers with conventional water management match or exceed the mean of 44 kg grain per kg N fertilizer, obtained for irrigated rice from about 700 observations in farmers' fields in six Asian countries (Witt et al. 1999; Dobermann and Fairhurst 2000). Estimates of PFP for N fertilizer in ecosystems with aerobic rice at IRRI are not available yet. Current knowledge, however, does not provide a conceptual basis for anticipating higher PFP for N fertilizer with aerobic rice than with rice grown under conventional water management, using recently developed, N-efficient, real-time N management practices.

The efficiency of N-fertilizer use in lowland rice by farmers in Asia is typically very low. A mean agronomic efficiency of only 10 kg increase in grain production per kg applied N fertilizer was obtained from an extensive survey in farmers' fields in six Asian countries (Witt et al. 1999). Recent research has demonstrated considerable opportunity to improve the efficiency of N-fertilizer use in farmers' fields, through improved timing of N applications and reduced quantities of N fertilizer applied (Dobermann and Fairhurst 2000). With optimum crop and fertilizer management, an efficiency of 20–25 kg grain per kg N fertilizer can be achieved.

Recently, an important and costly pathogen of rice was controlled in a large region of China by planting alternating rows of two rice varieties (Zhu et al. 2000). This practice increased profitability and reduced the use of a potent pesticide. The mixed planting of crop genotypes that have different disease-resistance profiles, can also reduce or even effectively eliminate a pathogen.

Ultimately, productivity and environmental quality of a rice field are both linked to processes in the organic phase. Here, microorganisms are the main carriers of biocatalytic functions. They affect nutrient supply to the crop, as well as the cycling of bioelements, which is a crucial function in any ecosystem. This linkage allows us to look for promising combined measures that promote yield- and ecosystem-related sustainability.

Rapid progress in microbial ecology provides a number of options for short-term assays to quantify the sustainability of biocatalytic functions in the soil environment. One such category of assays targets certain enzymatic processes as potential indicators of functional imbalances in an ecosystem (Reichardt 1996). The relatively new discipline of ecotoxicology that focuses on man-made damage to ecosystem functions and environmental health has adopted biochemical techniques that were designed for holistic analyses of ecosystem functions.

Another measure to promote environmental sustainability may be based on the concept of functional microbial diversity and richness (Atlas 1984; Coleman et al. 1994). It partly requires advanced techniques such as biomarker analysis (Tunlid and White 1992; Reichardt et al. 1997). An alternative methodology is based on substrate mineralisation patterns (Zak et al. 1994; Reichardt et al. 1996, 1997). Notwithstanding the potential role of biodiversity as an indicator of an agro-ecosystem's sustainability, intensive lowland rice production systems are composed of an extremely large number of very diverse microbial subhabitats in space and time. The floodwater compartment with its primary production in particular is viewed as a major supporter of the system's sustainability (Roger 1996).

CONCLUSION

Based on current projections, lowland rice could increasingly be produced within a political, economic, social, cultural, and environmental framework of less irrigation water, lower labor availability, more mechanization, greater environmental considerations, and alternative income opportunities. This will lead to changes in water and nutrient management, soil tillage, and cropping patterns—which include increased aeration of soil and less burning of crop residues. A change from soil submergence to greater soil aeration in lowland rice ecosystems, while increasing the efficiency of water use, could have huge—and largely unknown—effects on the productivity, sustainability, nutrient dynamics, greenhouse gas emissions, carbon storage, and weed and pest dynamics of the ecosystem.

The production of lowland rice with less water, resulting in reduced duration of soil submergence and perhaps even growth of lowland rice in unsaturated soil, can lead to increased water productivity (grain yield per unit water input), but this should not come at the expense of rice yields or risks compromising food security. Increased soil aeration can, however, have significant negative impacts, such as loss of soil organic matter, increased emissions of nitrous oxide, and a more competitive weed flora. It could also present challenges for managing crop residues. Critical management issues in the future may, for example, include how much water and what frequency of soil submergence is essential for sustaining the productivity and services of rice ecosystems. Precision management of water through intermittent irrigation and the control of outflows can be challenging to implement at the field scale. Water management, unlike fertilizer and pesticide management, is influenced by the performance of higher scale infrastructures and institutional organization. Consequently, a prerequisite for on-farm water saving technology is a reliable and flexible supply system, which is presently lacking in most of the large irrigation systems in Asia. We are challenged to ensure that critical factors accounting for the sustained productivity of submerged rice soils are maintained in rice ecosystems as they evolve. This will necessitate management of natural resources at high and increasing levels of efficiency (output per unit input) to ensure profitable rice production, while minimizing or reducing negative environmental impacts within and beyond the rice ecosystems.

BIBLIOGRAPHY

Anderson, J. P. E. and K. H. Domsch. "Quantities of Plant Nutrients in the Microbial Biomass of Selected Soils." *Soil Science* 130 (1980): 211–16.

Atlas, R. M. "Use of Microbial Diversity Measurements to Assess Environmental Stress." Pp. 540–45 in *Current Perspectives in Microbial Ecology*, edited by J. M. Klug, M. and C. A. Reddy. Washington, D.C.: American Society of Microbiology (ASM), 1984.

Bajwa, M. I. "Soil Potassium Status, Potash Fertilizer Usage and Recommendations in Pakistan." *Potash Review*, no. 3, Subject 1, 20th suite. International Potash Institute, Basel, 1994.

Bambaradeniya, C. N. B., J. P Edirisinghe, D. N. De Silva, C. V. S Ginatilleke, K. B. Ranawana, and S. Wijekoon. "Biodiversity Associated with an Irrigated Rice Agroecosystem in Sri Lanka." *Biodiversity Conservation* 13 (2004): 1715–53.

Bastiaans, L., M. J. Kropff, N. Kempuchetty, A. Rajan, and T. R. Migo. "Can Simulation Models Help Design Rice Cultivars That Are More Competitive Against Weeds?" *Field Crops Research* 51 (1997): 101–11.

Becker, M., J. K. Ladha, and J. C. C. Ottow. "Nitrogen Losses and Lowland Rice Yield as Affected by Residue Nitrogen Release." *Soil Science Society of America Journal* 58 (1994a): 1660–65.

Becker, M., J. K. Ladha, I. C. Simpson, and J. C. C. Ottow. "Parameters Affecting Residue Nitrogen Mineralization in Flooded Soils." *Soil Science Society of America Journal* 58 (1994b): 1666–71.

Blair, G. J., C. P. Mamaril, and E. Momuat. "Sulfur Nutrition of Wetland Rice." *IRRI Research Paper Series* 21, International Rice Research Institute, Manila, 1978.

Bouman, B. A. M. and T. P. Tuong. "Field Water Management to Save Water and Increase its Productivity in Irrigated Lowland Rice." *Agricultural Water Management* 49 (2001): 11–30.

Bouman, B. A. M., H. Hengsdijk, B. Hardy, P. S. Bindraban, T. P. Tuong, and J. K. Ladha, eds. *Water-wise Rice Production*. Los Baños: IRRI and Wageningen: Plant Research International, 2002.

Bray, F. *The Rice Economies: Technology and Development in Asian Societies*. Oxford: Blackwell Publishers, 1986.

Broadbent, F. E. "Plant Use of Soil Nitrogen." Pp. 171–82 in *Nitrogen in Crop Production*. Madison, WI: ASA-SSSA, 1984.

Cantrell, R. P. "IRRI's Perspective on Integrated Natural Resource Management for Sustainable Production: Regional Efforts and Concerns—Opportunities and Challenges." Presentation at the Sixth General Assembly of APAARI and the Expert Consultation on Strategies for Implementing APAARI "Vision 2025: Strengthening Agricultural Research for Development in the Asia-Pacific Region." Chiang Rai, 8 November 2000.

Cassman, K. G. and P. L. Pingali. "Intensification of Irrigated Rice Systems: Learning from the Past to Meet Future Challenges." *GeoJournal* 35 (1995a): 229–305.

Cassman, K. G. and P. L. Pingali. "Extrapolating Trends from Long-term Experiments to Farmer's Fields: The Case of Irrigated Rice Systems in Asia." Pp. 64–84 in *Agricultural Sustainability in Economic, Environmental, and Statistical Terms*, edited by V. Barnett, R. Payne, and R. Steiner. London: John Wiley & Sons, 1995b.

Cassman, K. G., G. C. Gines, M. Dizon, M. I. Samson, and J. M. Alcantara. "Nitrogen-use Efficiency in Tropical Lowland Rice Systems: Contributions from Indigenous and Applied Nitrogen." *Field Crops Research* 47 (1996): 1–12.

Cassman, K. G., S. K. De Datta, D. C. Olk, J. Alcantara, M. I. Samson, J. P. Descalsota, and M. Dizon. "Yield Decline and the Nitrogen Economy of Long-term Experiments on Continuous, Irrigated Rice Systems in the Tropics." Pp. 181–222 in *Sustainable Management of Soils*, edited by R. Lal and B. A. Stewart. Michigan: Lewiston Publishers, CRC Press Inc. 1995.

Chen, F. X., Y. A. Chen, and C. M. Zou. "Potassium Balance in Red Earth Paddy Soil of Southern Hunan and Its Application." Pp. 193–94 in *Proceedings of the International Symposium on Paddy Soils*. Nanjing: Chinese Academy of Sciences, 1992.

Clarholm, M. "The Microbial Loop in Soil." Pp. 221–30 in *Beyond the Biomass*, edited by K. Ritz, J. Dighton, and K. E. Giller. Chichester: J. Wiley & Sons, 1994.

Coleman, D. C., J. Dighton, K. Ritz, and K. E. Giller. "Perspectives on the Compositional and Functional Analysis of Soil Communities." Pp. 261–70 in *Beyond the Biomass*, edited by K. Ritz, J. Dighton, and K. E. Giller. Chichester: J. Wiley & Sons, 1994.

Dawe. D. "Water Productivity in Rice-Based Systems in Asia—Variability in Space and Time." In *New Directions for a Diverse Planet*, edited by T. Fischer et al. *Proceedings for the 4th International Crop Science Congress*, Brisbane, Australia, 26 September–1 October 2004. www.cropscience.org.au.

Dawe, D. and A. Dobermann. "Defining Productivity and Yield." *IRRI Discussion Paper Series*, no. 33. Los Baños: International Rice Research Institute, 1999.

Dawe, D., A. Dobermann, P. Moya, S. Abdulrachman, B. Singh, P. Lal, S. Y. Li, B. Lin, G. Panaullah, O. Sariam, Y. Singh, A. Swarup, and P. S. Tan. "How Widespread are Yield Declines in Long-term Rice Experiments in Asia?" *Field Crops Research* 66 (2000): 175–93.

De Datta, S. K. and D. S. Mikkelsen. "Potassium Nutrition of Rice." Pp. 665–99 in *Potassium in agriculture*, edited by R. D. Munson, M. E. Summer, and W. D. Bishop, Madison, WI: ASA, CSSA, SSSA, 1985.

De Datta, S. K., K. A. Gomez, and J.P. Descalsota. "Changes in Yield Response to Major Nutrients and in Soil Fertility Under Intensive Rice Cropping." *Soil Science* 146: 350–58.

Dobermann, A. and T. Fairhurst. *Rice: Nutrient Disorders and Nutrient Management*. Singapore: Potash and Phosphate Institute (PPI), Potash and Phosphate Institute of Canada (PPIC), Los Baños: International Rice Research Institute, 2000.

Dobermann, A., C. Witt, and D. Dawe, eds. *Increasing Productivity of Intensive Rice Systems Through Site-Specific Nutrient Management*. Enfield, UK: Science Publishers, and Los Baños: International Rice Research Institute (IRRI), 2004.

Dobermann, A., P. C. Sta Cruz, and K. G. Cassman. "Potassium Balance and Soil Potassium Supplying Power in Intensive, Irrigated Rice Ecosystems." Pp. 199–234 in: *Potassium in Asia. Proceedings of the 24th International Colloquium of the International Potash Institute*, 21–24 February 1995, Chiang Mai and Basel: IPI.

Dobermann, A., D. Dawe, R. P. Rötter, and K. G. Cassman. "Reversal of Rice Yield Decline in a Long-Term Continuous Cropping Experiment." *Agronomy Journal* 92 (2000): 633–43.

Dobermann, A., P. C. Sta Cruz, and K. G. Cassman. "Fertilizer Inputs, Nutrient Balance, and Soil Nutrient-Supplying Power in Intensive, Irrigated Rice Systems. I. Potassium Uptake and K Balance." *Nutrient Cycling in Agroecosystems* 46 (1996a), 1–10.

Dobermann, A., K. G. Cassman, C. P. Mamaril, and J. E. Sheehy. "Management of Phosphorus, Potassium and Sulfur in Intensive, Irrigated Lowland Rice." Paper presented at the workshop on Nutrient Use Efficiency of the Cropping System(s). 13–15 Dec. 1995, International Rice Research Institute, Manila, (1996b).

Dobermann, A., K. G. Cassman, P. C. Sta Cruz, M. A. A. Adviento, and M. F. Pampolino. "Fertilizer Inputs, Nutrient Balance, and Soil Nutrient-Supplying Power in Intensive, Irrigated Rice Systems. III. Phosphorus." *Nutrient Cycling in Agroecosystems* 46 (1996c): 111–25.

Dobermann, A., K. G. Cassman, S. Peng, P. S. Tan, C. V. Phung, P. C. Sta Cruz, J. B. Bajita, M. A. A. Adviento, and D. C. Olk. "Precision Nutrient Management in Intensive Irrigated Rice Systems." Pp. 133–54 in *Proceedings of the International Symposium on Maximizing Sustainable Rice Yields through Improved Soil and Environmental Management*, edited by T. Attanandana, I. Kheoruenromne, P. Pongsakul, and T. Vearasilp. Khon Kaen, 1996d.

Flinn, J. C. and S. K. De Datta. "Trends in Irrigated Rice Yields Under Intensive Cropping at Philippine Research Stations." *Field Crops Research* 9 (1984): 1–15.

Geng, S., Y. Zhou, M. Zhang, and K. S. Smallwood. "A Sustainable Agro-ecological Solution to Water Shortage in the North China Plain (Huabei Plain)." *Journal of Environmental Planning and Management* 44 (2001): 345–55.

George, T., J. K. Ladha, R. J. Buresh, and D. P. Garrity. "Managing Native and Legume-fixed Nitrogen in Lowland Rice-based Cropping Systems." *Plant and Soil* 141 (1992): 69–91.

George, T., J. K. Ladha, R. J. Buresh, and D. P. Garrity. "Nitrate Dynamics During the Aerobic Soil Phase in Lowland Rice-based Cropping Systems." *Soil Science Society of America Journal* 57 (1993): 1526–32.

George, T., J. K. Ladha, D. P. Garrity, and R. J. Buresh. "Legumes as 'Nitrate Catch' Crops During the Dry-to-Wet Transition in Lowland Rice Cropping Systems." *Agronomy Journal* 86 (1994): 267–73.

George, T., J. K. Ladha, D. P. Garrity, and R. O. Torres. "Nitrogen Dynamics of Grain Legume-Weedy Fallow-flooded Rice Sequences in the Tropics." *Agronomy Journal* 87 (1995): 1–6.

George, T., R. J. Buresh, J. K. Ladha, and G. Punzalan. "Use and Dissipation of *In situ* Recycled Soil and Legume-Fixed Nitrogen in Tropical Lowland Rice Production." *Agronomy Journal* 90 (1998): 429–37.

Huke, R. E. and E. H. Huke. "Rice Area by Type of Culture: South, Southeast, and East Asia; A Revised and Updated Database." Manila: International Rice Research Institute, 1997.

Inubushi, K. and I. Watanabe. "Dynamics of Available Nitrogen in Paddy Soils. II. Mineralized N of Chloroform-Fumigated Soil as a Nutrient Source for Rice." *Soil Science and Plant Nutrition* 32 (1986): 561–77.

Ismunadji, M. "Rice Diseases and Physiological Disorders Related to Potassium Deficiency." *Proceedings of the 12th IPI-Colloquium*, Izmir, 1976.

Ismunadji, M., I. Zulkarnaini, and M. Miyake. "Sulfur Deficiency in Lowland Rice in Java." *Contr. Centr. Res. Inst. Agric. Bogor.* 14 (1975).

Kawaguchi, K. and K. Kyuma. *Paddy Soils in Tropical Asia. Their Material Nature and Fertility.* Honolulu: University Press of Hawaii, 1977.

Kemmler, G. "Potassium Deficiency in Soils of the Tropics as a Constraint to Food Production." Pp. 253–75 in *Priorities for Alleviating Soil-related Constraints to Food Production in the Tropics.* Manila: International Rice Research Institute, Manila, 1980.

Kimura, M., I. Watanabe, P. Patcharapreetcha, S. Panichsapatana, H. Wada, and Y. Takai. "Quantitative Estimation of Decomposition Process of Plant Debris in Paddy Field." *Soil Science and Plant Nutrition* 36 (1990): 33–42.

Kon, K. F. "Weed Management: Towards Tomorrow (With Emphasis on the Asia-Pacific Region)." Pp. 1–9 in *Proceedings of the 10th Australia and 14th Asia-Pacific Weed Science Society Conference.* Brisbane, 1993.

Kropff, M. J. and K. M. Moody. "Weed Impact on Rice and Other Tropical Crops." Pp. 123–26 in *Proceedings of the First International Weed Control Congress.* Melbourne, 1992.

Kropff, M. J., H. H. van Laar, and H. F. M. ten Berge. *ORYZA1: A Basic Model for Lowland Rice Production.* Los Baños: IRRI, 1993.

Kundu, D. K. and J. K. Ladha. "Enhancing Soil Nitrogen Use and Biological Nitrogen Fixation in Wetland Rice." *Experimental Agriculture* 31 (1995): 261–77.

Labrada, R. "The Need for Improved Weed Management in Rice." In *Sustainable Rice Production for Food Security: Proceedings of the 20th Session of the International Rice Commission,* Bangkok, 23–26 July 2002. Rome: FAO, 2003.

Ladha, J. K., D. Dawe, H. Pathak, A. T. Padre, R. L. Yadav, B. Singh, Yadvinder Singh, Y. Singh, P. Singh, A. L. Kundu, R. Sakal, N. Ram, A. P. Regmi, S. K. Gami, A. L. Bhandari, R. Amin, C. R. Yadav, E. M. Bhattarai, R. K. Gupta, and R. P. Hobbs. "How Extensive are Yield Declines in Long-term Rice-wheat Experiments in Asia?" *Field Crops Research* 81 (2003): 159–80.

Lee, K. E. "The diversity of Soil Organisms." Pp. 73–87 in *The Biodiversity of Microorganisms and Invertebrates: Its Role in Sustainable Agriculture,* edited by D. L. Hawksworth. Wallingford: CAB International, 1994.

Marx, D. B., J. T. Gilmour, H. D. Scott, and J. A. Ferguson. "Effects of Long-term Water Management in a Humid Region on Spatial Variability of Soil Chemical Status." *Soil Science* 145 (1988): 188–93.

Mohanty, S. K. and L. N. Mandal. "Transformation and Budgeting of N, P, and K in Soils for Rice Cultivation." *Oryza* 26 (1989): 213–31.

Morris, R. A., R. E. Furoc, and M. A. Dizon. "Rice Responses to a Short-duration Green Manure. I. Grain Yield." *Agronomy Journal* 78 (1986): 409–12.

Nambiar, K. K. M. and I. P. Abrol. "Long-term Fertilizer Experiments in India—An Overview." *Fertilizer News* 34 (1989): 11–20.

Nannipieri, P., L. Badalucco, and L. Landi. "Holistic Approaches to the Study of Populations, Nutrient Pools and Fluxes: Limitations and Future Research Needs." Pp. 231–46 in *Beyond the Biomass,* edited by K. Ritz, J. Dighton, and K. E. Giller. Chichester: John Wiley & Sons, 1994.

Oberthuer, T., A. Dobermann, and H. U. Neue. "How Good is a Reconnaissance Soil Map for Agronomic Purposes?" *Soil Use and Management* 12 (1996): 33–43.

Olk, D. C. and K. G. Cassman. 1995. "Characterization of Two Chemically Extracted Humic Acid Fractions in Relation to Nutrient Availability." In *Soil Organic Matter*

Management for Sustainable Agriculture, edited by R. D. B. Lefroy, G. J. Blair, and E. T. Craswell. ACIAR *Proceedings,* no. 56. Canberra: ACIAR.

Olk, D. C., K. G. Cassman, E. W. Randall, P. Kinchesh, L. J. Sangers, and J. M. Anderson. "Changes in Chemical Properties of Organic Matter with Intensified Rice Cropping in Tropical Lowland Soil." *European Journal of Soil Science* 47 (1996): 293-303.

Ottow, J. C. G., G. Benckiser, and I. Watanabe. "Iron Toxicity as a Multiple Nutritional Stress of Rice." *Giessener Beitrage zur Entwicklungsforschung* I, no. 7 (1981): 203-15.

Palm, C. A. and P. A. Sanchez. "Nitrogen Release From the Leaves of Some Tropical Legumes as Affected by their Lignin and Polyphenolic Contents." *Soil Biology and Biochemistry* 23 (1991): 83-88.

Peng, K. L. and F. X. Li. "The Selection and Application of Potassium-rich Green Manure." Pp. 66-69 in *Proceedings of Studies on Green Manure in China,* edited by L. Hen. Shinying: Lining University Press, 1992.

Peng, S., F. V. Garcia, R. C. Laza, A. L. Sonic, R. M. Vespers, and K. G. Caspian. "Increased N-use Efficiency Using Chlorophyll Meter on High-yielding Irrigated Rice." *Field Crops Research* 47 (1996): 243-52.

Pingali, P. L., M. Hossain, S. Pandey, and L. Price. *Economics of Nutrient Management in Asian Rice Systems: Toward Increasing Knowledge Intensity.* Los Baños: International Rice Research Institute, 1995.

Ponnamperuma, F. N. "The Chemistry of Submerged Soils." *Advances in Agronomy* 24 (1972): 29-96.

Reichardt, W. "Enzymatic Potential for Decomposition of Detrital Biopolymers in Sediments from Kiel Bay." *Ophelia* 28 (1986): 369-84.

Reichardt, W. "Ecotoxicity of Certain Heavy Metals Affecting Bacteria-mediated Biogeochemical Pathways in Sediments." Pp. 159-78 in *Sediments and Toxic Substances—Environmental Effects and Ecotoxicity,* edited by W. Calmano and U. Foerstner. Berlin: Springer Verlag, 1996.

Reichardt, W., G. Mascaria, B. Padre, and J. Doll. "Microbial Communities of Continuously Cropped Irrigated Rice Fields." *Applied and Environmental Microbiology* 63 (1997): 233-38.

Reichardt, W., A. Briones, B. Padre, R. de Jesus, and G. Mascaria. "Dynamics of Soil Microbial Communities and Sustainable Nutrient Supply in Highly Intensified Rice Cultivation." Pp. 887-96 in *Proceedings of the International Symposium on Maximizing Sustainable Rice Yields through Improved Soil and Environmental Management,* edited by T. Attanandana, I. Kheoruenromne, P. Pongsakul, and T. Vearasilp. Khon Kaen, 1996.

Rijsberman, F. R. "Water Scarcity: Fact or Fiction?" In *New Directions for a Diverse Planet,* edited by T. Fischer et al. *Proceedings of the 4th International Crop Science Congress,* Brisbane, 26 September-1 October 2004. The proceedings are available online at: www.cropscience.org.au/.

Roger, P. A. *Biology and Management of the Floodwater Ecosystem in Rice Fields.* Manila: International Rice Research Institute and Paris: ORSTOM, 1996.

Saito, M. and I. Watanabe. "Organic Matter Production in Rice Field Floodwater." *Soil Science and Plant Nutrition* 28 (1978): 427-40.

Schink, B. "Principles and Limits of Anaerobic Degradation: Environmental and

Technological Aspects." Pp. 771–846 in *Biology of Anaerobic Microorganisms*, edited by A. J. B. Zehnder. New York: John Wiley & Sons, 1988.

Schoenly, K. G., J. E. Cohen, K. L. Heong, A. S. Arida, A. T. Barrion, and J. A. Litsinger. "Quantifying the Impact of Insecticides on Food Web Structure of Rice-arthropod Populations in a Philippine Farmer's Irrigated Field: A Case Study." Pp. 343–51 in *Foodwebs: Integration of Patterns and Dynamics*, edited by G. Polis and K. Winemiller. New York: Chapman and Hall, 1996.

Singh, Y., C. S. Khind, and B. Singh. "Efficient Management of Leguminous Green Manures in Wetland Rice." *Advances in Agronomy* 45 (1991): 135–89.

Sri Adiningsih, J., D. Santoso, and M. Sudjadi. "The Status of N, P, K and S of Lowland Rice Soils in Java." Pp. 68–76 in *Sulfur Fertilizer Policy for Lowland and Upland Rice Cropping Systems in Indonesia*, edited by G. Blair. Melbourne: ACIAR, 1991.

Tiwari, K. N. "Changes in Potassium Status of Alluvial Soils under Intensive Cropping." *Fertilizer News* 30 (1985): 17–24.

Torsvik, V., K. Salte, R. Sorheim, and J. Goksoyr. "Comparison of Phenotypic Diversity and DNA Heterogeneity in a Population of Soil Bacteria." *Applied Environmental Microbiology* 56 (1990): 776–81.

Tsutsuki, K. and F. N. Ponnamperuma. "Behavior of Anaerobic Decomposition Products in Submerged Soils." *Soil Science and Plant Nutrition* 33 (1987): 13–33.

Tunlid, A. and D. C. White. "Biochemical Analysis of Biomass, Community Structure, Nutritional Status, and Metabolic Activity of Microbial Communities in Soil." Pp. 229–62 in *Soil Biochemistry*, vol. 7, edited by G. Stotzky and J. M. Bollag. New York: Marcel Dekker, 1992.

Tuong, T. P. "Productive Water Use in Rice Production: Opportunities and Limitations." *Journal of Crop Production* 2 (1999): 241–64.

Tuong, T. P., B. A. M. Bouman, and M. Mortimer. 2004. "More Rice, Less Water— Integrated Approaches for Increasing Water Productivity in Irrigated Rice-Based Systems in Asia." In *New Directions for a Diverse Planet*, edited by T. Fischer et al. *Proceedings of the 4th International Crop Science Congress*, Brisbane, 26 September–1 October 2004. The proceedings are available online at: www.cropscience.org.au/.

Uexkuell, H. R. von. "Availability and Management of Potassium in Wetland Rice Soils." Pp. 293–305 in *Wetland Soils: Characterization, Classification and Utilization*. Manila: International Rice Research Institute, 1985.

Uexkuell, H. R. von. and J. D. Beaton. "A Review of Fertility Management of Rice Soils." Pp. 288–300 in *Characterization, Classification, and Utilization of Wet Soils*, edited by J. M. Kimble. *Proceedings of the 8th International Soil Correlation Meeting* (VIII. ISCOM). Lincoln, NE: USDA, Soil Conservation Service, 1992.

Ventura, W. and I. Watanabe. "Green Manure Production of *Azolla Microphylla* and *Sesbania Rostrata* and Their Long-term Effects on Rice Yields and Soil Fertility." *Biology and Fertility of Soils* 15 (1993): 241–48.

Witt, C., A. Dobermann, S. Abdulrachman, H. C. Gines, G. H. Wang, R. Nagarajan, S. Satawathananont, T. T. Son, P. S. Tan, L. V. Tiem, G. C. Simbahan, and D. C. Olk. "Internal Nutrient Efficiencies of Irrigated Lowland Rice in Tropical and Subtropical Asia." *Field Crops Research* 63 (1999): 113–38.

Ye, W. and Q. X. Wen. "Characteristics of Humic Substances in Paddy Fields." *Pedosphere* 1 (1991): 229–39.

Yoshida, S. *Fundamentals of Rice Crop Science.* Manila: International Rice Research Institute, 1981.

Zak, J. C., M. R. Willig, D. L. Moorhead, and H. G. Wildman. "Functional Diversity of Microbial Communities: A Quantitative Approach." *Soil Biology and Biochemistry* 26 (1994): 1101–8.

Zhu, Y., H. Chen, J. Fan, Y. Wang, Y. Li, Y. Chen, J. Fan, S. Yang, L. Hu, H. Leung, T. S. Mew, P. S. Teng, Z. Wang, and C. C. Mundt. "Genetic Diversity and Disease Control in Rice." *Nature* 406 (2000): 718–22.

15

Agricultural Technology and Nitrogen Pollution in Southeast China

Marrit van den Berg, Guanghuo Wang, and Reimund Roetter

China's recent agricultural growth is impressive. Between 1965 and 2002, the production of cereals has increased by 148 percent and the production of vegetables and melon by 703 percent (FAOSTAT database, 2004). Research-induced technical change accounts for 20 percent of agricultural growth between 1993 and 1995 (Fan and Pardey 1997). Important aspects of the new technologies are improved seeds and an intensive use of agro-chemicals. Total fertilizer use, for example, increased 50 times from 1962 to 2001, and China presently consumes about 30 percent of the world's nitrogen fertilizer (Fan and Pardey 1997).

The intensive use of agro-chemicals associated with production growth has had a negative impact on environmental quality. Gaseous nitrogen losses from paddy fields contribute significantly to global warming (Bodegom et al. 2000). On a local scale, losses of fertilizers from agricultural fields have resulted in eutrophication of surface water in the intensively cultivated East (Ellis and Wang 1997). Furthermore, a recent study on water quality in vegetable-production areas in northern China demonstrates that nitrate pollution in ground and drinking water has become a serious problem (Zhang et al. 1996). Nitrate contents exceeded the European limit of 50 mg/liter for drinking water in over half of the sixty-nine locations investigated, and in the worst case the nitrate content of the groundwater reached 3.5 times the drinking water limit. In all locations, high amounts of nitrogen fertilizer were applied: recorded N rates amounted to up to 1,900 kg/ha/annum. Similar levels of fertilizer input and emissions have been observed in vegetable production in eastern and southern China (Sheldrick et al. 2003).

Agricultural research has recognized the downside of the rapid rise in agro-chemical use and focuses increasingly on the development of more sustainable technologies (Smil 1998, 2002; Zhu and Chen 2002). The question is how large the potential impact of research on decreasing agricultural pollution is. That is, to what extent can new technological developments help reduce pollution and to what extent can government policies aimed at changing cropping patterns and stimulating adoption of new technologies play a role?

The present chapter analyzes these issues for *Pujiang*, a county in the province of *Zhejiang* in Eastern China. On June 2, 2002, the local government declared that "after three years' effort, the unit amount of chemical fertilizer and pesticide application should be reduced one-third, to ensure. . . agro-products to be green, safe and environmental(ly) health(y)" (People's Government of *Jinhua* City, 2002). This declaration reflects a general concern for the environmental consequences of current agricultural practices. The government had not elaborated a concrete plan of action to achieve the set goals but stressed the importance of innovation and science. In this study we assess whether science and innovation can indeed play an important role in increasing the sustainability of agricultural production in the area.

The case study approach used below compares farmer practices with data from local research and technologies developed for other regions. This approach deviates from the standard production function approach used in economic studies to assess the impact of agricultural research. The latter approach involves estimation of aggregate agricultural production or total factor productivity at the regional or national level as a function of inputs or resources and research expenditures (e.g., Fan 2000). This approach has the advantage of producing results with validity for large areas but cannot be used to assess the impact of research on agricultural pollution, as the required aggregate data on nutrient losses are not available. Moreover, the case study approach has the advantage to be forward looking. We do not assess the past impact of research as in the production function approach but the potential future impact. Two different questions are taken into account: (i) which are the suitable technologies available or developed at research stations, and (ii) which other technologies could be developed given the availability of similar technologies elsewhere—thus identifying gaps in local and region-specific research.

The structure of the chapter is as follows. The second section presents the background theory on agricultural technology and fertilizer use. In the third section we give a description of the case study area and estimate nitrogen losses based on farm survey data. In the fourth section, the potential impact of research on nitrogen pollution is analyzed for both on-the-shelf technologies and additional research, while the last section presents the conclusion.

AGRICULTURAL TECHNOLOGY AND FERTILIZER USE

When farmers decide on fertilizer application, they do not simply balance fertilizer prices and expected marginal returns. Farmers often use more than the profit-maximizing level of fertilizers (Babcock 1992). They use decision criteria such as "fertilizing for the good years" or "applying a little extra fertilizer just in case it is needed." This behavior weakens the direct link between prices and fertilizer use, but is not necessarily inconsistent with expected profit maximization. If ex-post optimal fertilizer rates are positively correlated with yield, fertilizing for average conditions leads to relatively high levels of foregone income in good years, while the cost of some additional fertilizers in normal years are relatively low (Babcock and Blackmer 1994).

Typical fertilizer recommendations from agricultural research institutes are independent of prices and reflect "optimal" fertilizer rates, that is, the minimum amount of fertilizer needed to reach maximum yield under optimal conditions. This type of recommendation disregards economic optimization, but is not inconsistent with the general farmer strategy of fertilizing "for the good years." Optimizing farmers have no reason to apply more than this amount of fertilizers, unless the rates are only optimal with a costly technology or farmers have large surpluses of organic fertilizers from intensive livestock production and the cheapest means of disposal is on-farm application.

Not fertilizer use per se but its impact on the quality of soils, water, and other natural resources are relevant for sustainability. The intensive agriculture in eastern China results in large emissions of nutrients to the environment. These emissions depend not only on the amount of fertilizer applied but also on the method of fertilizer application. Agronomic research has aimed to increase both crop yields and nutrient uptake, through, for example, fine-tuning the timing and doses of fertilizer applications with respect to crop requirements under given local conditions. That is, improved technologies not only involve higher yields but also higher use efficiency—and thus lower nutrient losses—per kg of product.

Figure 15.1 illustrates the relation between technology, fertilizer use, and losses. The positive y-axis pictures yield. There are two production functions; one for the farmer practice and one for the improved technology. Both functions increase to plateaus that lie below the potential yield (Van Ittersum et al. 2003) as determined by crop characteristics, temperature, and radiation. The improved technology produces higher yields at lower levels of fertilization due to better timing of applications. Hence, shifting from current farmer practice to improved technology only requires additional knowledge/skills and no change in inputs other than fertilizers. The consequences of fertilizer use for the environment are pictured below the x-

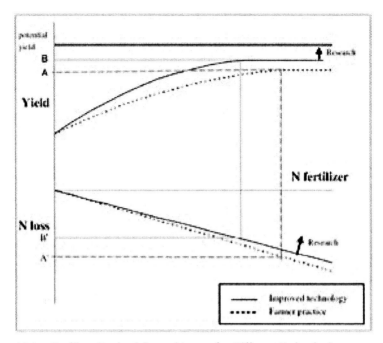

Figure 15.1.   Fertilizer Productivity and Losses for Different Technologies

axis. Fertilizer uptake rates are highest for the improved technology. That is, for the improved technology the same fertilizer application leads to lower losses to the environment than for the current farmer practice. Additional research could result in a further increase in fertilizer uptake rates, and the yield plateau (but not higher than the potential yield).

Using farmer technology, a knowledgeable farmer aiming for maximum yield at minimum fertilizer use will produce at point A with fertilizer loss A'. After being informed about the new technology, the farmer will shift to slightly higher yield B with lower fertilizer use. Hence, the improved technology results in a reduction in fertilizer losses from A' to B' through both decreasing total fertilizer use and the share of fertilizers lost to the environment.

FERTILIZER USE AND THE ENVIRONMENT IN *PUJIANG* COUNTY, *ZHEJIANG* PROVINCE

Fertile soils and abundant water resources make *Zhejiang*, and the other parts of China's greater *Yangtze* River Delta, one of the world's most productive rice growing regions. With low per capita land availability, farmers have traditionally generated some of Asia's highest rice yields through

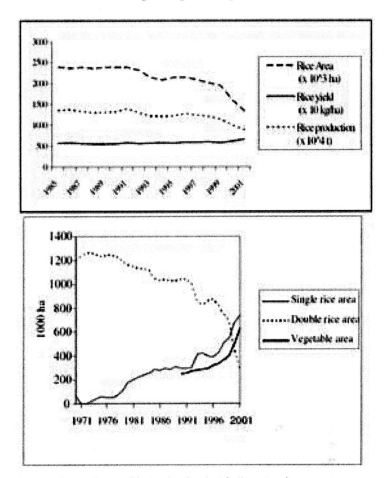

Figure 15.2.   Rice and Vegetable Production in *Zhejiang* Province

intensive use of labor. In the era of the P. R. China, continuous population growth and a concern for rice self-sufficiency stimulated technological development, which increased yields even further. The modern technologies involve dwarf varieties and hybrids that are highly responsive to fertilizers and have a climate-adjusted, genetic yield potential of 10–12 ton/ha/crop in *Zhejiang* Province (Huang and Rozelle 1996; Wang et al. 2004; Widawsky et al. 1998). *Zhejiang* farmers have adopted these varieties at a large scale and have increased their use of chemical fertilizers at about 5 percent per year during the 1980s and early 1990s (Widawsky et al. 1998).

Despite these large technological developments, farmer yields have stagnated at 5.5–6 ton/ha/crops since 1985, and total rice production has decreased dramatically, especially since the late 1990s (see figure 15.2). Industrialization and urbanization have caused a decline in rice area of about 2 percent per year between 1980 and 2000, which has resulted in a loss of about 500,000 hectare of rice cultivation area (Wang et al. 2001). Moreover, *Zhejiang* was the first province where farmers became completely free in their choice of crops. Rice prices have been low (Dawe 2002) and many farmers have replaced double rice for single rice production or for alternative crops such as fruits and vegetables. These crops accounted for about half of all fertilizers applied in the province in 2001 (estimate based on data from Zhejiang Statistical Bureau 2001).

At a national level, grain production showed a declining trend for four consecutive years since 2000. Out of concern for the nation's food security, the government promulgated a series of measures in 2004 to rejuvenate promotion of grain production (see also chapter 8). Major policies include setting minimum product prices, lowering agricultural taxes, giving subsidies to grain producers directly, and revoking industrial development zones (People's Daily Online 2004a, b). Preliminary statistics indicate that these policies have resulted in a favorable turn for grain output from decline to growth at the national level (People's Daily Online 2004c). In *Zhejiang*, rice prices increased by about 30 percent. This has resulted in an increase in rice production from 6.47 million tons in 2003 to 6.87 million tons in 2004 (NBS 2004, 2005). The future will show whether the shift in policy can structurally reverse the decline in rice production in *Zhejiang* Province.

Zhejiang faces severe water quality problems, which are partly caused by agriculture. One of the three most polluted lakes in China, Lake *Taihu*, lies on the border between *Zhejiang* and *Jiangsu* provinces. The lake is highly eutrophicated by nutrients, and the water of the lake is no longer suitable for drinking. During summer, excessive growth of algae causes a foul smelling layer on the surface of the lake. The water quality of tributary rivers, canals, and ditches is even worse (Luijkx 2002). Chemical fertilizers are responsible for 23 percent of the discharge of nitrogen in the Taihu Basin (NIES 1996, cited by Luijkx 2002).

In November and December 2002, we held a survey among 107 farm households in *Pujiang*, a county in the center of *Zhejiang* (see figure 15.3). In the lowlands at the heart of *Pujiang*, which are the focus of the present study, the main cropping systems are single rice, double rice and vegetables. In the surrounding hills and mountain areas mainly perennial crops are grown. As the main purpose of the survey was to get insight in the diversity of agricultural technologies in the area, the design was such to cover all important farm types. For the larger farm types, e.g., fruit plantations on the hillsides, households were randomly selected from a (complete) list of large farms from the Agricultural Bureau. As smallholder farms were not formally listed, the survey group drafted lists with local officials of representative villages of each township. Survey households were randomly selected from these lists. The number of households to be interviewed per agro-ecological zone was determined according to population proportions, and the minimum number of households in each of the townships was set to five.

During the survey, data on crop production and input use were collected for each farm plot separately. This has resulted in input and output data for 101 fields with single rice, double rice, or horticultural crops, the major cropping systems of the lowlands. In our analysis we focus on the use and losses on nitrogen, the nutrient that has caused most problems to the environment. Virtually all fertilizers applied are inorganic, as organic manure is produced mainly on specialized farms and not traded with crop farmers.

Average fertilizer use was especially high for horticultural crops: 743 kg N/ha/yr, compared to 298 kg N/ha/yr for double rice and 150 kg N/ha/yr for single rice (see table 15.1). Previous research suggests that these high values are realistic. The average fertilization rate of field-grown vegetables was found to be 781 kg N/ha in Beijing suburbs and 1,894 kg N/ha in *Fanzuhuang, Yutian* (Zhang and Liu 2002, cited by Härdter and Fairhurst 2003; Zhang et al. 1996). Fertilization of greenhouse vegetables is even more excessive, with measured averages of 2,388 kg N/ha in *Shouguang, Shandong* (Zhang and Liu 2002, cited by Härdter and Fairhurst 2003).

Nitrogen losses were computed for each cropping system at average fertilizer applications for all fields using TechnoGIN (Ponsioen et al. 2003), a generic expert tool for integrating different types of information on crop production (Hengsdijk and van Ittersum 2003). Based on soil, crop, and technology characteristics, TechnoGIN allows characterization of cropping systems in terms of inputs and outputs, including the amount of nutrients lost to the environment. TechnoGIN has been calibrated for conditions in *Pujiang* on the basis of the 2002 survey and local experimental data (Wang et al. 2001). TechnoGIN computes per hectare nitrogen leaching and gaseous nitrogen losses, which we aggregated to total nitrogen losses for *Puji-*

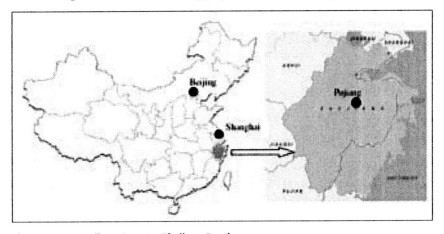

Figure 15.3. *Pujiang* County, *Zhejiang* Province

Table 15.1. Nitrogen Use and Losses in Pujiang County

	Double Rice	*Single Rice*	*Annual Horticulture*
Field-level averages			
1. Number of observations	17	45	39
2. N-fertilizer (kg/ha/yr)[a]	298	150	743
3. N-leaching (kg/ha/yr)[b]	29	18	391[e]
4. Gaseous N losses (kg/ha/yr)[b]	222	116	201[e]
5. Apparent N-recovery efficiency[b,d]	0.17	0.21	0.14[e]
County-level estimates			
6. Total area under cropping system (ha)[c]	5,231	12,205	1,762
7. Total N-leaching (10³ kg/yr)[f]	152	220	826
8. Total gaseous N-losses (103 kg/yr)[g]	1,161	1,416	423

Sources: [a] Own survey in December 2002; [b] TechnoGIN computations; [c] *Zhejiang* Statistical Bureau (2001).
Notes: [d] ANRE = (crop uptake–uptake on 0 N plot)/N fertilizer.
   [e] Computed for the common cropping system greens-celery-radish with average fertilizer use of 869 kg N/ha. This is higher than average, since not all systems have triple crops. Single or double cropping systems have lower fertilizer gifts, but also lower crop uptake.
   [f] Row 6 × row 2.
   [g] Row 6 × row 4.

ang County using data from the *Zhejiang* Statistical Bureau on county-level acreage per cropping system.

As nitrogen uptake efficiency was low for all crops, the greater part of all nitrogen fertilizer ended up in the environment. The estimated county-level total losses of nitrogen are highest for single rice, the dominant cropping system in the county: 1,636 tons/yr. Estimated total losses were of similar magnitude for double rice and vegetables; 1,313 tons/yr for double rice and 1,249 tons/yr for annual horticulture (table 15.1: rows 7 and 8). The differences in total estimated nitrogen losses between the systems are small relative to the differences in acreage: on a per hectare basis, single rice has the smallest nitrogen losses and annual horticulture the largest (table 15.1: rows 2 and 3). This is not only the result of a larger cropping intensity, but also a less efficient fertilizer use as reflected by a lower apparent nitrogen-recovery efficiency (table 15.1: row 5).

Due to the relatively large share of leaching in total nitrogen losses for horticultural crops, horticulture already contributed more to water pollution than rice did (table 15.1: row 7). If the trend of substitution of rice for horticulture continues, this will cause a further deterioration of ground and surface water quality. Research in northern China indicates that the nitrate content in groundwater exceeds the European limits for drinking water when more than 500 kg N/ha is applied and less than 40 percent of applied N is taken up (Zhang et al. 1996). In *Pujiang* County, fertilization of vegetables largely exceeds these limits. On the other hand, if China's new policies to stimulate grain production succeed in offsetting the trend of increased cultivation of horticultural crops, water quality is likely to improve.

While leaching from rice cultivation is limited, gaseous nitrogen losses from paddy fields are relatively high. These gasses contribute to global warming and are thus harmful for the global environment. The market-induced shift from double to single rice is good news in this respect: it involves a significant reduction in nitrogen losses.

AGRICULTURAL TECHNOLOGY FOR REDUCING NITROGEN POLLUTION IN *PUJIANG*

The local government recognizes the damage that intensive fertilization can do to the environment and has declared the intention to reduce fertilizer use (People's Government of *Jinhua* City 2002). No specific measures have been announced, but the proclamation stresses the role of research and technology. Below, we analyze the potential impact of recently developed technologies and expected development of more advanced future technologies.

In recent experiments on twenty-one farms in seven villages in *Jinhua* district, which also comprises *Pujiang* County, researchers have succeeded in increasing the apparent nitrogen-recovery efficiency (ANRE) in double rice from 0.2 kg plant N per kg fertilizer N in the farmer's fertilizer practice to 0.3 kg/kg using site-specific nutrient management (SSNM) (Wang et al. 2001). The SSNM approach optimizes the use of nutrients from soil, crop residues, and fertilizers. Whereas most farmers apply all fertilizers within the first 10–15 days of the crop cycle, SSNM involves a basal dressing (before transplanting) and three top dressings at different development stages of the crop. Moreover, the new technology involves abolishing the common practice of midseason drainage, which causes large losses of nitrogen to the atmosphere. The result is an 8 percent higher yield at a 12 percent lower N-fertilizer use (see table 15.2). This is associated with a reduction in gaseous losses of 10 percent and leaching of 14 percent.

We translated the experimental results for SSNM in double rice to single rice. We assumed the yield increase from farmer practice to SSNM and the ANRE of SSNM to be the same for single rice as for double rice. As the ANRE in farmer practice was higher in single rice than in double rice, the SSNM as defined has relatively little impact on nitrogen losses: Nitrogen leaching decreased by 6 percent and gaseous nitrogen losses by 3 percent. The newly developed technology of SSNM seems attractive for farmers, especially for double rice. It involves an increase in yields at lower (double rice) or identical (single rice) fertilizer use. The additional costs are low. Planting

Table 15.2.   Comparison of N Use and Losses between Farmer Practice and Improved Technologies for Rice

	Double Rice	*Single Rice*
Farmer practice		
Yield (10³ kg/ha/yr)	10.7	7.1
N-fertilizer (kg/ha/yr)	298	150
N-leaching (kg/ha/yr)	29	18
Gaseous N losses (kg/ha/yr)	222	116
Site-specific nutrient management (SSNM)		
Yields	11.5 (8%)	7.7 (8%)
N-fertilizer (kg/ha/yr)	263 (−12%)	150 (0%)
N-leaching (kg/ha/yr)	26 (−10%)	17 (−6%)
Gaseous N-losses (kg/ha/yr)	192 (−14%)	112 (−3%)
Future-oriented technology: apparent N-recovery = 0.40		
Yields	11.5 (0%)	7.7 (0%)
N-fertilizer (kg/ha/yr)	165 (−37%)	105 (−30%)
N-leaching (kg/ha/yr)	16 (−38%)	12 (−29%)
Gaseous N-losses (kg/ha/yr)	110 (−43%)	71 (−37%)
Future-oriented technology: apparent N-recovery = 0.50		
Yields	11.5 (0%)	7.7 (0%)
N-fertilizer (kg/ha/yr)	132 (−50%)	84 (−44%)
N-leaching (kg/ha/yr)	12 (−54%)	10 (−41%)
Gaseous N-losses (kg/ha/yr)	81 (−58%)	52 (−54%)

Sources: Farmer yields and fertilizer use are survey averages. Yields for SSNM are 8 percent higher than average farmer yields (Wang et al. 2004). Fertilizer use for SSNM is computed based on an apparent N recovery of 0.27 and indigenous nutrient supply of 0.7 kg/day (Wang et al. 2004). All nitrogen losses are computed by TechnoGIN.

Note: Numbers in parentheses are changes with respect to farmer practice for SSNM and changes with respect to SSNM for future technologies.

density is somewhat higher, which requires more seeds and labor. But these additional costs are more than compensated by the lower fertilizer costs and higher yields. There are also additional labor costs for real-time N management. These costs are low, but do involve some daily work over a longer period. This could be problematic in a region where many farmers are involved in non-farm employment, but this problem can be overcome through a community-oriented program with one person doing it for many (say about 20) fields per day (Wang et al. 2004). Alternatively, the task could be assigned to those household members who are not involved in non-farm employment or who are involved in local non-farm employment and thus live on the farm.

Despite these recent technological developments, there is still scope for further improving N-use efficiency in rice. Apparent recovery rates of SSNM are 0.3 kg plant N per kg fertilizer N, while with knowledge-intensive management it is possible to achieve rates of 0.5–0.6 in irrigated rice (Dobermann et al. 2000; Fischer 1998; Peng and Cassman 1998). These rates could be reached through better monitoring and matching crop N demand and supply—that is, by using a chlorophyll meter (Peng et al. 1996). However, it should be borne in mind that these upper thresholds of nitrogen uptake and fertilizer use efficiencies in rice were all obtained in high-yielding situations at experimental stations. Moreover, unlike in the SSNM approach for *Zhejiang,* in most of these high-yielding situations, basal N applications had been replaced by delaying the first application until mid tillering to avoid high losses during early growth stages (Peng et al. 1996). We computed nitrogen requirements and losses for apparent N-recovery rates of 0.4 and 0.5 to gain insight into the possible benefits of the development and adoption of such technologies for environmental sustainability (see table 15.2). The results show that a lot can be gained from these technologies. An increase of the apparent N-recovery of 0.3 to 0.4 results in a decrease in total nitrogen losses of 42 percent for double rice and 36 percent for single rice. A further increase to 0.5 would result in another decrease of 15 and 16 percent points for double and single rice, respectively.

Fertilizer recovery rates are even lower for horticultural production than for rice (see table 15.1). Extension for non-rice crops is relatively weak, and there is limited site-specific research on annual vegetables and fruits. Hence, we could not find data on improved non-rice technologies. To compensate for this absence of field data, we used TechnoGIN to compute potential technologies for the common rotation of greens-celery-radish (see table 15.3). We increased the ANRE from 0.14 in farmer practice to 0.20 and 0.25 assuming that yields remain the same. Like SSNM techniques in rice, these changes could be achieved by fine-tuning fertilizer gifts to crop requirements. The results of our computations are promising: An increase

Table 15.3. Comparison of N Use and Losses Between Farmer Practice and Improved Technologies for Vegetables

	Greens-Celery-Radish
Farmer practice	
Yield (103 kg/ha/yr)	42.2 - 50 - 30
N-fertilizer (kg/ha/yr)	869
N-leaching (kg/ha/yr)	434
Gaseous N losses (kg/ha/yr)	221
Future-oriented technology: apparent N-recovery = 0.20	
N-fertilizer (kg/ha/yr)	625 (–28%)
N-leaching (kg/ha/yr)	330 (–24%)
Gaseous N-losses (kg/ha/yr)	168 (–24%)
Future-oriented technology: apparent N-recovery = 0.25	
N-fertilizer (kg/ha/yr)	500 (–42%)
N-leaching (kg/ha/yr)	259 (–40%)
Gaseous N-losses (kg/ha/yr)	132 (–40%)

Sources: Farmer yields and fertilizer use are survey averages. Future technologies and nitrogen losses are computed by TechnoGIN as described in table 2. Yields are assumed identical for all technologies.
Note: Numbers in parentheses are changes with respect to the farmer practice.

of the ANRE from 0.14 to 0.20 would result in a decrease in nitrogen fertilizer of 28 percent and a decrease in nitrogen losses of 24 percent. A consecutive increase of the ANRE to 0.25 would imply another decrease in nitrogen costs by 14 percent points and nitrogen losses by 18 percent points. These computations indicate that there is ample scope for the introduction of new technologies that are beneficial for farmer income as well as the environment.

CONCLUSION

China's agricultural growth has been impressive but has had a negative impact on environmental quality. In the intensively cultivated eastern areas, leaching of fertilizers from agricultural fields has resulted in eutrophication of surface water and has made groundwater unsuitable for drinking. Moreover, rice production contributes significantly to the emission of greenhouse gasses to the atmosphere. Agricultural researchers have recognized the downsides of past developments and are designing more sustainable technologies. Policy makers consider these new technologies an important solution to pollution problems. However, it is open to debate how large the role of technology can be and to what extent government policies aimed at changing cropping patterns and stimulating adoption of new technologies can play a role.

This chapter analyzed the potential role of agricultural technology for *Pujiang*, a county in the province of *Zhejiang* in southeastern China. Contrary to the common practice of introducing research as an explanatory variable in an aggregate production function, we use a case study approach involving a comparison of farmer practices and improved technologies. This approach allows assessing the impact of technology on pollution and is forward looking: it can inform about the potential of on-the-shelf technology and further opportunities for technology development. In our analysis we consider the following aspects of agricultural technology: the choice of cropping pattern, the choice between available technologies for a given crop, and the potential to develop improved technologies.

The choice of cropping system is found to be a major determinant of environmental sustainability. A single hectare of annual horticulture produced the same amount of nitrogen leaching as thirteen hectares of double rice or twenty-two hectares of single rice. As a result, annual horticulture contributed more to water pollution than rice in 2002, despite the still limited area under horticultural crops. On the other hand, rice production contributed six times more to global warming through gaseous N-losses. This is mainly the result of the large area under rice, as differences in gaseous N-losses per hectare were relatively small.

Until recently, observed changes in land use were associated with rising pollution of water resources: the share of vegetables increased at the cost of rice production. Recent policies have offset this trend at least temporarily. Declining grain production since 1998 induced the Chinese government to stimulate the production of rice, wheat, and maize through direct subsidies and other policies since 2004. At the same time, rice prices also increased rapidly since the end of 2003 (see also chapter 8). As a result, farmers in *Zhejiang* increased their area planted with rice by 48,700 ha and reduced the area planted with vegetables by 39,800 ha (NBS 2004, 2005). Our computations indicate that a side-effect of this shift in cropping pattern is a decrease in nitrogen pollution. Hence, stimulating self-sufficiency in grain production and maintaining or improving the quality of water resources are complementary objectives for *Zhejiang* Province.

Given the choice of cropping pattern, farmers can affect nitrogen losses through changes in choice of technology. We could not find information on on-the-shelf technologies for annual horticulture. On the other hand, there is a newly developed rice technology, developed in collaboration with farmers that can markedly decrease fertilizer pollution, especially for double rice. Compared to average farmer technology, SSNM decreases N-leaching with 10 percent and gaseous N-losses with 14 percent. The monetary costs for the new technology are lower than the benefits, but implementation requires knowledge and small amounts of daily labor over a longer period. Extension, supported by enabling policies, could cover these factors to promote adoption.

We have shown that there are large potential benefits for additional research on increasing fertilizer efficiency through crop management. The apparent nitrogen recovery rates of the recently developed SSNM technology are 0.3 kg plant N per kg fertilizer N, while with knowledge-intensive management it is feasible to achieve rates of 0.5–0.6. A question that requires further examination is as to why certain features of these advanced technologies were not even considered in developing the SSNM approach. Irrespective of the question of current non-adoption of such technologies, we analyzed the potential ecological consequences. Simulations show that a further increase in the apparent N-recovery from 0.3 to 0.5 would result in a cut in nitrogen losses by more than 50 percent. For vegetable production, the apparent N-recovery is currently as low as 0.14. An increase to the still low level of 0.20 would already result in a decrease of nitrogen losses of 24 percent. Hence, for both rice and vegetable production, there are clear possibilities to develop new technologies with substantially lower nitrogen losses.

In conclusion, we want to stress the role of the government in creating an enabling environment for technology adoption. The availability of more sustainable technologies is a necessary but not a sufficient condition for decreasing agricultural pollution. An important factor in this respect is the capacity and structure of the extension service. At present, the motivation of the extension service to disseminate fertilizer-saving technologies is low, as extension workers earn money from selling fertilizers. A first important step of the government in decreasing agricultural pollution would therefore be increasing basic funds for extension and severing the link between extension and fertilizer sales. Only then will sustainable technologies get a fair chance.

BIBLIOGRAPHY

Babcock, B. A. "The Effects of Uncertainty on Optimal Nitrogen Applications." *Review of Agricultural Economics* 14, no. 2 (1992): 271–80.

Babcock, B. A. and A. M. Blackmer. "The Ex-Post Relationship between Growing Conditions and Optimal Fertilizer Levels." *Review of Agricultural Economics* 16, no. 3 (1994): 353–62.

Bodegom, P. M., P. A. van Leffelaar, A. J. M. Stams, and R. B. Wassman. "Modeling Methane Emissions from Rice Fields: Variability, Uncertainty, and Sensitivity Analysis of Process Involved [Review]. *Nutrient Cycling in Agroecosystems* 58 (2000): 231–48.

Dawe, D. "The Changing Structure of the World Rice Market, 1950–2000." *Food Policy* 27, no. 4 (2002): 355–70.

Dobermann, A., D. Dawe, R. P. Roetter, and K. G. Cassman. "Reversal of Rice Yield Decline in a Long-Term Continuous Cropping Experiment." *Agronomy Journal* 92, no. 4 (2000): 633–43.

Ellis, E. C. and S. M. Wang. "Sustainable Traditional Agriculture in the Tai Lake Region in China." *Agriculture, Ecosystems and Environment* 61 (1997): 177–93.

Fan, S. G. "Research Investment and the Economic Returns to Chinese Agricultural Research." *Journal of Productivity Analysis* 14, no. 2 (2000): 163–82.

Fan, S. G. and P. G. Pardey. "Research, Productivity, and Output Growth in Chinese Agriculture." *Journal of Development Economics* 53, no. 1 (1997): 115–37.

FAOSTAT. *Statistical Database*. Rome: Food and Agricultural Organization, 2001. www.fao.org.

Fischer, K. S. "Toward Increasing Nutrient-Use Efficiency in Rice Cropping Systems: The Next Generation of Technology." *Field Crops Research* 56, nos. 1–2 (1988): 1–6.

Haerdter, R. and T. Fairhurst. *Nutrient Use Efficiency in Upland Cropping Systems of Asia. Proceedings of the 2003 IFA Regional Conference for Asia and the Pacific*, Cheju Island, Republic of Korea, 6–8 October 2003.

Hengsdijk, H. and M. K. van Ittersum. "Formalizing Agro-Ecological Engineering for Future-Oriented Land Use Studies." *European Journal of Agronomy* 19, no. 4 (2003): 549–62.

Huang, J. and S. Rozelle. "Technological Change: Rediscovering the Engine of Productivity Growth in China's Rural Economy." *Journal of Development Economics* 49 (1996): 337–69.

Luijkx, T. *Troubled Water: Agricultural Non-Point Pollution in the Catchment Area of Taihu Lake.* MSc thesis. Wageningen: Wageningen University, 2002.

National Bureau of Statistics of China (NBS). *China Statistical Yearbook 2003.* Beijing: China Statistics Press, 2004.

National Bureau of Statistics of China (NBS). *China Statistical Yearbook 2004.* Beijing: China Statistics Press, 2005.

NIES. *An Analysis of the Pollution Load of the Taihu Area.* Nanjing: Nanjing Environmental Science Institute under the National Environment Protection Bureau (NEPA), 1996.

Peng, S., F. V. Garcia, R. C. Laza, A. L. Sanico, R. M. Visperas, and K. G. Cassman. "Increased N-Use Efficiency Using a Chlorophyll Meter on High-Yielding Irrigated Rice." *Field Crops Research* 47, nos. 2–3 (1996): 243–52.

Peng, S. and K. G. Cassman. "Upper Thresholds of Nitrogen Uptake Rates and Associated Nitrogen Fertilizer Efficiencies in Irrigated Rice." *Agronomy Journal* 90 (1988): 178–85.

People's Daily Online. "China Clears Development Zones to Make Way for Grain Production." Available online at: english.people.com.cn/200406/25/eng20040625_147568.html, accessed June 25, 2004a.

People's Daily Online. "China Revives Farmers' Enthusiasm for Grain Production." Available online at: english.people.com.cn/200406/25/eng20040623_147295.html, accessed June 23, 2004b.

People's Daily Online. "China's Grain Production This Year Expected to Exceed 455 Billion Kg." Available online at: english.people.com.cn/200406/25/eng20041109_163226.html, accessed November 8, 2004c.

Ponsioen, T. C., A. G. Laborte, R. P. Roetter, H. Hengsdijk, and J. Wolf. "Technogin-3: A Technical Coefficient Generator for Cropping Systems in Eats and Southeast Asia." *Quantitative Approaches in Systems Analysis report* no. 26, 2003.

Sheldrick, W. F., J. K. Syers, and J. Lingard. "Soil Nutrient Audits for China to Estimate Nutrient Balances and Output/Input Relationships." *Agriculture, Ecosystems and Environment* 94, no. 3 (2003): 341–54.

Smil, V. "Food, Energy, and the Environment: Implications for Asia's Rice Agriculture." Pp. 321–34 in *Sustainability of Rice in the Global Food System*, edited by Dowling, N. G., S. M. Greenfield, and K. S. Fischer. Davis, CA: Pacific Basin Study Centre and Manila International Rice Research Institute, 1998.

Smil, V. "Nitrogen and Food Production: Proteins for Human Diets." *Ambio* 31, no. 2 (2002): 126–30.

Van Ittersum, M. K., P. A. Leffelaar, H. van Keulen, M. J. Kropff, L. Bastiaans, and J. Goudriaan. "On Approaches and Applications of the Wageningen Crop Models." *European Journal of Agronomy* 18, nos. 3–4 (2003): 201–34.

Wang, G., A. Dobermann, C. Witt, Q. Su, R. Fu, R. Simbahan, and M. A. A. Adviento. "Reducing the Gap between Attainable and Potential Yield in Double Rice-Cropping Systems of Zhejiang Province, China." Pp. 479–87 in *Rice Research for Food Security and Poverty Alleviation. Proceedings of the International Rice Research Conference*, edited by S. Peng and B. Hardy. Los Baños: International Rice Research Institute, 2001.

Wang, G., Q. Sun, R. Fu, X. Huang, X. Ding, J. Wu, Y. He, A. Dobermann, and C. Witt. "Site-Specific Nutrient Management in Irrigated Rice Systems of Zhejiang

Province, China." Pp. 243–63 in *Increasing Productivity of Intensive Rice Systems through Site-Specific Nutrient Management*, edited by A. Dobermann, C. Witt, and D. Dawe. Enfield: Science Publishers, Inc. and Los Baños: International Rice Research Institute, 2004.

Widawsky, D., S. Rozelle, S. Jin, and J. Huang. "Pesticide Productivity, Host-Plant Resistance and Productivity in China." *Agricultural Economics* 19, nos. 1–2 (1998): 203–17.

Zhang, F. and Z. Liu. "Effect of Potassium Fertilizer Application on Nitrogen Uptake by Vegetable and Nitrate Leaching from Vegetable Soil in Shandong Province." Annual report CAU/SAAS/IPI cooperative project on N:K interaction in vegetables, 2002.

Zhang, W. L., Z. X. Tian, N. Zhang, and X. Q. Li. "Nitrate Pollution of Groundwater in Northern China." *Agriculture, Ecosystems and Environment* 59, no. 3 (1996): 223–31.

Zhejiang Statistical Bureau. *Rural Statistical Yearbook of Zhejiang*. Chinese Statistics Press, Beijing, 2001.

Zhu, Z. L. and D. L. Chen. "Nitrogen Fertilizer Use in China—Contributions to Food Production, Impacts on the Environment and Best Management Strategies." *Nutrient Cycling in Agroecosystems* 63, nos. 2–3 (2002): 117–27.

16

China's Farmland Use

A Scenario Analysis of Changes and Trends

Changhe Lu, Xiubin Li, and Minghong Tan

The P.R. China is a country with a large population and relatively scarce arable land resources. In 2000, the population was 1.26 billion, about one-fifth of the world's total, while the farmland area was just more than 128 million ha. As a result farmland per capita is only 0.11 ha on average, less than half of the world average of 0.24 ha (Lin et al. 2003). In several provinces in the coastal zone, average farmland availability per capita is below 0.05 ha. This problem has been exacerbated by recent farmland losses and by degradation problems. During the past decade, a total of more than 8 million ha of farmland has been converted into urbanized areas for construction, or for orchards, forests, and pastures. Of the total farmland, an estimated area of about 38 percent is threatened by water and wind erosion.

Due to the large size of its population and the relatively small size of the world market for grain as compared to China's domestic market, food self-sufficiency plays an important role in China's policy making. Maintaining a sufficiently large area of land for crop production plays an important role in this respect. The purpose of this chapter is to review the changes in China's land use over the past fifty years and to explore possible changes in farmland availability in the future using a scenario analysis approach. The chapter proceeds in the following manner. The second section provides an overview of China's current land use and its spatial distribution as well as the major changes in land use patterns since the founding of the People's Republic in 1949. In the third section, a detailed analysis is provided of China's arable land resources and the factors responsible for the changes in farmland area since 1987. The methodology used for analyzing future trends in farmland availability is explained in the fourth section. It is followed by a discussion of the assumptions made and the definition of scenarios in the fifth section. Results of the scenario analysis for the period 2000–2050 are presented in the sixth section. The chapter concludes with some policy suggestions for improving the current policy of balancing farmland loss and gains at the provincial level.

CHINA'S LAND USE AND ITS CHANGES

Land Use and Its Spatial Distribution

Based on the 1996 national land use survey, mainland China has a total land area of 950.7 million ha, of which 66 percent is mountainous and hilly land, and nearly half (45 percent) has a dry climate with an annual rainfall below 400 mm. In 2000, the largest part (69 percent) of the land was used for agricultural activities, including 14 percent for crop cultivation, 1.1 percent for orchards, 24 percent for forestry (forest/shrubs), 28 percent for pasture, and 2.3 percent for fishery (water area). Nonagricultural land use (settlements, industry and mining sites, and transportation) covered an area of 3.8 percent (MLR 2001). The remaining part of 28 percent was unused or unusable land, including desert, wasteland, swamps, saline land, and mountain areas, as seen in figure 16.1.

China's land use varies greatly due to effects of climatic and terrain conditions. Figure 16.1 shows that the western part of China (particularly Inner Mongolia, *Gansu, Ningxia, Xinjiang, Qinghai,* and Tibet) is predominately covered by grassland and unused lands. This is very different from the other parts, where land is mostly used for forestry and crop cultivation. The spatial

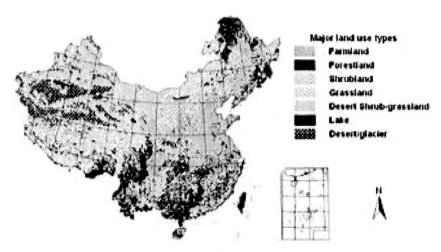

Figure 16.1.   Schematic Representation of Spatial Distribution of China's Major Land Use Types

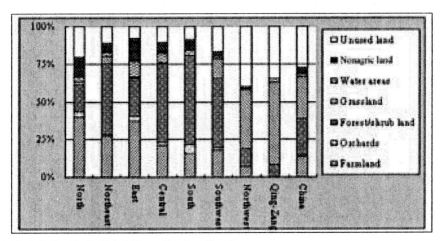

North: *Beijing, Tianjin, Hebei, Henan, Shandong, Shanxi;*
Northeast: *Liaoning, Jilin, Heilongjiang;*
East: *Shanghai, Jiangsu, Anhui, Zhejiang;*
Central: *Hubei, Hunan;*
South: *Guangdong, Guangxi, Fujian, Hainan;*
Southwest: *Yunnan, Guizhou, Jiangxi, Sichuan, Chongqing;*
Northwest: *Inner Mongolia, Shaanxi, Gansu, Ningxia, Xinjiang;*
Qing-Zang: *Tibet, Qinghai.*
Data source: MLR (2001).

Figure 16.2.   Regional Land Use Structure of China (2000).

variation of land use among different regions is shown in more detail in
figure 16.2. Clearly, among the various regions, North China and East China
have markedly higher proportions of farmland (38–40 percent) and nonagricultural land (13–15 percent), while Northwest China and the *Qing-
Zang* Plateau comprise mainly of grassland (39–55 percent) and unused
land (35–40 percent), with a very low share of land devoted to crop cultivation and nonagricultural use. The other regions have an area of 46–55 percent of their land used for forestry and 12–27 percent for crop cultivation.

Changes in Land Use

Since the founding of the People's Republic of China in 1949, China's
land use has changed significantly. As shown in table 16.1, the area of
cropland, orchard, forest and nonagricultural land use has dramatically
expanded, while the grassland and unused land substantially shrank. In
particular, the land used for orchards and nonagricultural purposes has
increased by 892 and 227 percent since 1949. This was largely realized by
the conversion of cropland, which caused a profound change of agricultural structure, and rapid industrialization and urbanization. Forestland
and grassland demonstrated the greatest changes in absolute terms (table

Table 16.1. China's Land Use in 1949 and 2000

× 1,000 ha	Area		Change	
	1949[a]	*2000*[b]	*Area*	*%*
Farmland	97,881.3	128,243.1	30,361.8	31.0
Orchards	1,066.7	10,576.0	9,509.3	891.5
Forest/shrubs	125,000.0	228,789.2	103,789.2	83.0
Grassland	391,918.7	263,768.8	−128,149.9	−32.7
Built-up area	6,733.3	21,985.2	15,251.9	226.5
Water area	22,533.3	36,206.5	13,673.2	60.7
Unused land	305,542.9	261,107.3	−44,435.6	−14.5

Notes:
[a] Based on data of Li (2000).
[b] Based on MLR (2001).

16.1). However, this change seems exaggerated, probably due to the use of different criteria for land use classification. In 2000, shrubland is included in the category of forestland, but in 1949 it may have been largely classified as grassland.

In the past decades, land reclamation played a key role in China's land use change, and resulted in an impressive conversion of grassland, unused land, wetland, and forestland into farmland. It is estimated that a land area of 72 million ha was reclaimed between 1949 and 2000 (Sun and Shi 2003; Chen 2001; and unpublished MLR land monitoring data). The land reclamation was largely carried out in the 1950s and early 1960s in relatively low populated provinces of *Xinjiang, Heilongjiang,* Inner Mongolia, *Qinghai, Gansu, Yunnan,* and *Guizhou,* in particular converting marginal lands (e.g., sandy land and hilly areas) into crop land. In the end, around 30 percent of the total reclaimed land is estimated to be useless due to serious land degradation (soil salinity, water, and wind erosion), and thus 70 percent or about 50 million ha was actually added to the existing farmland area. Meanwhile, some 20 million ha of farmland was transformed to built-up (construction) areas, orchards, and fishponds. On balance, the area of China's farmland had increased in 2000 by 30 million ha or 31 percent of the area in 1949. As the newly added farmland was mainly obtained in dry or mountainous areas of western China, its quality is generally poorer than the lost fertile farmland largely in the humid areas of eastern China (Lin et al. 2003).

CHINA'S ARABLE LAND RESOURCE AND ITS CHANGES SINCE 1987

Arable Land Resources

Arable land resources include two types, farmland that is currently used for growing crops and vegetables, and cultivable "wasteland" (*Yinong*

Table 16.2. Area of Different Farmland Types, 1996 (% of the total farmland)

Region[a]	Flat Farmland[b]		Sloped Farmland				
	Total	*Irrigated*	*Total*	*2–6°*	*6–15°*	*15–25°*	*>25°*
North	88.2	49.1	11.8	5.1	4.3	1.9	0.4
Northeast	71.5	12.3	28.5	19.8	7.0	1.4	0.3
East	87.4	63.4	12.6	7.3	2.8	1.6	0.9
Central	78.0	60.7	22.0	6.2	6.8	5.1	3.9
South	77.1	67.5	22.9	12.0	7.3	3.2	0.5
Southwest	41.4	29.7	58.6	8.3	18.7	21.0	10.5
Northwest	56.7	33.6	43.3	13.7	13.7	9.7	6.3
Qing-Zang	50.4	46.3	49.6	12.6	21.8	14.1	1.1
China	69.1	39.8	30.9	10.4	9.5	7.2	3.8

Source: Based on the 1996 national land survey of the Ministry of Land Resources (MLR).
Notes:
[a] See footnote of figure 2.
[b] Flat farmland also includes the terraced farmland in hilly and mountain areas.

huangdi) that is potentially suitable for reclamation. Both land resources, in general, are very limited. In the year 2000, total farmland in mainland China was 128.2 million ha or not more than 0.11 ha per capita, comprised of 40 percent irrigated land and 60 percent dry land. According to the information collected in the 1996 national land survey, 31 percent of the farmland in China is sloped land, with shares of sloped land ranging from 43 to 59 percent in the Southwest, Northwest, and *Qing-Zang* regions (table 16.2). In several provinces of western China, more than half of the farmland is steeply sloped land on mountains and hills, which is marginally suitable for crop cultivation. For instance, steeply sloped farmland (with a slope above 15°) covers a share of 59 percent in *Guizhou*, 50 percent in *Shaanxi*, and 46 percent in *Yunnan* Province.

Potentially cultivable wasteland can be found mainly in western and northeastern China. According to the agricultural resource survey of the Ministry of Agriculture undertaken during the period 1991–1993, the total potentially cultivable land was 9.5 million hectare (Chen 2001). It was estimated, based on land monitoring data of the Ministry of Land Resources (MLR), that a total area of 2.8 million ha was reclaimed during the period 1993–2000. Taking this addition into account, the potentially cultivable land in 2000 was estimated at 6.7 million ha. More than half of the area (3.5 million ha) is located in the three provinces of *Heilongjiang*, Inner Mongolia, and *Xinjiang*. There is no land available for reclamation in most provinces of eastern and central China.

Farmland Change since 1987

With the implementation of the open-door policy and the economic reform since 1978, China has experienced a rapid socioeconomic develop-

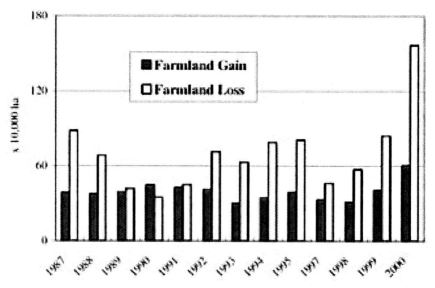

Source: Based on land monitoring data of the Ministry of Land Resources (MLR).

Figure 16.3.   Changes in the Farmland Area between 1987 and 2000

ment that stimulated a profound structural change of agriculture (and vice-versa) and a rapid expansion of construction land. Meanwhile, China has strategically shifted the focus from land reclamation to the rehabilitation of seriously degraded land that was largely caused by the overcultivation of marginal land and the destruction of natural vegetation. The agricultural sector change, urban expansion, and environmental rehabilitation transformed large amounts of farmland into orchards and fishponds, built-up area, and forests or pastures. In the years between 1987 and 2000, a farmland area of between 0.34 and 1.57 million ha was converted each year, with an annual average of 0.7 million ha (see figure 16.3). In total, more than 9 million ha of farmland was lost during this period. In the same period, about 5 million ha of farmland was gained by land reclamation, and land consolidation or rehabilitation. On balance, China therefore lost a net farmland area of more than 4 million ha, or 0.3 million ha per year, between 1987 and 2000.

Four factors contribute to farmland losses, namely government-promoted ecological conversion of marginal farmland (the so-called *Tuigeng huanlin huancao*), restructuring of agricultural land use (particularly into orchards and fishponds), built-up area expansion, and destruction by natural hazards and soil degradation. Of the total reduction in farmland area between 1987 and 2000, 38 percent was transformed from marginal land into forestland and grassland, 25 percent into orchards and fishponds, and 22 percent was transformed into built-up (construction) areas. The remaining 15 percent suffered serious damage from natural hazards and soil degradation of farmland and could no longer be used.

The conversion of steeply sloped farmland to forestland or grassland took place mainly in the northwestern and southwestern regions, where cultivation of marginal lands is widespread. From 1987 to 2000, this conversion totaled 3.6 million ha (annually between 0.12 and 0.76 million ha), of which 71 percent was located in the above-indicated regions. Moreover, a total area of 2.45 million ha (annually 0.06–0.33 million ha) was changed from crop cultivation to fruit and fishery production. Destruction by natural hazards affecting farmland occurred every year, and the resulting farmland loss was reported at between 0.05 and 0.16 million ha per year during the same period. Of the total hazard-induced farmland loss, nearly 80 percent was located in the northeast, northwest, and southwest.

The encroachment of construction into farmland occurred in particular in highly populated urban areas. It has been perceived as the most important factor reducing farmland and thus food self-sufficiency, because the lost farmland was fertile, high-productive land. In addition, urban expansion and the resulting farmland loss have deprived a great number of farmers of their land and income source, and therefore aroused great concerns by Chinese government recently.

From 1987 to 2000, farmland conversion to built-up land was reported (by the MLR) to equal 2.05 million ha (on average, 0.16 million ha per year). It shows an increasing trend in the recent past (see figure 16.4). It has been estimated that in this conversion more than 1.5 million rural inhabitants per year have lost their land.

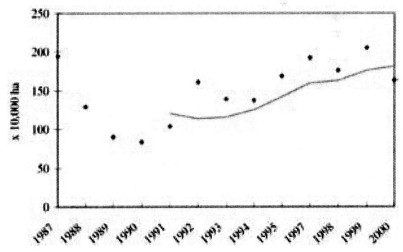

Note: Fitted curve shows five-year moving averages.

Figure 16.4. Farmland Loss Due to Built-up Land Expansion (1987–2000)

TREND ANALYSIS OF FUTURE FARMLAND AVAILABILITY

With socioeconomic development, population growth, and urbanization, it is anticipated that China's arable land area will continue to decrease in the next decades. This decrease will depend on land demand to support economic development and urbanization, and availability of cultivable marginal land to compensate for unavoidable farmland losses in the future. This section outlines the procedure and assumptions of an explorative analysis on possible changes of China's farmland until 2050.

General Procedure

For simplicity, this study estimates the farmland area only for the years 2005, 2010, 2015 . . . and 2050. For a specific year, the farmland area is determined by two factors, the loss and gain. This estimation is simply presented as:

$$F_y = F_{y-5} - 5 \cdot (L_T - R_T) \tag{1}$$

in which F_y is farmland area in the year y, F_{y-5} is farmland area in the year five years before the year y, and L_T and R_T are mean annual farmland losses and farmland gains in the period T, respectively. T represents the five-year periods 2001–2005, 2006–2010 . . . and 2046–2050, respectively. The year 2000 is considered as the reference year, that is, farmland area in 2000 is used for estimating the farmland area in 2005.

For the estimation of farmland losses, the four major factors listed above are considered: government-promoted ecological conversion of marginal farmlands, farmland conversion to built-up land, farmland destruction by natural disasters and land degradation, and farmland transformation to orchards and fishponds. The ecological conversion generally concerns steeply sloping and seriously degraded farmland. In this study, only farmland with a slope above 25° is assumed to be returned to its original use, as the China Water and Soil Conservation Law prescribes that land used for crop cultivation should not exceed 25°. For simplicity, this steep farmland is excluded from the farmland area at the beginning of the calculations. Thus, the farmland area for the reference year 2000 is 123.89 million ha (table 16.3), which excludes the steep farmland area of 4.35 million ha (i.e., 3.4 percent). Excluding the ecological conversion of marginal farmlands, the total farmland loss (L_T) in equation (1) is a summation over the lost areas due to the above three factors:

$$L_T = [B_T \cdot b_T + F_{y-5} \cdot (d_T + o_T)]/100 \tag{2}$$

in which B_T is mean annual increase of built-up area in the period T, F_{y-5} is total farmland area in the year $y-5$, b_T is the percentage of annual built-up land increase consisting of transformed farmland, d_T is the relative rate of farmland loss due to natural disasters (as a percentage of total farmland),

and o_T is the relative rate of farmland conversion to orchards and fishponds (as a percentage of total farmland).

Farmland conversion to orchards and fishponds, and hazards' induced farmland losses are simply estimated as a percentage of the farmland area in the year y–5, respectively. For factors of b_T, d_T, and o_T, the values are estimated using historical data for the past ten years, while for the built-up land expansion (B_T), a regression equation is estimated. Table 16.3 presents the basic data to be used for the analysis.

Table 16.3. Basic Data Used for the Estimation of Farmland Area in the Scenario Analysis

Province	Area (1,000 ha) in 2000		Reference Values (%) for Coefficients in Equation (2)[b]			Relative Annual Growth Rate (%)	
	Farmland[a]	Cultivable Land	d_t	o_t	b_t	GDP[c]	Population[d]
Beijing	332.1	0.0	0.088	0.296	57.0	9.97	2.194
Hebei	6,842.2	156.2	0.030	0.061	60.3	11.07	0.800
Henan	8,081.3	129.7	0.012	0.095	73.8	10.09	0.910
Shandong	7,672.0	143.8	0.017	0.131	58.5	10.97	0.608
Shanxi	4,385.3	193.8	0.062	0.329	67.3	8.68	1.087
Tianjin	483.4	23.2	0.002	0.117	54.1	11.31	1.190
Heilongjiang	11,739.4	1,105.1	0.144	0.027	44.7	8.89	0.783
Jilin	5,558.1	252.0	0.106	0.022	60.6	9.85	0.780
Liaoning	4,155.3	51.8	0.117	0.191	63.7	8.58	0.599
Anhui	5,898.5	0.0	0.026	0.108	75.1	10.40	1.040
Jiangsu	5,016.3	53.3	0.016	0.279	82.0	11.18	0.746
Shanghai	289.0	0.0	0.146	0.399	73.1	11.36	1.119
Zhejiang	2,026.5	21.2	0.081	0.221	74.7	10.98	0.639
Hebei	4,535.0	83.5	0.030	0.187	69.0	10.81	1.003
Hunan	3,858.4	166.5	0.088	0.208	35.1	9.95	0.741
Jiangxi	2,816.5	152.7	0.081	0.560	46.2	9.78	1.143
Fujian	1,381.5	0.0	0.096	0.089	57.2	12.12	1.004
Guangdong	3,127.8	0.0	0.045	0.484	42.8	10.35	1.405
Guangxi	4,202.6	56.2	0.054	0.190	48.6	8.89	1.117
Hainan	727.9	55.3	0.023	0.704	42.9	8.17	1.507
Guizhou	4,115.5	190.0	0.090	0.025	65.5	9.56	1.341
Sichuan	8,244.4	264.8	0.068	0.150	58.2	9.85	0.860
Yunnan	5,551.6	227.2	0.139	0.200	54.4	9.27	1.255
Gansu	4,948.9	348.5	0.009	0.094	53.2	10.17	1.326
Inner Mongolia	7,590.3	1,075.5	0.152	0.004	42.5	10.88	0.982
Ningxia	1,265.9	282.3	0.030	0.006	63.4	9.92	1.691
Shaanxi	4,035.6	42.5	0.085	0.251	75.9	9.87	0.985
Xinjiang	3,985.7	1,452.4	0.134	0.050	41.9	8.79	1.577
Qinghai	669.2	191.8	0.018	0.001	41.1	9.63	1.380
Xizang (Tibet)	356.3	21.8	0.031	0.000	45.7	11.82	1.720
China	123,892.3	6,741.1	0.076	0.157	60.0	10.31	0.965

Notes:
[a] Farmland area excludes the farmland with a slope above 25°.
[b] Average values for 1990–2000, based on the MLR land monitoring data (unpublished).
[c] Average values for 1996–2000 based on the China Yearbooks 1996–2000.
[d] Average values for 1990–1999 based on the China Yearbooks 1990–1999.

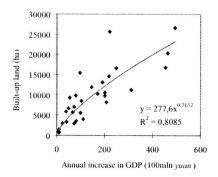

Figure 16.5a. Scatter Diagrams of Mean Annual Expansion of Built-up Land and Annual Growth Rate of GDP and Population, Thirty Provinces (1993–1999)

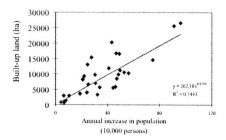

Figure 16.5b. Scatter Diagrams of Mean Annual Expansion of Built-up Land and Annual Growth Rate of GDP and Population, Thirty Provinces (1993–1999)

Built-up land expansion is driven by increasing land requirements for settlements, urban development, industries, services, and infrastructure, which are heavily affected by economic development and population growth. Considering the quality and availability of data, we use the yearly averages for 1993–1999 for the thirty provinces in mainland China to conduct a regression analysis of the factors driving the expansion of built-up (construction) land. Simple scattergrams, shown in figures 16.5a and b, indicate that the annual increase of built-up land area is highly related to the annual growth of GDP as well as population.

In addition to these two variables, we added the efficiency of built-up land use to the regression equation. In general, the efficiency is often expressed as the ratio of total built-up land area to total GDP or the ratio of total built-up land area to total population. Here, we combine these two factors into one, called the *utilization efficiency index* of built-up land, expressed as the ratio of total built-up area (ha) to the product of total GDP (in million *yuan*) and total population (in million inhabitants). Using these three factors as independent variables, we estimated the following multivariate equation by a linear regression through the origin:

$$B_T = 218.89 \cdot G_T^{0.7132} + 148.32 \cdot P_T - 1168.02 \cdot e_T^{-0.5956} \qquad (3)$$

in which B_T is the mean annual increase of built-up land area (ha) in period T, G_T is the mean annual growth rate of GDP (in 10^8 *yuan*) at 1992 constant price in period T, P_T is the mean annual growth rate of population (in 10^4 persons) in period T, and e_T is the mean utilization efficiency index of built-up land in period T.[1]

In its quest for economic development, China is rapidly developing its infrastructure. To meet housing demand for increasing population, it needs land to improve its transportation and communication systems, to develop industries, and to build new houses, which all require land. This land demand is expected to be positively related to the growth of GDP and population. However, the land demand would be reduced by an improvement of land use efficiency, and thus a negative relationship is expected between built-up land expansion and land use efficiency. The above regression equation is in accordance with this expectation, indicating that annual expansion of built-up land increases with annual growth rate of GDP and population, but decreases with an increase of the use efficiency of built-up land.

Assumptions and Scenario Definitions

To estimate the farmland area in the future, various assumptions are defined and incorporated into model (1)–(3). This section will give a brief description of the assumptions.

Land Reclamation

Available land area for reclamation ("cultivable land") is presented in table 16.3. It is the major source to compensate the cropland losses. In recent years, China has implemented several policy measures to control the conversion of farmland into built-up land, and issued a regulation that farmland conversion to built-up land should be fully compensated by land reclamation. Realization of this aim is largely dependent on the availability of cultivable lands. In several provinces in the coastal zone, there is actually no cultivable land to compensate the farmland loss (see table 16.3). Considering this fact and the land management regulation, we assume that the lost farmland in a province is compensated by land reclamation within the same province, provided that cultivable land is available, and is not compensated otherwise. With this assumption, possible farmland gain can be simply related to farmland loss and the available area of cultivable land. For a province, if the cultivable land area is more than the lost farmland in the period T, the annual farmland gain in that period (R_T) equals the annual farmland loss (L_T), otherwise, it equals the available area of the cultivable land.

Farmland Conversion to Orchards and Fishponds

According to the policy of strict protection for the farmland base (delineated by the government for growing vegetables and food crops, covering around 80 percent of the farmland area), the farmland base cannot be converted into orchards, forests, and fishponds. With decreasing availability of farmland in the future, the farmland conversion to orchards and fishponds is therefore expected to slow down. We assume in this study that the rate of farmland conversion to orchards and fishponds (o_T) is 20 percent lower in 2001–2010 as compared to the average during 1990–2000 (see table 16.3). For 2011–2030, we assume it to be 35 percent lower and for 2030–2050 we assume it to be 50 percent lower than in 1990-2000. Increasing requirements for fruit and fish products are assumed to be met by an increase in land productivity.

Farmland Loss Caused by Natural Hazards

According to the MLR land monitoring data, the mean annual farmland loss caused by natural hazards and land degradation in the 1990s was 0.076 percent of the farmland area in mainland China (table 16.3). Part of the lost farmland can be restored by appropriate measures. In recent years, the government has increased investment in ecological protection as an important part of the Western Region Development Program. This may stimulate soil conservation and improvement of cropping and soil management in the future, and thus reduce farmland loss due to degradation. Considering these aspects, it is assumed that the relative rate of hazard-induced farmland loss (d_T) in 2001–2050 is 20 percent lower than the average in 1990–2000.

Built-up Land Expansion

This study assumes that the built-up land expansion for each of the thirty provincial units is fully determined by equation (3). Regarding China's GDP growth, the World Bank (1997, cited by Hubacek and Sun 2001) projected that it would be slowing down over time, from some 8 percent today to 5 percent in 2020. The State Department Center for Economic Development (Li et al. 2001; Wang et al. 2002) predicted that the GDP growth rates would be at 6.6–7.9 percent in 2001–2010, 5.5–6.6 percent in 2011–2020, and at 5.4, 4.5, and 3.4 percent in the next three decades from 2031 to 2050,

Table 16.4. Definitions of Three GDP Growth Scenarios, 2001–2050 (Percentage growth rates)

	2001–2010	2011–2020	2021–2030	2031–2040	2041–2050
High GDP growth	7.6	6.6	5.4	4.5	3.4
Medium GDP growth	7.0	6.0	4.8	3.8	2.8
Low GDP growth	6.5	5.5	4.2	3.2	2.2

respectively. Zheng (2002) estimated that the GDP growth rate could be at 7 percent in 2001–2010 and 6 percent in 2011–2020. This study defines three GDP growth scenarios for the years 2001–2050 (Table 16.4). The GDP growth rate for each province is estimated using its mean value for the years 1996–2000 (see table 16.3).[2, 3]

Population growth rates for China as a whole are based on the data obtained from the China Population Information Research Centre (table 16.5). To capture regional variation, the mean population growth rate in 1990–1999 taken from the statistical yearbooks of China is used to derive the growth rate for each province. For the factor b_T in equation (2), the same value as the average in the past years (table 16.3) is used.

ESTIMATED FARMLAND AREA IN 2050

Table 16.6 presents farmland area for each province in the years 2010, 2030, and 2050 for the three GDP growth scenarios. The results show that the farmland area decreases at a much faster speed in the coastal provinces such as *Shanghai, Beijing, Tianjin, Gangdong, Jiangsu,* and *Zhejiang* than in the other areas. Since there is land available for reclamation to compensate lost farmland in most provinces, total farmland area in 2010 is rather similar

Table 16.5. Mean Population Growth Rates Assumed and Predicted Population Size in Different Periods

Period	Mean Growth Rate (%)	Predicted Population Size at the End of Period (million persons)
2001–2005	0.8470	1,306.64
2006–2010	0.7696	1,358.22
2011–2015	0.7148	1,408.12
2016–2020	0.5630	1,448.86
2021–2025	0.4155	1,479.78
2026–2030	0.2975	1,502.36
2031–2040	0.1875	1,516.80
2041–2045	−0.0585	1,522.78
2046–2050	−0.1930	1,518.23

Source: Obtained from China Population Information Research Centre, 1999.
    http://www.cpirc.org.cn/est2.htm

for the three scenarios (figure 16.6), and not much different from that in 2000. With decreased availability of cultivable land, farmland loss cannot be compensated by land reclamation, and therefore a considerably lower farmland area is found for all scenarios in 2030 and 2050.

In figure 16.7, the net annual farmland loss per five-year period is presented. It shows that a high reduction in the farmland area is expected to occur in the period 2015–2035 because of reduced availability of cultivable land and the high demands for built-up land.

Farmland conversion into built-up land in the next thirty years will be at a similar level (slightly higher) as in the late 1990s for the three scenarios (figure 16.8). This rate is expected to be maintained for 2030–2050 in the *high GDP growth* scenario. All scenarios suggest that most of the cultivable land will be reclaimed in the first decade, to compensate the farmland loss. The abrupt decrease in annual farmland reclamation (figure 16.9) during 2011–2050 reflects that only very limited area of cultivable land is available after reclamation in the first decade. In general, the national aim to maintain a balance in farmland area gains and losses is probably only possible in the provinces of *Heilongjiang, Xinjiang,* Inner Mongolia, *Ningxia, Qinghai,* Tibet, and *Gansu,* where the population density is relatively low and cultivable land is available to compensate for farmland losses. In other provinces, this aim can hardly be achieved, particularly with a high GDP growth rate after the year 2030. In *Shanghai,* there will probably be no farmland available anymore in 2050 in all three scenarios (see table 16.6).

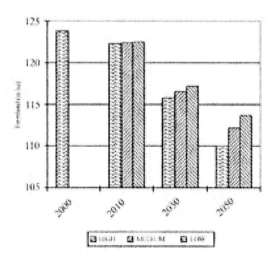

Figure 16.6. Farmland Area in the Years 2010, 2030, and 2050 for the Three GDP Growth Scenarios

Table 16.6.   Farmland Area (10³ ha) in the Years 2010, 2030, and 2050 for the Three GDP Growth Scenarios

Province	High GDP Growth			Medium GDP Growth			Low GDP Growth		
	2010	2030	2050	2010	2030	2050	2010	2030	2050
Beijing	280.5	166.2	66.8	283.7	188.2	136.5	286.3	205.8	181.6
Hebei	6,842.2	6,518.2	6,169.0	6,842.2	6,566.3	6,325.0	6,842.2	6,604.6	6,439.2
Henan	8,004.7	7,602.8	7,293.1	8,011.8	7,650.7	7,437.6	8,017.6	7,689.3	7,537.9
Shandong	7,608.1	7,134.7	6,630.1	7,616.8	7,199.5	6,839.8	7,623.8	7,251.2	6,993.9
Shanxi	4,385.3	4,078.8	3,861.6	4,385.3	4,093.9	3,899.8	4,385.3	4,106.3	3,922.8
Tianjin	473.6	379.4	257.6	476.0	398.4	321.8	478.0	413.5	368.7
Heilongjiang	11,739.4	11,739.4	11,739.4	11,739.4	11,739.4	11,739.4	11,739.4	11,739.4	11,739.4
Jilin	5,558.1	5,493.7	5,280.5	5,558.1	5,514.3	5,342.3	5,558.1	5,531.0	5,388.9
Liaoning	4,041.5	3,716.7	3,433.3	4,046.6	3,747.8	3,521.0	4,050.8	3,773.4	3,588.7
Anhui	5,731.6	5,408.4	5,156.1	5,736.6	5,443.1	5,262.2	5,740.6	5,471.0	5,337.0
Jiangsu	4,776.6	4,091.0	3,326.4	4,789.9	4,192.4	3,660.3	4,800.7	4,273.0	3,904.6
Shanghai	217.4	0.0	0.0	224.0	49.3	0.0	229.4	92.0	0.0
Zhejiang	1,924.8	1,636.3	1,322.8	1,932.0	1,689.9	1,501.3	1,937.8	1,732.4	1,626.2
Hubei	4,433.6	4,051.3	3,703.2	4,440.3	4,099.8	3,858.2	4,445.8	4,138.6	3,971.8
Hunan	3,858.4	3,642.0	3,439.7	3,858.4	3,661.0	3,496.6	3,858.4	3,676.3	3,534.2
Jiangxi	2,781.8	2,470.9	2,260.7	2,784.0	2,484.7	2,297.9	2,785.8	2,495.8	2,318.9
Fujian	1,303.6	1,099.8	847.7	1,308.4	1,140.1	991.5	1,312.4	1,171.7	1,093.5
Guangdong	2,900.3	2,489.9	2,176.0	2,905.8	2,527.7	2,293.7	2,910.3	2,558.0	2,373.8
Guangxi	4,134.1	3,928.6	3,802.6	4,136.1	3,940.0	3,824.1	4,137.7	3,949.4	3,838.4
Hainan	727.9	646.1	593.2	727.9	649.4	599.8	727.9	652.1	604.2
Guizhou	4,115.5	4,073.4	3,992.5	4,115.5	4,084.1	4,010.9	4,115.5	4,092.7	4,022.2
Sichuan	8,244.4	7,788.7	7,457.9	8,244.4	7,825.3	7,562.6	8,244.4	7,854.8	7,623.6
Yunnan	5,551.6	5,229.3	4,981.9	5,551.6	5,242.5	5,010.1	5,551.6	5,253.2	5,028.2
Gansu	4,948.9	4,948.9	4,948.9	4,948.9	4,948.9	4,948.9	4,948.9	4,948.9	4,948.9
Inner Mongolia	7,590.3	7,590.3	7,590.3	7,590.3	7,590.3	7,590.3	7,590.3	7,590.3	7,590.3
Ningxia	1,265.9	1,265.9	1,265.9	1,265.9	1,265.9	1,265.9	1,265.9	1,265.9	1,265.9
Shaanxi	3,915.2	3,629.7	3,424.3	3,918.2	3,649.3	3,479.9	3,920.7	3,665.1	3,511.7
Xinjiang	3,985.7	3,985.7	3,985.7	3,985.7	3,985.7	3,985.7	3,985.7	3,985.7	3,985.7
Qinghai	669.2	669.2	669.2	669.2	669.2	669.2	669.2	669.2	669.2
Xizang (Tibet)	356.3	356.3	346.4	356.3	356.3	355.4	356.3	356.3	356.3
China	122,366.4	115,831.4	110,022.6	122,449.2	116,593.1	112,227.7	122,516.7	117,206.6	113,765.5

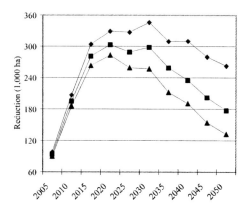

Figure 16.7. Mean Annual Net Farmland Loss per Five-Year Period for the Three GDP Growth Scenarios

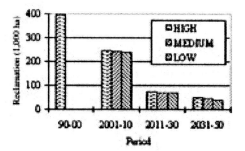

Figure 16.8. Mean Annual Farmland Loss Due to Built-up Land Expansion (1993–2000) and Scenario Results (2001–2050)

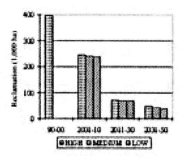

Figure 16.9. Mean Annual Farmland Reclamation Area (1990–2000) and Scenario Results (2001–2050)

CONCLUSION

Increasing requirements of built-up land, which are largely driven by economic and population growth will have a great impact on farmland availability in the future. With decreasing availability of marginal land that is potentially cultivable, complete compensation of farmland losses caused by expansion of built-up areas and restructuring of land use seems hardly achievable in most provinces, particularly the coastal ones. It is therefore expected that farmland area will continuously decrease, especially after 2010.

Farmland conversion to built-up land is anticipated to be at a similar level to that in the 1990s for the years 2001–2030, because continued economic growth, increasing income, and a growing population will imply a steadily increasing demand for nonagricultural land. Achievement of the compulsory aim to balance the loss (due to built-up land expansion) and gain of farmland in each province seems not realistic for most provinces of China. Hence, this policy should be amended. It could be an option to introduce a specific tax for farmland conversion to nonagricultural use. The tax can be used to improve the quality and productivity of current farmland by improving the irrigation systems, building terraces (of slope farmland), and alleviating land degradation, through public investment. This optional policy measure can be applied to all provinces, even for provinces with sufficient cultivable land to compensate the farmland loss. Investment in farmland improvement is more economic than investment in land reclamation (to compensate the farmland losses), as cultivable land is mostly located in remote areas with poor accessibility and low land qualities. In addition, development of land markets and appropriate policies/laws for land transactions are crucial for the control of illegal land transactions and for an improvement of land use efficiency to reduce farmland losses.

ACKNOWLEDGMENTS

We gratefully acknowledge the anonymous referees for their valuable comments. This work is supported by the EU project (ICA4-2001-10085), *Policy decision support for sustainable adaptation of China's agriculture to globalization.*

NOTES

1. The value of the R-square is 0.941, while the t-values equal 3.293 for variable G_t, 4.177 for P_t, and -1.100 for e_t, respectively.

2. *Sichuan* and *Chongqing* are considered as one unit, because base data for the two provinces separately are lacking.

3. GDP growth rate for each province (R_i) is determined by: $R_i = r_i (1 + f_i) R_{GDP}$, in which r_i = mean GDP growth rate for province i in 1996–2000 (table 16.3), and f_i = a factor to account for the effect of the Western Region Development Program on GDP growth in the relevant provinces (which equals 0.1 for the twelve concerned provinces in the west region, and 0 otherwise). R_{GDP} = ratio of the growth rate assumed for the scenario analysis (table 16.4) to the mean GDP growth rate of China in 1996–2000 (table 16.3).

BIBLIOGRAPHY

Chen, B. M. ed. *Production and Population-supporting Capacity of Agricultural Resources in China.* Beijing: Meteorological Press (*in Chinese*), 2001.

Hubacek, K. and L. Sun. 2001. "A Scenario Analysis of China's Land Use and Land Cover Change: Incorporating Biophysical Information into Input-output Modeling." *Structural Change and Economic Dynamics* 12 (2001): 367–97.

Li, S. T., Y. Z. Hou, and F. W. Zhai. "An Analysis of China's Economic Development Potentials in the Forthcoming Ten Years. Information Network of Centre for China Development." Available online at: drcnet.com.cn/New_product/Expert (*in Chinese*), 2001.

Lin, G., C. S. Ho, and S. P. S. Ho. "China's Land Resources and Land-use Change: Insights from the 1996 Land Survey." *Land Use Policy* 20 (2003): 87–107.

MLR (Ministry of Land and Resources of P. R. China). *China Land and Resources Almanac.* Beijing: China Statistics Press (*in Chinese*), 2001.

Sun, H. and Y. L. Shi. eds. *Agricultural Land Use in China.* Nanjing: Jiangsu Science & Technology Press (*in Chinese with English abstract*), 2003.

Wang, M. K., B. P. Lu, and Z. Y. Lu. "Three Steps to Achieve Modernization of China Economy. Information Network of Centre for China Development." Available online at: drcnet.com.cn/New_product/Expert (*in Chinese*), 2002.

Zheng, M. Q. "Prediction on Trend of RMB Exchange Rate in 2001–2020." *Economy Dynamics* 4 (2002): 34–36 (*in Chinese*).

Appendix

Table 10.A.1: Descriptive Statistics for Educational Level (schooling years) of All Labor Force Members in Four Household Groups

Schooling years	Group 1	Group 2	Group 3	Group 4	Total
Mean	2.25	5.76	4.05	5.61	4.83
Maximum	4	13	12	11	13
Standard Deviation	1.56	3.15	3.04	2.40	2.85
Number of Cases	34	40	93	158	325

Notes:
Group 1: Households with no persons educated more than 4 years.
Group 2: Households with no oxen, at least 1 person educated more than 4 years.
Group 3: Households with oxen, 1–2 persons educated more than 4 years.
Group 4: Households with oxen, 3 or more persons educated more than 4 years.

Table 10.A.2: Results of t-Tests for Differences of Household Size and Number of Laborers Among Four Household Groups

		Group 1	Group 2	Group 3	Group 4
			Labor Force Size		
Household Size	Group 1		−2.74***	−1.64**	−6.82***
	Group 2	2.71***		1.15	−2.92***
	Group 3	3.60***	0.70		−5.01***
	Group 4	6.17***	2.90***	2.53***	

Notes: Group definition is the same as in table 10.A.1.
All t-values presented below the diagonal refer to differences in household size between the two groups in question. The t-values presented above the diagonal refer to differences in labor force sizes. A positive sign of the t-value indicates that the mean value for the group listed in the column group exceeds the mean value of the group listed in the row. If the value is negative, the reverse is the case.
***, **, and * indicates statistical significance at 1% level, 5% level, and 10% level, respectively.

Table 10.A.3: Results of t-Tests for Differences of Per Capita Irrigated Contracted Land and Per Capita Contracted Forestland among Four Household Groups

		Group 1	*Group 2*	*Group 3*	*Group 4*
		Per Capita Contracted Irrigated Land			
Per Capita	Group 1		0.22	1.59*	0.17
Contracted	Group 2	1.48*		0.72	0.79
Forest Land	Group 3	0.03	−1.90**		0.04
	Group 4	1.62*	−2.05**	0.16	

Notes: Group definition is the same as in table 10.A.1.
All t-values presented below the diagonal refer to differences in per capita forestland between the two groups in question. The t-values presented above the diagonal refer to differences in per capita irrigated land. A positive sign of the t-value indicates that the mean value for the group listed in the column group exceeds the mean value of the group listed in the row. If the value is negative, the reverse is the case.
***, **, and * indicates statistical significance at 1% level, 5% level, and 10% level, respectively.

Table 10.A.4: List of All Entries within Each Account

Activities		Commodities			Factors	Institutions
Production	Transaction	Products	Services	Rented Factors		
One-season rice	Irrigated land rent in	Rice	Draught power	Irrigated land	Irrigated land	Household Group 1
One-season rice with green manure	Irrigated land rent out	Vegetable	Agricultural labor	Oxen	Dry land	Household Group 2
Two-season rice	Oxen rent in	Bamboo	Nonagricultural labor	Agricultural labor	Forestland	Household Group 3
Two-season rice with green manure	Oxen rent out	Straw	Self-employment			Household Group 4
Vegetable		Livestock	Migration		Low-educated labor	Household Group 5
Perennial crops		Livestock manure	Leisure		High-educated labor	
Livestock production		Processed manure			Capital	
Manure activity		Fuel wood				
Fuel wood collection		Livestock feed				
Agricultural works by low-educated labor		Other inputs of livestock				
Agricultural works by high-educated labor		Food				
Nonagricultural works by low-educated labor		Non-food				
Nonagricultural works by high-educated labor		Durable goods				
Self-employment by low-educated labor		Transaction goods				
Self-employment by high-educated labor						
Low-educated labor Migration						
High-educated labor Migration						

Table 11.A.1. Comparison of Functional Forms

Cost Categories	Total	Labor	Fertilizers	Seed	Herbicides and Pesticides	Oxen
Semi-log						
Normality test	6.17	1.02	1.65	4.99	5.72	3.31
(prob.)	(0.05)	(0.60)	(0.44)	(0.08)	(0.06)	(0.19)
Ramsey test	0.97	1.80	6.39	0.44	0.22	0.24
(prob.)	(0.32)	(0.16)	(0.01)	(0.51)	(0.64)	(0.62)
R^2	0.34	0.39	0.30	0.33	0.13	0.33
R^2-adjusted	0.31	0.37	0.28	0.30	0.09	0.31
F-statistic	12.13	15.39	10.39	11.24	3.451	1.92
(prob.)	(0.00)	(0.00)	(0.00)	(0.00)	(0.00)	(0.00)
Linear						
Normality test	3,472	5,340	255	9,121	854	814
(prob.)	(0.00)	(0.00)	(0.00)	(0.00)	(0.00)	(0.00)
Ramsey test	7.17	9.18	0.25	0.93	2.11	8.40
(prob.)	(0.01)	(0.00)	(0.62)	(0.34)	(0.00)	(0.15)
R^2	0.30	0.33	0.22	0.27	0.13	0.26
R^2-adjusted	0.27	0.31	0.19	0.24	0.09	0.23
F-statistic	10.08	11.88	6.62	8.69	3.52	8.45
(prob.)	(0.00)	(0.00)	(0.00)	(0.00)	(0.00)	(0.00)
Double log						
Normality test	3.63	0.924	3.4	69.5	22.07	0.50
(prob.)	(0.16)	(0.63)	(0.00)	(0.00)	(0.00)	(0.78)
Ramsey test	1.08	2.90	0.56	0.02	0.00	0.57
(prob.)	(0.30)	(0.09)	(0.46)	(0.89)	(0.97)	(0.45)
R^2	0.35	0.40	0.26	0.25	0.17	0.36
R^2-adjusted	0.33	0.37	0.23	0.22	0.13	0.33
F-statistic	12.81	15.6	8.28	7.89	4.75	13.1
(prob.)	(0.00)	(0.00)	(0.00)	(0.00)	(0.00)	(0.00)

Note: In the double-log functional form, dummy variables are used to replace variables that have one or more zero observations (i.e., education, share of good irrigated land area, forest area, savings, and credit).

Table 11.A.2: Results for Rice Production Cost Category (yuan/ton)

			Regression Results		
	Labor Coefficient	Fertilizer Coefficient	Seed Coefficient	Herbicides and Pesticides Coefficient	Oxen and Tractor Coefficient
Land fragmentation					
Simpson index	0.429***	-0.351**	-12.07**	-0.123	-0.386**
Farm size (mu)	-0.021***	0.0011	-0.3447**	-0.014**	-0.004
Average distance (minute)	0.0089***	0.008***	0.0897	0.0067*	0.007**
Household characteristics					
Age (year)	0.005**	0.003	-0.146	0.0023	0.0016
Education (year)	-0.021**	-0.03***	-0.794**	-0.0047	-0.005
Household size (person)	-0.010	-0.002	0.035	0.0374*	-0.015
Farm characteristics					
Share good irrigated land	-0.001	-0.0006	-0.027	0.00058	-0.001
Forestland (mu)	0.0068	-0.010	0.330	-0.0122	-0.008
Oxen ownership	-0.098**	-0.065	0.828	-0.016	0.0357
Available savings (yuan)	-8.64E-07	-7.30E-06*	0.0003*	-7.99E-06	-2.00E-06
Available credit (yuan)	-7.42E-06	-1.31E-05**	-0.0002	-6.92E-06	-1.27E-06
Village dummies					
Shangzhu village	0.623***	-0.607***	22.50***	-0.391***	0.847***
Gangyan village	0.196***	-0.242***	6.916***	-0.283***	0.477***
Constant	5.89***	5.56***	42.13***	3.849***	4.247***
Number of observations	322	322	315	319	322

Notes: [a] Except for seed, all the dependent variables are in logarithms.
[b] *Significant at 10% level; **Significant at 5% level; and ***Significant at 1% level.

Index

About the Contributors

Chan-Kang, Connie Research Assistant, International Food Policy Research Institute (IFPRI), Washington, D.C., The United States.

Chen, Zhigang Assistant Professor, Department of City and Regional Planning, Nanjing University (NU), Nanjing, China.

Dung, Nguyen Huu Researcher, Environmental Economics Unit, University of Economics (UOE), Ho Chi Minh City, Vietnam.

Fan, Shenggen Senior Research Fellow and Division Director, Development Strategies and Governance Division, International Food Policy Research Institute (IFPRI), Washington, D.C., The United States.

Heerink, Nico Research Fellow and Office Coordinator, International Food Policy Research Institute (IFPRI), Beijing Office, Beijing, China; Associate Professor, Wageningen University, Wageningen, The Netherlands.

Holden, Stein Professor, The Department of Economics and Resource Management (IØR), at the Norwegian University of Life Sciences (UMB, formerly Agricultural University of Norway), Oslo, Norway.

Kruseman, Gideon Researcher, Agricultural Economics Institute (LEI), The Hague, The Netherlands.

Kuiper, Erno Assistant Professor, Wageningen University (WUR), Wageningen, The Netherlands.

Kuiper, Marijke Researcher, Agricultural Economics Institute (LEI), The Hague, The Netherlands.

Li, Xiubin Researcher, Institute of Geographic Sciences and Natural Resources Research, Chinese Academy of Sciences (CAS), Beijing, China.

Li, Yousheng Associate Professor, Nanjing Agricultural University (NAU), Nanjing, China.

Lu, Changhe Researcher, Institute of Geographic Sciences and Natural Resources Research, Chinese Academy of Sciences (CAS), Beijing, China.

Lu, Hualiang PhD Researcher, Wageningen University (WUR), Wageningen, The Netherlands.

Qu, Futian Professor, College of Land Management and Vice-President, Nanjing Agricultural University (NAU), Nanjing, China.

Roetter, Reimund Researcher, Alterra, Wageningen, The Netherlands.

Ruben, Ruerd Professor and Director of Centre for International Development Issues Nijmegen (CIDIN), Radboud University, Nijmegen, The Netherlands.

Shi, Xiaoping Associate Professor, College of Land Management, Nanjing Agricultural University (NAU), Nanjing, China.

Spoor, Max Associate Professor, Institute of Social Studies (ISS), and Coordinator of the Centre for the Study of Transition and Development (CESTRAD), The Hague, The Netherlands.

Tan, Minghong Researcher, Institute of Geographic Sciences and Natural Resources Research, Chinese Academy of Sciences (CAS), Beijing, China.

Tan, Shuhao Associate Professor, College of Land Management, Nanjing Agricultural University (NAU), Nanjing, China.

Tuan, Nguyen Do Anh Lecturer, National Economic University (NEU), Hanoi, Vietnam.

van den Berg, Marrit Assistant Professor, Wageningen University (WUR), Wageningen, The Netherlands.

van Keulen, Herman Professor, Plant Research International (PRI), Wageningen, The Netherlands.

Wang, Guanghuo Professor, Zhejiang University, Hangzhou, China.

Zhang, Xiaobo Research Fellow, International Food Policy Research Institute (IFPRI), Washington, D.C., The United States.

Zhong, Funing Professor, College of Economics and Management, Nanjing Agricultural University (NAU), Nanjing, China.

Zhu, Jing Professor, College of Economics and Management, Nanjing Agricultural University (NAU), Nanjing, China.

Zhu, Peixin Associate Professor, College of Land Management, Nanjing Agricultural University (NAU), Nanjing, China.